CHAOS AND
INTEGRABILITY
IN NONLINEAR
DYNAMICS

CHAOS AND INTEGRABILITY IN NONLINEAR DYNAMICS

AN INTRODUCTION

MICHAEL TABOR

Department of Applied Physics
Columbia University
New York, New York

WILEY

A Wiley-Interscience Publication

JOHN WILEY & SONS

New York / Chichester / Brisbane / Toronto / Singapore

Library of Congress Cataloging in Publication Data:

Tabor, Michael, 1951–
   Chaos and integrability in nonlinear dynamics : an introduction /
Michael Tabor.
      p.   cm.
   "A Wiley-Interscience publication."
   Bibliography: p.
   Includes index.   1 0 0 2 4 8 7 7 0 5
   ISBN 0-471-82728-2
   1. Dynamics.   2. Nonlinear theories.   3. Chaotic behavior in
systems.   I. Title.
QA871.T33   1989
531'.11—dc19   88-15516
                                                         CIP

Printed in the United States of America

10 9 8 7 6 5

Printed and bound by Quinn - Woodbine, Inc.

To the memory of
Ruth Braham

PREFACE

The aim of this introductory text is to provide a transition from traditional courses in differential equations and classical mechanics to the rapidly growing areas of nonlinear dynamics and chaos and to present these "old" and "new" ideas in a broad and unified perspective.

The reason for such an approach is that many students in the physical and engineering sciences (with typical backgrounds in differential equations and classical mechanics), eager to learn about the new and exciting results concerning chaos in dynamical systems, sometimes tend to see this "new" field as distinct from the "old" courses. I believe that the former can be more fully appreciated when seen as a natural extension of the latter. One way of emphasizing this relationship is through Hamilton's equations of motion. These fundamental equations of classical mechanics provide a natural framework in which to discuss the phase-space dynamics of systems of differential equations that are capable of displaying both integrable and chaotic behavior. Furthermore, considerable emphasis is placed on the concept of integrability, since a firm grasp of this concept considerably enhances one's appreciation of nonintegrable dynamics and the significance of the fundamental KAM theorem. This also provides the background for a later chapter on integrable partial differential equations and soliton dynamics.

The first four chapters of this book, namely, "The Dynamics of Differential Equations," "Hamiltonian Dynamics," "Classical Perturbation Theory," and "Chaos in Hamiltonian Systems and Area-Preserving Mappings," form the core of the book and are written at a level suitable for advanced undergraduates and beginning graduate students. Here the prerequisites need be little more than the standard courses on differential equations and classical mechanics. Although it might be argued that some of the material contained in these chapters can be found scattered among other, existing texts and reviews, the aim (and hope) is that these "old" and "new" ideas are here drawn together in a unified way. In addition, Chapter 2 on "Hamiltonian Dynamics" is presented in a "traditional" way—but with a definite "geometric" slant. This is achieved by means of an appendix, "Geometric concepts in classical mechanics," which is linked to the main text through a series of footnotes and asides. The aim here is to help students make a relatively painless transition (should they wish to do so) from the "traditional" to the "geometric" presentations of classical mechanics.

Many of the excellent texts now appearing on chaos and nonlinear dynamics concentrate on dissipative dynamical systems. Here, for the reasons stated above, our main emphasis is on chaos in Hamiltonian systems. Nonetheless, in order to keep a broad view, one introductory chapter (Chapter 5) on dissipative dynamics is included. Where possible, an attempt is made to emphasize the connections with the dynamical concepts of the preceding chapters, as well as the relevance of these ideas to real fluid dynamical problems.

The remaining three chapters are an attempt to show the breadth and diversity of topics in nonlinear dynamics. Although written at a slightly more advanced level than the preceding chapters, they really only require the additional prerequisites of a course in elementary quantum mechanics (for Chapters 6 and 7) and complex variable theory (for Chapter 8). Chapter 6 introduces a currently active research area exploring possible connections between classical chaos and the corresponding quantum mechanical behavior of systems in the "semiclassical limit." Chapter 7 is concerned with the dynamics of nonlinear partial equations and, in particular, with those that have the property of being integrable and exhibiting solitons. At first sight it might appear that studies of chaos and solitons are mutually exclusive, since the former is concerned with the properties of nonintegrable systems and the latter with the properties of integrable systems. However, I believe it is very important that these concepts be learned together. It will only be through an appreciation of chaos *and* integrability in *both* ordinary and partial differential equations that significant breakthroughs will be achieved in the understanding of real physical problems involving spatiotemporal chaos such as fluid dynamical turbulence.

The final chapter, on "Analytic Structure of Dynamical Systems," is an introduction to another recent research topic that attempts to unify the enormously diverse properties of ordinary and partial differential equations, be they integrable or nonintegrable, through the study of the singularities that their solutions can exhibit in the complex domain.

This text could be used as the basis of a two-semester course with the first four or five chapters providing a first semester of general background and the second semester being devoted to more specialized topics such as those introduced in the last three chapters. Nonetheless, the text is very much designed to be read as a whole. Sections or subsections marked by an asterisk (*) tend to contain slightly more advanced or technical information and can be omitted at a first reading without loss of continuity.

It should be emphasized that this book is *not* meant to be an up-to-the-minute review on chaos. Accordingly, I have tended to draw only on well-established physical models and their associated references. The growth of the literature has been enormous in recent years, and no attempt has been made to provide a complete bibliography (now almost an impossible task). This being so, I apologize in advance to all those many researchers whose excellent works have not been cited here.

I would particularly like to thank my graduate students Frank Cariello, Joel Chaiken, and Gregory Levine for their considerable assistance in preparing this manuscript. I could consider no chapter complete until it had been brutally dissected and constructively criticized by them. If this text has any merit, aimed as it is at a student audience, it will be in no small part due to their efforts.

Finally, I would like to thank William G. Lewis for his help in preparing the diagrams and Lois Winter for her heroic efforts in typing this manuscript.

MICHAEL TABOR

New York, New York
November 1988

CONTENTS

CHAOS AND INTEGRABILITY IN NONLINEAR DYNAMICS

1

THE DYNAMICS OF DIFFERENTIAL EQUATIONS

1.1 INTEGRATION OF LINEAR SECOND-ORDER EQUATIONS

Integrating even a simple differential equation should be regarded as much more than an automatic exercise in mathematical manipulation. It can, especially for those equations pertaining to dynamical systems, cast light on a system's deepest geometric properties.

The ubiquitous second order, linear, ordinary differential equation

$$\ddot{x} + \omega^2 x = 0 \tag{1.1.1}$$

(where "·" denotes time derivative d/dt) can describe the motion of a particle under a linear restoring force, that is, small displacement oscillations. Here the standard notation prevails, that is, $\omega = \sqrt{k/m}$, where k is the force constant and m is the particle mass. For such a simple equation, one "automatically" makes the substitution $x(t) = e^{\alpha t}$ and obtains the solution

$$x(t) = a \sin(\omega t + \delta) \tag{1.1.2}$$

The two constants of integration, namely, the amplitude a and phase shift δ, can be obtained from the initial value data $x(0)$ and $\dot{x}(0)$, that is,

$$a = (\dot{x}^2(0) + \omega^2 x^2(0))^{1/2}/\omega, \quad \delta = \tan^{-1}(\dot{x}(0)/\omega x(0)) \tag{1.1.3}$$

1.1.a Integration by Quadrature

We now go back to (1.1.1) and solve it via, at first sight, a more convoluted route. Although not strictly necessary, it is, as a first step, convenient to cast (1.1.1) in the form of a pair of coupled first-order equations, that is,

$$\dot{x} = y \tag{1.1.4a}$$

$$\dot{y} = -\omega^2 x \tag{1.1.4b}$$

As we shall soon see, this first-order representation, especially for Hamiltonian systems (described in Chapter 2), is particularly convenient and leads to a much more "geometrical" picture of the motion. If we multiply both sides of (1.1.4a) by $\omega^2 x$ and both sides of (1.1.4b) by y and add the two together, we obtain the identity

$$y\dot{y} + \omega^2 x\dot{x} = 0 \tag{1.1.5}$$

The left-hand side is simply the time derivative $(d/dt)(\frac{1}{2}y^2 + \frac{1}{2}\omega^2 x^2)$, which immediately identifies the bracketed quantity as a constant, that is,

$$\tfrac{1}{2}(y^2 + \omega^2 x^2) = I_1 \tag{1.1.6}$$

The constant (in time) function I_1 is usually termed a *constant of motion* or a *first integral*. Since $y = \dot{x}$, I_1 can be immediately identified as the mechanical energy of the system. This quantity can now be used to reduce Eq. (1.1.4) from a second-order to a first-order system. From (1.1.6) one explicitly expresses y in terms of x and I_1, that is, $y = \sqrt{2(I_1 - \frac{1}{2}\omega^2 x^2)}$, and hence (1.1.4a) becomes

$$\frac{dx}{dt} = \sqrt{2(I_1 - \tfrac{1}{2}\omega^2 x^2)} \tag{1.1.7}$$

This equation can now be "solved" as an explicit integral or *quadrature*, that is,

$$\int dt = \int \frac{dx}{\sqrt{2(I_1 - \tfrac{1}{2}\omega^2 x^2)}} \tag{1.1.8}$$

Of course, having integrated both sides of (1.1.7), we now have a second constant of integration (denoted as I_2) so we can write

$$t + I_2 = \int \frac{dx}{\sqrt{2(I_1 - \tfrac{1}{2}\omega^2 x^2)}} \tag{1.1.9}$$

The right-hand side of (1.1.9) is easily identified as the arcsin integral, that is,

$$t + I_2 = \frac{1}{\omega} \sin^{-1}\left(\frac{x\omega}{\sqrt{2 I_1}}\right) \qquad (1.1.10)$$

At this point, t is expressed as a (multivalued) function of x, but here the inversion is trivial and the final result is

$$x(t) = \frac{\sqrt{2 I_1}}{\omega} \sin(\omega t + I_2 \omega) \qquad (1.1.11)$$

which is, of course, identically equivalent to the result originally obtained in Eq. (1.1.2). Notice that there have been essentially four steps in this procedure:

  (i) Identification of a first integral.
 (ii) Use of the integral I_1 to reduce the order of the differential equation by one.
(iii) An explicit "integration" or "quadrature".
 (iv) An inversion to obtain a single-valued solution.

For such a simple equation as (1.1.1), this (apparently tortuous) route would appear to have no advantage over the solution obtained by the simple substitution $x(t) = e^{\alpha t}$. However, as soon as nonlinear terms appear in the differential equation, that simple method is no longer applicable and the "integration by quadratures" becomes a more natural procedure. However, as will soon become strikingly clear in the course of this book, there are deep dynamical reasons why many, if not most, equations *cannot* be integrated by quadratures—especially when higher than second order.†
We again emphasize that our ability to integrate an equation is more than a mathematical nicety but touches right at the heart of the deepest properties of a system.

---

†In fact, even for lower-order equations there can sometimes be technical difficulties. For example, the nonlinear first-order equation

$$\dot{x} = x^2$$

is trivially integrated (do it!), whereas its nonautonomous version

$$\dot{x} = x^2 - t$$

cannot be solved by quadratures (try it!). For this particular case, however, the substitution $x = \dot{y}/y$ transforms it to the linear second-order equation

$$\ddot{y} + ty = 0$$

for which a convergent power-series solution, called the *Airy function*, is known. The substitution that we have used is an example of a "linearizing" transformation. Such a linearization is tantamount to completely solving the problem at hand. We will return to such exact linearizations in Chapter 7.

Historically, it is worth noting that the development of methods for integrating differential equations (the first person to attempt this was Isaac Newton†) was a major activity of eighteenth- and nineteenth-century mathematics. This reached considerable levels of virtuosity in the hands of such great mathematicians as Jacobi, who was instrumental in developing the theory of elliptic functions, which can be used to integrate certain classes of nonlinear differential equations. Before we consider such equations, some further aspects of the linear system (1.1.1) and its modifications, are worthy of discussion.

For the linear system of (1.1.1) the *period* of the motion is obvious from the explicit form of the solution (1.1.2); that is, a complete cycle of the motion is completed in a time $T = 2\pi/\omega$. This can also be deduced from the explicit quadrature (1.1.8) when evaluated as a definite integral between the limits of the motion. Denoting the first integral I_1 explicitly as the mechanical energy, $E = \frac{1}{2}(\dot{x} + \omega^2 x^2)$, the limits of integration are simply $x = \pm\sqrt{2E}/\omega$ (the square root in (1.1.8) vanishes at these two points). The time to complete a full cycle of the motion is thus:

$$T = 2 \int_{-\sqrt{2E}/\omega}^{\sqrt{2E}/\omega} \frac{dx}{\sqrt{2(E - \frac{1}{2}\omega^2 x^2)}} = \frac{2\pi}{\omega} \qquad (1.1.12)$$

Obviously, for this linear system, the period is independent of the energy (or, equivalently, independent of the initial conditions). Now compare with a simple nonlinear system, that is,

$$\ddot{x} + \beta x^3 = 0 \qquad (1.1.13)$$

which might correspond to the motion of a particle under a nonlinear restoring force (with "force constant" β). The first integral of (1.1.13) is easily deduced to be $I_1 = \frac{1}{2}\dot{x}^2 + \frac{1}{4}\beta x^4$, which again may be interpreted as the mechanical energy, E, of the system. The definite integral equivalent to (1.1.12) is now

$$T = 2 \int_{-(4E/\beta)^{1/4}}^{+(4E/\beta)^{1/4}} \frac{dx}{\sqrt{2(E - \frac{1}{4}\beta x^4)}} \qquad (1.1.14)$$

which fortunately can be evaluated explicitly in terms of the gamma function (Γ), that is,

$$T = \frac{1}{2} \sqrt{\frac{2}{\pi\beta^{1/2}}} \, \Gamma^2(\tfrac{1}{4}) E^{-1/4} \qquad (1.1.15)$$

Now the period is explicitly dependent on the energy. In this particular case it decreases as a function of energy, that is, the more excited the particle the

---

†However, the first integration, in today's notation, is due to Leibniz—apparently performed on November 11, 1675. (A brief historical survey is given by Ince (1956).)

quicker it completes a cycle of motion. Equivalently we can associate with the motion a characteristic, and now energy dependent, frequency $\omega(E)$ given by $\omega(E) = 2\pi/T(E)$.

This simple example illustrates an all-important difference between linear and nonlinear systems—namely, for the latter the characteristic frequency is a function of the energy (or, equivalently, a function of the initial conditions). This becomes particularly significant when the system is subjected to a perturbation or to an external force.

1.1.b The Damped Oscillator

Returning to the linear equation (1.1.1), it can easily be modified to include the effect of a frictional force, that is,

$$\ddot{x} + \lambda\dot{x} + \omega^2 x = 0 \tag{1.1.16}$$

where λ is the "damping" or "frictional" coefficient. The solution of (1.1.16) is, again, most easily found by making the substitution $x(t) = e^{\alpha t}$ leading to

$$x(t) = ae^{-\lambda t/2}\sin(\nu t + \delta) \tag{1.1.17}$$

where now the frequency ν is given by $\nu = \frac{1}{2}\sqrt{4\omega^2 - \lambda^2}$ (which presupposes that $4\omega^2 > \lambda^2$ since for $4\omega^2 < \lambda^2$ the solution is purely decaying) and the amplitude a and phase δ can be determined (although not the same as for (1.1.1)) from the initial data $x(0)$ and $\dot{x}(0)$. Clearly, the presence of friction retards motion and the solution (1.1.17) corresponds to that of damped oscillations. For $\lambda \ll \omega$ the characteristic period $2\pi/\nu$ is little different from the undamped period $2\pi/\omega$. In this limit it is still meaningful to think of a mean mechanical energy constructed from the sum of the mean (i.e., averaged over a period) squared velocity $\overline{(\dot{x})^2}$ and displacement $\overline{(x)^2}$, namely, $\bar{E} = \frac{1}{2}[\overline{(\dot{x})^2} + \omega^2\overline{(x)^2}]$. However, this is no longer a conserved quantity and one finds that

$$\bar{E}(t) = E(0)e^{-\lambda t} \tag{1.1.18}$$

where $E(0)$ is the initial value of the "energy." Having emphasized the idea of integration by quadratures, one might ask whether the system (1.1.16), now written as a first-order pair,

$$\dot{x} = y \tag{1.1.19a}$$

$$\dot{y} = -\omega^2 x - \lambda y \tag{1.1.19b}$$

can be similarly treated. In the undamped case the energy provided a time-independent first integral with which to reduce the order of the system.

Now, as just illustrated, such a quantity is no longer conserved. It might appear, then, that a quadrature is no longer possible. However, as we shall see in Section 1.5, for one value of λ one can introduce a *time-dependent integral* which can be used to effect an explicit reduction of order. These time-dependent integrals are rather rare beasts, and the few cases in which they can be found are consequently rather interesting.

1.2 INTEGRATION OF NONLINEAR SECOND-ORDER EQUATIONS

For differential equations describing physical processes (vibrations, chemical reactions, population growth, etc.), nonlinearity is the rule rather than the exception. A fairly wide class of second-order nonlinear equations can be cast in the form

$$\frac{d^2x}{dt^2} = F(x) \tag{1.2.1}$$

where $F(x)$ might be some polynomial, rational, or transcendental function of x. In the case of polynomial functions, for example,

$$F(x) = a_0 + a_1 x + a_2 x^2 + \cdots + a_n x^n \tag{1.2.2}$$

we can easily effect a quadrature, that is,

$$\int dt = \int \frac{dx}{[2(a_0 x + \frac{1}{2}a_1 x^2 + \cdots + (1/n+1)a_n x^{n+1} + I_1)]^{1/2}} \tag{1.2.3}$$

where I_1 is the integral of motion. The problem lies in evaluating the integral on the right-hand side of (1.2.3). For those $F(x)$ with a highest nonlinearity of x^3, this can be achieved "explicitly" and the solution to (1.2.1) can be expressed in terms of what are known as *elliptic functions*. These are the natural generalization of the standard circular functions (sine, cosine, etc.). The *elliptic integrals* are the corresponding generalization of the inverse circular functions (arcsin, arccos, etc.). In a simple minded way, it is not too hard to see how such functions might arise. For the linear problem (1.1.1), the key to its solution was the integration (1.1.8)—the integral being identified as the arcsin function. For integrals involving higher powers of x (i.e., $\sqrt{1-x^3}$, $\sqrt{1-x^4}$, etc.), it seems (in hindsight!) not unreasonable to expect the corresponding integral to be the inverse of some more complicated periodic function. The theory of elliptic functions was independently developed by Abel and Jacobi, the latter published a monumental treatise *Fundamenta Nova Theoriae Functionum Ellipticarum* on the subject in 1829. As was often the case at that time, Gauss (who was extremely cautious about publishing results—a welcome change from today) had probably obtained a number of these results previously. Although elliptic functions only enable us to solve a relatively small class of equations of the form (1.2.1), these include (suitably transformed) important problems such as the pendulum; and for that case alone, it is worth learning a little more about them. The subject is enormous, so here we will just sketch out some of the basic ideas.† (Extra, technical details are given in Appendix 1.1.)

Consider now some general second-order equation of the form

$$\frac{d^2x}{dt^2} = A + Bx + Cx^2 + Dx^3 \tag{1.2.4}$$

which might, for example, correspond to the equation of motion for a particle moving under a force function expanded to third order in the displacement. As we shall soon learn, the solutions to this equation can be expressed in terms of *Jacobi* elliptic functions. If the right-hand side of (1.2.4) only includes terms up to second order in x, the solution can then be expressed in terms of *Weierstrass* elliptic functions. (As might be imagined, there are a variety of relationships between these two functions.) Furthermore, other right-hand sides to (1.2.4), including certain rational functions of x and simple transcendental functions such as $\sin(x)$ and $\cos(x)$ (which arise in the all-important pendulum equation), can also be integrated in terms of one or other of the elliptic functions.

1.2.a Jacobi Elliptic Functions

The first integral of (1.2.4) is easily identified as

$$\left(\frac{dx}{dt}\right)^2 - \left(Ax + \frac{1}{2}Bx^2 + \frac{1}{3}Cx^3 + \frac{1}{4}Dx^4\right) = I_1 \tag{1.2.5}$$

and for convenience we rewrite (1.2.5) in the form

$$\left(\frac{dx}{dt}\right)^2 = a + bx + cx^2 + dx^3 + ex^4 \tag{1.2.6}$$

The quartic expression on the right-hand side can be factorized, that is,

$$\left(\frac{dx}{dt}\right)^2 = e(x - \alpha)(x - \beta)(x - \gamma)(x - \delta) \tag{1.2.7}$$

---

†A discussion of elliptic functions might appear to be somewhat "old-fashioned." However, at a deeper level their algebraic geometry gives important insights into the meaning of "integrability." These ideas will be discussed further in later chapters and Chapter 8 in particular.

which, in turn, can be transformed (see Appendix 1.1) into the canonical form (due to Legendre)

$$\left(\frac{dx}{dt}\right)^2 = (1 - x^2)(1 - k^2 x^2) \tag{1.2.8}$$

The right-hand side of the resulting integration, that is,

$$\int dt = \int \frac{dx}{\sqrt{(1 - x^2)(1 - k^2 x^2)}} \tag{1.2.9}$$

is termed an *elliptic integral of the first kind* and is often denoted as $F(x, k)$, that is,

$$F(x, k) = \int_0^x \frac{dx'}{\sqrt{(1 - x'^2)(1 - k^2 x'^2)}} \tag{1.2.10}$$

(clearly, in the limit $k = 0$ this just reduces to the arcsin function). An equivalent form to (1.2.10) is obtained by making the transformation $x = \sin \theta$, that is,

$$F(\theta, k) = \int_0^\theta \frac{d\theta'}{\sqrt{1 - k^2 \sin^2 \theta'}} \tag{1.2.11}$$

There is a whole bagful of terminology associated with elliptic integrals (to avoid clutter, elliptic integrals of the second and third kind are summarized in Appendix 1.1), and some of the more important terms are:

(i) k—"modulus"
(ii) $k' = \sqrt{1 - k^2}$—"complementary modulus"
(iii) $K = F(1, k) = F(\pi/2, k)$—"complete elliptic integral of the first kind"
(iv) $K' = F(1, k') = F(\pi/2, k')$—"complementary complete integral of the first kind"

The Jacobi elliptic functions are the inverses of the elliptic integral (1.2.11) (or 1.2.10). Writing

$$u = \int_0^\theta \frac{d\theta'}{\sqrt{1 - k^2 \sin^2 \theta'}} \tag{1.2.12}$$

the elliptic function, denoted by the symbol sn, is defined as†

$$\text{sn}(u, k) = \sin \theta \qquad (1.2.13)$$

There are, in fact, other elliptic functions—for example, the one denoted as cn is defined as

$$\text{cn}(u, k) = \cos \theta \qquad (1.2.14)$$

The analogy with the circular functions (i.e., sine and cosine), should be clear, as are the definitions of the inverse elliptic functions, that is,

$$u = \text{sn}^{-1}(\sin \theta, k) = \text{cn}^{-1}(\cos \theta, k) \qquad (1.2.15)$$

There is a bewildering array of properties and identities involving the elliptic functions, and some of them are given in the appendix. Here we just draw attention to their periodicity properties. The periods are determined by the complete elliptic integrals defined earlier, and the integral K plays a role analogous to that of π for the circular functions, that is,

$$\text{sn}(u \pm 2K, k) = -\text{sn}(u, k), \qquad \text{cn}(u \pm 2K, k) = -\text{cn}(u, k) \quad (1.2.16)$$

The complementary complete integral K'—or to be more precise, iK'—plays a role analogous to that of $i\pi$ as the period of the hyperbolic functions sinh and cosh, that is,

$$\text{sn}(u \pm iK', k) = \frac{1}{k \sin(u, k)}, \qquad \text{cn}(u \pm iK', k) = \frac{i \, \text{dn}(u, k)}{k \, \text{sn}(u, k)} \quad (1.2.17)$$

The elliptic functions are thus *doubly periodic* functions; that is, they have both a real and imaginary period.

---

†At a first reading, the definition (1.2.13) can be a bit confusing. To understand how it comes about, consider evaluating the elliptic integral (1.2.12) (or (1.2.10)) for the case $k = 0$. This is just the arcsin integral, so one has

$$u = \theta \equiv \sin^{-1}(x) \qquad (i)$$

Thus taking the sin of both sides, one obtains

$$\sin u = \sin \theta (\equiv x) \qquad (ii)$$

Now consider the case $k \neq 0$. The right-hand side of (i) is no longer arcsin but, instead, is the inverse of some more complicated function which we denote as sn. Thus (ii) becomes

$$\text{sn} \, u = \sin \theta (\equiv x) \qquad (iii)$$

1.2.b Weierstrass Elliptic Functions

As mentioned earlier, equations of the form (1.2.4) terminating at second order in x can be solved in terms of the Weierstrass elliptic functions. The form of equation analogous to (1.2.6) now involves a cubic right-hand side and is written in standard form as†

$$\left(\frac{dx}{dt}\right)^2 = 4x^3 - g_2 x - g_3 \qquad (1.2.18)$$

(where now the coefficients 4, g_2, and g_3 have been chosen for consistency with standard notation). The quadrature is thus

$$t = \int \frac{dx}{\sqrt{4(x - e_1)(x - e_2)(x - e_3)}} \qquad (1.2.19)$$

where the cubic form has been factorized. The e_i and g_i are related by the standard formulae

$$e_1 + e_2 + e_3 = 0, \qquad e_1 e_2 + e_1 e_3 + e_2 e_3 = \tfrac{1}{4} g_2, \qquad e_1 e_2 e_3 = \tfrac{1}{4} g_3 \quad (1.2.20)$$

The inverse of (1.2.19) is the Weierstrass elliptic function $\mathcal{P}(t)$, that is,

$$x = \mathcal{P}(t) \qquad (1.2.21)$$

This Weierstrass elliptic function is also doubly periodic. Denoting the two periods as Ω and Ω' one finds that

$$\mathcal{P}(t + 2\Omega) = \mathcal{P}(t), \qquad \mathcal{P}(t + 2\Omega') = \mathcal{P}(t) \qquad (1.2.22)$$

where there is a series of relationships between the Ω and Ω' and the K and K' of the Jacobi functions.

1.2.c Periodic Structure of Elliptic Functions

The doubly periodic structure (i.e., the presence of both a real and imaginary period) of the elliptic functions suggests that they have a nontrivial structure when extended into the complex plane, that is, when regarded as a function of a complex independent variable. The circular functions, which are singly periodic, such as $\sin(z)$ and $\cos(z)$, have a very simple structure when extended into the complex z-plane. They are just *entire* functions; that is, they have no singularities in the finite complex plane. On the other

†Terms quadratic in x have been removed by the simple change of variables $x \to x - c/3d$.

hand, the elliptic functions of both Jacobi and Weierstrass are *meromorphic* functions; that is, they have (isolated) pole singularities in the complex plane. In fact, these poles are organized in a regular, latticelike structure, which reflects the doubly periodic nature of these functions. Indeed it would appear that the behavior of these functions in the complex plane (i.e., their *analytic structure*) is a fundamental factor in determining our ability to "integrate" the associated differential equations. This deep, and as yet not fully understood, issue is discussed further in Chapter 8.

1.2.d The Pendulum Equation

A most important nonlinear equation, which can be integrated exactly in terms of the Jacobi elliptic functions, is the classical pendulum equation. Taking the angle of libration to be θ (see Figure 1.1), the equation of motion for an oscillating particle of unit mass is easily shown to be

$$\frac{d^2\theta}{dt^2} + \frac{g}{L}\sin\theta = 0 \qquad (1.2.23)$$

where L is the length of the string and g is the gravitational constant. In the limit of small displacement, the sin can be expanded to first order to yield the standard equation of motion for a simple harmonic oscillator (cf. (1.1.1))

$$\frac{d^2\theta}{dt^2} + \frac{g}{L}\theta = 0 \qquad (1.2.24)$$

the solution of which has the characteristic frequency $\omega = \sqrt{g/L}$. The first integral of (1.2.23) can again be identified as the (scaled) mechanical energy E' of the system, that is,

$$\frac{1}{2}\left(\frac{d\theta}{dt}\right)^2 - \frac{g}{L}\cos\theta = E' \qquad (1.2.25)$$

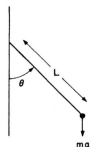

mg

Figure 1.1 The classical pendulum executing a swing of angle θ under the gravitational force mg.

(Here $E' = E/mL^2$, where E is the actual energy of the system (with mass m)). The quadrature becomes

$$t = \int \frac{d\theta'}{\sqrt{2(E' + (g/L)\cos\theta')}} \qquad (1.2.26)$$

(for convenience we simply drop the second constant of integration, which here merely plays the role of a phase shift). The integral in (1.2.26) can be transformed into the form of an integral of the first kind, that is, (1.2.11). The first step is to introduce the variable $\cos\omega = -E'(L/g)$, thereby changing (1.2.26) into

$$t = \sqrt{\frac{L}{2g}} \int_0^\theta \frac{d\theta'}{\sqrt{\cos\theta' - \cos\omega}} \qquad (1.2.27)$$

The next step is to introduce the transformations $\cos\theta = 1 - 2k^2\sin^2\varphi$ and $k = \sin(\omega/2)$, which, with a little manipulation, casts (1.2.27) into the desired form, that is,

$$t = \sqrt{\frac{L}{g}} \int_0^\varphi \frac{d\varphi'}{\sqrt{1 - k^2\sin^2\varphi'}}, \qquad (1.2.28)$$

where the modulus is given explicitly by $k = \sqrt{\frac{1}{2}(1 + E'L/g)} = \sin(\omega/2)$. The inversion in terms of the Jacobi elliptic function is immediate, namely,

$$\text{sn}\left(t\sqrt{\frac{g}{L}}, k\right) = \sin\varphi = \frac{1}{k}\sin\left(\frac{\theta}{2}\right) \qquad (1.2.29)$$

and hence the desired result of expressing θ as an explicit function of time is thus

$$\theta(t) = 2\sin^{-1}\left[k\,\text{sn}\left(t\sqrt{\frac{g}{L}}, k\right)\right] \qquad (1.2.30)$$

The period of the motion can be calculated, as we did for the linear problem, as the definite integral of (1.2.26). Here the range of integration is from the equilibrium position $\theta = 0$ to the classical "turning point" $\theta_0 = \cos^{-1}(-E'L/g)$, which we have defined to be the variable ω. Using the transformations $\cos\theta = 1 - 2k^2\sin^2\varphi$ and $k = \sin(\omega/2)$, it is easy to show that in terms of the angle φ, this turning point is just $\pi/2$. Since the period is defined as the time to complete a full cycle of the motion (i.e., from $\theta = 0$ to θ_0 to $-\theta_0$ and back again to $\theta = 0$), the integral (1.2.28), evaluated between the limits 0 and $\pi/2$, must be multiplied by 4, that is,

$$T(k) = 4\sqrt{\frac{L}{g}} \int_0^{\pi/2} \frac{d\varphi}{\sqrt{1 - k^2 \sin^2 \varphi}} = 4\sqrt{\frac{L}{g}} \, F\left(\frac{\pi}{2}, k\right) \qquad (1.2.31)$$

Thus the period is just the complete elliptic integral times a simple factor. Since $F(\pi/2, 0) = \pi/2$, the limiting behavior of (1.2.29) is

$$T(0) = 2\pi \sqrt{\frac{L}{g}} = \frac{2\pi}{\omega} \qquad (1.2.32)$$

which is merely the period of simple harmonic motion described by Eq. (1.2.24). If the modulus in (1.2.31) is replaced by the complementary modulus $k' = \sqrt{1 - k^2}$, the result would be the complex period

$$T(k') = 4\sqrt{\frac{L}{g}} \, F\left(\frac{\pi}{2}, k'\right) \qquad (1.2.33)$$

of the (doubly) periodic motion. At first sight, the idea of a complex period for a pendulum would appear to be devoid of physical meaning. However, it can be thought of as just the period the pendulum would have if the sign of gravity were changed; that is, as it were, if the world was turned on its head (this is easily seen from the definition of k, that is, $k = \sin(\omega/2) = \sqrt{\frac{1}{2}(1 + E'L/g)}$).

1.3 DYNAMICS IN THE PHASE PLANE

So far, little attempt has been made to visualize the nature of a solution to the various differential equations discussed above—other than a vague mental picture of the "position variable" $x = x(t)$ oscillating periodically in time. A most valuable description of a solution is obtained by examining its behavior in the *phase plane*. As we shall see, this can even be achieved, in some cases, without having to integrate the equations of motion. Returning to the first-order representation of the linear problem (1.1.1) (i.e., the pair of equations (1.1.4)), the two (independent) variables x and y define a space in which the solution moves. This is the *phase space* of the system; and since it is (strictly) two-dimensional, we also refer to it as the *phase plane*. For a system of n first-order equations of the form

$$\dot{x}_1 = f_1(x_1, \ldots, x_n)$$
$$\dot{x}_2 = f_2(x_1, \ldots, x_n)$$
$$\vdots$$
$$\dot{x}_n = f_n(x_1, \ldots, x_n)$$

each variable $x_1, x_2, \ldots, x_n$, can be thought of as an independent phase-space coordinate in the associated n-dimensional phase space.

Although the notion of a phase space can be associated with any system of differential equations, we shall learn that for those systems of equations derived from a Hamiltonian system, the associated phase space has an especially rich geometric structure. The value of the phase-space coordinates (x and y in this simple case (1.1.4)), at any instant in time, completely defines the state of the system at that time. Typically, a given solution to the equations of motion will map out, as a function of time, a smooth curve in the phase plane. This is termed a *phase curve* (sometimes *level curve*), and the motion along it is termed the *phase flow*. Furthermore, owing to fundamental uniqueness properties of solutions to differential equations, different phase-space trajectories do not cross each other. When many different solutions (corresponding to different initial conditions) are plotted out on the same phase plane, a (sometimes quite complicated) picture of the overall behavior emerges. Such a picture, made up of sets of phase curves, is often referred to as a *phase portrait*.

At first sight, it is not necessarily obvious which regions of the phase plane the solutions to Eqs. (1.1.4) will explore. This is where the existence of a (constant) integral of motion plays a most significant role. In this particular case it is just the mechanical energy $E = \frac{1}{2}(y^2 + \omega^2 x^2)$. Clearly, the phase flow must be confined to ellipses whose major and minor axes are determined by the energy E (and hence the initial conditions $x(0)$ and $y(0)$). These ellipses cut the x-axis at $x = \pm\sqrt{2E}/\omega$ (which are, of course, just the classical "turning point" of motion used as the limits of integration in (1.1.12)) and the y-axis at $y = \pm\sqrt{2E}$ (see Figure 1.2). *Thus the existence of a (constant) first integral has provided a definite geometrical constraint on the phase flow,* that is, it has constrained the motion in the two-dimensional phase space to closed one-dimensional regions (the phase curves). The origin of the phase plane, $x = y = 0$, corresponds to an obvious

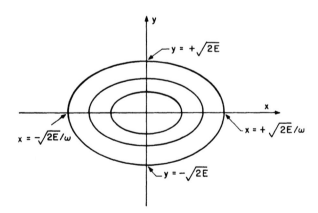

Figure 1.2 Phase curves for the simple harmonic oscillator (1.1.4).

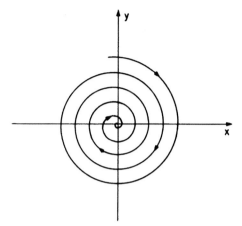

Figure 1.3 Phase curve for the damped harmonic oscillator (1.1.19).

equilibrium point of the motion and the overall phase portrait is just a set of concentric ellipses, growing smoothly as a function of energy, centered around this point.

By contrast, the phase portrait for the frictionally damped linear oscillator, Eq. (1.1.19), is quite different. For this case there is no constant first integral to constrain the motion. Now all solutions just spiral (at a rate depending on the damping coefficient) into the equilibrium point at the origin (see Figure 1.3).

For nonlinear equations (such as (1.1.13)), with a constant first integral and executing bounded motion, the phase portrait is again a set of concentric (quartic in this case) curves centered around the origin. For more general nonlinearities, such as in Eq. (1.2.4), the phase portrait can become more complicated, depending on the signs of the various coefficients. This will be discussed shortly.

1.3.a Phase Portrait of the Pendulum

First we consider the phase portrait for the pendulum. For notational consistency, we now denote the variables θ and $\dot\theta$ of Eq. (1.2.23) as x and y, respectively. As a first-order system, the equations of motion are just

$$\dot x = y \tag{1.3.1a}$$

$$\dot y = -\frac{g}{L}\sin x \tag{1.3.1b}$$

and the first integral is

$$\frac{1}{2}y^2 - \frac{g}{L}\cos x = E' \tag{1.3.2}$$

A number of purely physical considerations can provide a guide to how the phase portrait will turn out. For small energies, the pendulum will just oscillate about the equilibrium point $x = y = 0$ in nearly linear fashion. Clearly, in this regime the phase portrait will just be a set of curves centered about this point. As the energy increases, the pendulum will execute ever-larger librations until finally a point is reached when the pendulum (assuming a "stiff" string) goes right "over the top" and starts to execute unhindered rotational motion which gets ever faster as the energy increases further. This critical point is reached when the pendulum has sufficient energy to swing all the way from the rest angle $x = 0$ to $x = \pm \pi$—this energy is just $E' = g/L$. The point $x = \pm \pi$, with $y = 0$ (i.e., the pendulum "standing on its head"), is an equilibrium point—but clearly an unstable one. Furthermore, owing to the periodic nature of the restoring force, this pattern must be repeated at every multiple of 2π to the left and right of $x = 0$. Thus at every $x = \pm 2n\pi$ there is a *stable* equilibrium point, and at every $x = \pm(2n + 1)\pi$ there is an *unstable* equilibrium point. Furthermore, these latter points mark a transition from librational to rotational motion, and the associated phase curves change from closed to open (corresponding to unbounded rotational motion). Thus the overall phase portrait takes the form shown in Figure 1.4. The pair of phase curves that separate the librational and rotational motion and that meet at the unstable equilibrium points is termed the *separatrix*.

An analytical description of the motion along the separatrix is easily obtained. Here the energy is $E' = g/L$, so the modulus takes the value $k = \sqrt{\frac{1}{2}(1 + E'L/g)} = 1$; hence the period, evaluated according to (1.2.31), becomes infinite since

$$T(1) = 4\sqrt{\frac{L}{g}} \int_0^{\pi/2} \frac{d\varphi}{\sqrt{1 - \sin^2 \varphi}} = 4\sqrt{\frac{L}{g}} F\left(\frac{\pi}{2}, 1\right) = \infty \qquad (1.3.3)$$

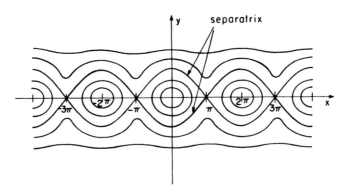

Figure 1.4 Phase curves for the pendulum (1.3.1). Stable equilibrium points are located at $x = \pm 2n\pi$, and unstable ones are located at $x = \pm(2n + 1)\pi$.

Furthermore, there is a convenient identity (see Appendix 1.1) $\text{sn}(u, 1) =$ $\tanh(u)$, so the actual solution (1.2.30) takes the form (where now we are using the notation $x \equiv \theta$)

$$x(t) = 2 \sin^{-1}[\tanh(t\sqrt{g/E})] \qquad (1.3.4)$$

Since $\lim_{x \to \pm\infty} \tanh(x) = \pm 1$, it clearly takes an infinite time for the displacement $x(t)$ to reach the unstable equilibrium point at $x = \pm \pi$ (i.e., the approach is exponentially slow). This result is, of course, completely consistent with the infinite period obtained in (1.3.3).

1.3.b Phase Portraits for Conservative Systems

For the systems considered so far, the existence of a constant first integral has enabled us to define the phase curves globally. That is, the explicit function $E = E(x, y)$, which is just the mechanical energy, determines the phase curves for all initial conditions. An explicit solution to the equations of motion does not appear to be required. These systems, for which the energy is a constant of motion, are termed *conservative* systems; and when the energy function can be written in the (usual) form of kinetic energy + potential energy, the construction of the phase curves (often called *level curves* for conservative systems) is particularly easy. For the sorts of system considered here, we may write

$$E = E(x, y) = \tfrac{1}{2}y^2 + V(x) \qquad (1.3.5)$$

where the "potential function" $V(x)$ is (typically) some nonlinear function of x, for example, $V(x) = -(g/L)\cos x$ for the pendulum or $V(x) = ax^2 + bx^4$ for the "quartic oscillator." By thinking of the corresponding equation of motion

$$\ddot{x} + \frac{\partial V}{\partial x}(x) = 0 \qquad (1.3.6)$$

as describing the motion of a particle in the potential well $V(x)$ (although the equation may not have a mechanical origin), physical intuition guides the drawing of the level curves. For example, for the simple linear system (1.1.1) the potential function $V(x)$ is just a parabola in which the motion is forever confined—bouncing backwards and forwards between the classical turning points determined by $E = V(x)$ (i.e., $x = \pm\sqrt{2E/\omega^2}$). Now there is a rather obvious "matching" between the level curves of Figure 1.2 and the classical motion in the potential (see Figure 1.5). For the pendulum case, $V(x) = -(g/L)\cos x$ is a periodic potential well. Below the periodically spaced maxima, the motion is clearly bounded libration and above the

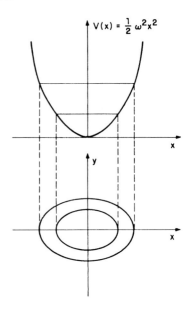

Figure 1.5 Matching of phase curves with classical motion in potential $V(x)$ for a simple harmonic oscillator.

maxima the motion is unbounded (rotation in this case). The maxima themselves provide obvious unstable equilibrium points (the slightest disturbance will send the particle rolling into one valley or another), whereas the minima are obviously stable equilibrium points. A plot of the potential energy functions $V(x)$ is again easily matched to the set of level curves (see Figure 1.6).

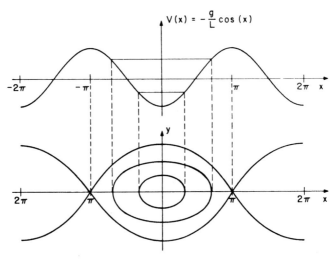

Figure 1.6 Matching of phase curves with classical motion for a pendulum.

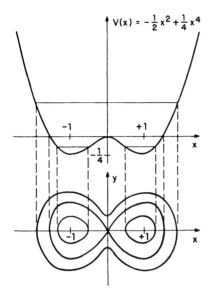

Figure 1.7 Matching of phase curves with classical motion for a quartic potential.

Now consider the level curves associated with the system

$$\ddot{x} - x + x^3 = 0 \tag{1.3.7}$$

The energy is just $E = \frac{1}{2}y^2 + V(x)$, where the potential energy function is $V(x) = -\frac{1}{2}x^2 + \frac{1}{4}x^4$ (again, $y = \dot{x}$). This function (plotted in Figure 1.7) has minima at $x = \pm 1$ (of depth $E = -\frac{1}{4}$), has a maximum at $x = 0$, and, for large x (positive or negative), increases steeply as a "quartic" parabola. For energies $-\frac{1}{4} < E < 0$, the particle will be trapped in one or other of the two minima in which it executes librational motion. The corresponding sets of level curves will obviously be closed curves centered about those stable equilibrium points. For energies above the maximum at $x = 0$, the motion will now be larger amplitude librations between the "outer walls" of the potential energy well. Again, the level curves will be a set of larger closed curves now centered about $x = 0$. The maximum of the potential at this point is clearly an unstable equilibrium point and separates the small amplitude motions (about $x = \pm 1$) from the large amplitude motions (about $x = 0$). Clearly, at the energy $E = 0$ the level curve is a (figure-eight) separatrix.

Another simple example is provided by the system

$$\ddot{x} + x - x^2 = 0 \tag{1.3.8}$$

Now the potential energy function $V(x) = \frac{1}{2}x^2 - \frac{1}{3}x^3$ has a single minimum at $x = 0$ and maximum at $x = 1$ (with height $E = \frac{1}{6}$). The set of level curves is easily deduced and is sketched in Figure 1.8.

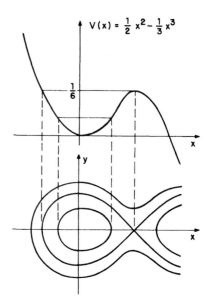

Figure 1.8 Matching of phase curves with classical motion for a cubic potential.

1.4 LINEAR STABILITY ANALYSIS

For conservative systems (in one dimension), a global phase portrait is particularly easy to construct. The equilibrium points play an important role and clearly have a characteristic *local* behavior, that is, sets of closed curves around the stable points and hyperbolic-looking regions in the neighborhood of the unstable points. For nonconservative systems, a global phase portrait is difficult to construct unless an explicit solution is known to the equations of motion. However, it is always possible to build up an approximate, local phase portrait by identifying the equilibrium points (henceforth referred to as the *fixed points*) and drawing the phase curves in their neighborhood, the nature of which will depend on their stability. Fixed points can be thought of as the "organizing centers" of a system's phase-space dynamics. Thus by identifying them and their stability, one can build up a fairly global picture of a system's behavior.

1.4.a The Stability Matrix

Consider some general second-order systems of the form

$$\dot{x} = f(x, y) \tag{1.4.1a}$$

$$\dot{y} = g(x, y) \tag{1.4.1b}$$

where f and g are any (typically nonlinear) smooth functions of x and y. The fixed points of the motion are those values of x and y, denoted as x_0 and y_0, for which the phase flow is stationary (i.e., those points for which $\dot{x} = \dot{y} = 0$), that is,

$$f(x_0, y_0) = 0 \qquad (1.4.2a)$$

$$g(x_0, y_0) = 0 \qquad (1.4.2b)$$

There can be any number of points (x_0, y_0) satisfying these conditions, depending on the precise functional form of f and g. Having identified these points, their stability can be determined by examining the evolution of some small displacement $(\delta x, \delta y)$ about (x_0, y_0). Expanding f and g in powers of δx, δy, one obtains

$$\delta \dot{x} = f_x(x_0, y_0)\,\delta x + f_y(x_0, y_0)\,\delta y + f_{xy}(x_0, y_0)\,\delta x\,\delta y + \cdots \qquad (1.4.3a)$$

$$\delta \dot{y} = g_x(x_0, y_0)\,\delta x + g_y(x_0, y_0)\,\delta y + g_{xy}(x_0, y_0)\,\delta x\,\delta y + \cdots \qquad (1.4.3b)$$

To first order, (1.4.3) can be written as the linear system (1.4.4), and we refer to them as the "linearized" equations (this first-order approximation is not to be confused with the type of exact linearization mentioned on p.3), that is,

$$\frac{d}{dt}\begin{bmatrix} \delta x \\ \delta y \end{bmatrix} = \begin{bmatrix} f_x(x_0, y_0) & f_y(x_0, y_0) \\ g_x(x_0, y_0) & g_y(x_0, y_0) \end{bmatrix}\begin{bmatrix} \delta x \\ \delta y \end{bmatrix} \qquad (1.4.4)$$

where the 2×2 matrix (to be denoted as **M**) is often referred to as the stability matrix. The system of first-order linear equations (1.4.4) (which is easily generalized for an nth-order system of the form $\dot{x}_i = f_i(x_1, \ldots, x_n)$, $i = 1, \ldots, n$) is easily solved with the eigenvalues of **M** determining the stability of the associated fixed point. Denoting the column vector $(\delta x, \delta y)$ as $\delta\mathbf{X}$ and denoting the two eigenvectors associated with the two eigenvalues λ_1 and λ_2 as $\mathbf{D}_1$ and $\mathbf{D}_2$, the general solution to (1.4.4) is the linear combination

$$\delta\mathbf{X} = c_1\mathbf{D}_1 e^{\lambda_1 t} + c_2\mathbf{D}_2 e^{\lambda_2 t} \qquad (1.4.5)$$

where c_1 and c_2 are arbitrary coefficients, and the eigenvalues are just the roots of the equation

$$\det|\mathbf{M} - \lambda\mathbf{I}| = 0 \qquad (1.4.6)$$

where **I** is the unit matrix. From our previous considerations, it is clear that if both λ_1 and λ_2 are pure imaginary numbers, the displacement will just rotate about the associated fixed point and locally the phase curves will be closed ellipses—a clear manifestation of a stable fixed point. On the other hand, if the λ_i ($i = 1, 2$) acquire real parts, the displacements will, depending on the sign, decay or grow exponentially, indicating some form of stability or instability respectively. In fact, there is a whole range of possibilities, each leading to a particular type of fixed point; this classification is listed below. Note, by the way, that although the stability type is determined by the eigenvalues λ_1, λ_2, the eigenvectors $\mathbf{D}_1$, $\mathbf{D}_2$ should not be ignored—their form determines the actual directions of the local phase flows.

1.4.b   Classification of Fixed Points

Case (i). $\lambda_1 < \lambda_2 < 0$—this is usually termed a *stable node*. The local flow decays in both directions, determined by D_1 and D_2, into the fixed point (Figure 1.9a).

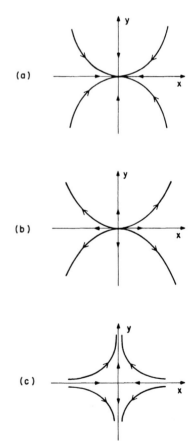

Figure 1.9 Local phase flows for (a) stable node, (b) unstable node, and (c) hyperbolic point.

Case (ii). $\lambda_1 > \lambda_2 > 0$—an *unstable node*. The local flow grows exponentially away from the fixed point in both directions (Figure 1.9*b*).

Case (iii). $\lambda_1 < 0 < \lambda_2$—a *hyperbolic point* (or "saddle" point). One direction grows exponentially, and the other decays exponentially (Figure 1.9*c*). This is, in fact, the type of pattern found for the unstable equilibrium points of the pendulum and the nonlinear oscillators (1.3.7) and (1.3.8). Often the incoming and outgoing directions are referred to as the *stable* and *unstable* manifolds (of the separatrix), respectively.

Case (iv). $\lambda_1 = -\alpha + i\beta$, $\lambda_2 = -\alpha - i\beta(\alpha, \beta > 0)$—a *stable spiral point*. Owing to the negative real part (since λ_1, λ_2 are the roots of a quadratic equation, they form a complex conjugate pair), the flow spirals *in* toward the fixed points (Figure 1.10*a*).

Case (v). $\lambda_1 = \alpha + i\beta$, $\lambda_2 = \alpha - i\beta$—an *unstable spiral point*. Now owing to the positive real parts, the flow spirals *away* from the fixed point (Figure 1.10*b*).

(a)

(b)

(c)

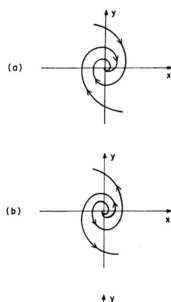

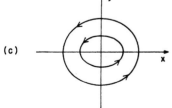

Figure 1.10 Local phase flows for (a) stable spiral point, (b) unstable spiral point, and (c) elliptic point.

Case (vi). $\lambda_1 = +i\omega$, $\lambda_2 = -i\omega$—an *elliptic point* (or "center"). The flow is locally pure rotation about the fixed point. This is the type of stable equilibrium point discussed in some of the earlier examples (Figure 1.10c).

For the spiral-point cases (cases (iv) and (v)) and the elliptic-point case (case (vi)), there is the question as to whether the flow (in or out) is clockwise or anticlockwise. This may be determined from the linearized equations (Eqs. (1.4.4)) by setting $\delta y = 0$ and $\delta x > 0$: If $\delta \dot{y} < 0$, this implies a "downward" motion (i.e., a locally clockwise motion), whereas if $\delta \dot{y} > 0$, the motion is "upwards" and hence anticlockwise.

So far, all the cases considered have nondegenerate roots λ_1, λ_2. In the case of degenerate roots, the general solution to (1.4.4) is of the form

$$\delta \mathbf{X} = (c_1 \mathbf{D}_1 + c_2 (\mathbf{D}_2 + \mathbf{D}_1 t)) e^{\lambda t} \qquad (1.4.7)$$

The nature of the fixed points will depend on (obviously) the sign of and the nature of the eigenvectors $\mathbf{D}_1$ and $\mathbf{D}_2$.

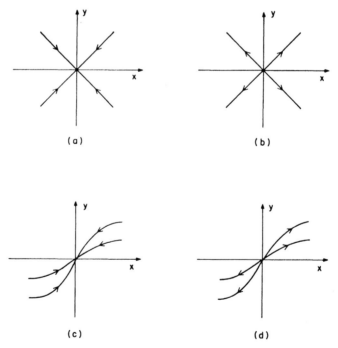

Figure 1.11 Local phase flows for (*a*) stable star, (*b*) unstable star, (*c*) stable improper node, and (*d*) unstable improper node.

Case (vii). If $\mathbf{D}_2 = (0,0)$ (i.e., a null vector and $\mathbf{D}_1$ arbitrary), then the flow lines will be independent, intersecting straight lines forming

(a) a stable star if $\lambda < 0$ (Figure 1.11a)
(b) an unstable star if $\lambda > 0$ (Figure 1.11b)

Case (viii). If $\mathbf{D}_2$ is not a null vector, then the flow lines will be curved, forming

(a) a stable improper node if $\lambda < 0$ (Figure 1.11c)
(b) an unstable improper node if $\lambda > 0$ (Figure 1.11d)

1.4.c Examples of Fixed-Point Analysis

We now work through a series of examples to illustrate the various types of fixed-point analysis. First of all, consider our old friend the damped linear oscillator (1.1.19); the only fixed point is clearly at $x_0 = y_0 = 0$. Obviously there is no need to linearize these equations, which we now write in the matrix form

$$\frac{d}{dt}\begin{bmatrix} x \\ y \end{bmatrix} = \begin{bmatrix} 0 & 1 \\ -\omega^2 & -\lambda \end{bmatrix}\begin{bmatrix} x \\ y \end{bmatrix} \tag{1.4.8}$$

and the two eigenvalues of $\mathbf{M}$ are easily determined to be

$$\lambda_1 = -\frac{\lambda}{2} + \frac{1}{2}\sqrt{\lambda^2 - 4\omega^2} \quad \text{and} \quad \lambda_2 = -\frac{\lambda}{2} - \frac{1}{2}\sqrt{\lambda^2 - 4\omega^2}$$

There are various possibilities depending on the relative sizes of λ and ω.

Case (a). For $\lambda^2 > 4\omega^2$ we obtain $\lambda_1 < \lambda_2 < 0$, and hence $(0,0)$ is a stable node.

Case (b). For $\lambda^2 < 4\omega^2$ we obtain $\lambda_1 = \lambda_2^* = -\lambda/2 + i\sqrt{4\omega^2 - \lambda^2}/2$, and hence $(0,0)$ is a stable spiral.
Setting $\delta y = 0$ and $\delta x > 0$ in Eqs. (1.4.8), we observe that $\delta\dot{y} < 0$ and hence deduce that the spiral motion is clockwise (see Figure 1.3).

Case (c). For $\lambda^2 = 4\omega^2$ we obtain $\lambda_1 = \lambda_2 = -\lambda/2$; hence for this degenerate case, $(0,0)$ is a stable improper node.
Another case worth considering again is that of the pendulum (Eqs. (1.3.1)). There is an infinite set of fixed points $(x_n, y_n) = (\pm n\pi, 0)$, $n =$

$0, 1, 2, \ldots$, and the linearized equations are

$$\frac{d}{dt}\begin{bmatrix} \delta x \\ \delta y \end{bmatrix} = \begin{bmatrix} 0 & 1 \\ -\dfrac{g}{L}\cos x_n & 0 \end{bmatrix}\begin{bmatrix} \delta x \\ \delta y \end{bmatrix} \tag{1.4.9}$$

The eigenvalues are $\lambda_{1,2} = \pm i\sqrt{(g/L)\cos x_n}$. If $x_n = 2n\pi$, they become $= \pm i\sqrt{g/L}$, corresponding to elliptic fixed points; if $x_n = (2n+1)\pi$, they become $= \pm \sqrt{g/L}$, corresponding to hyperbolic fixed points. Now consider the case of a damped pendulum, that is,

$$\dot{x} = y \tag{1.4.10a}$$

$$\dot{y} = -\frac{g}{L}\sin x - \lambda y \tag{1.4.10b}$$

The fixed points are still at the same positions, that is, $(x_n, y_n) = (\pm\, n\pi, 0)$, but now the linearized equations are

$$\frac{d}{dt}\begin{bmatrix} \delta x \\ \delta y \end{bmatrix} = \begin{bmatrix} 0 & 1 \\ -\dfrac{g}{L}\cos x_n & -\lambda \end{bmatrix}\begin{bmatrix} \delta x \\ \delta y \end{bmatrix} \tag{1.4.11}$$

with corresponding eigenvalues

$$\lambda_1 = -\frac{\lambda}{2} + \frac{1}{2}\sqrt{\lambda^2 - \frac{4g}{L}\cos x_n}, \quad \lambda_2 = -\frac{\lambda}{2} - \frac{1}{2}\sqrt{\lambda^2 - \frac{4g}{L}\cos x_n}$$

At the fixed points $(x_n, y_n) = (\pm(2n+1)\pi, 0)$, the situation is still $\lambda_2 < 0 < \lambda_1$, corresponding to hyperbolic behavior. At the points $(x_n, y_n) = (\pm\, 2n\pi, 0)$, a variety of cases now arise.

Case (a). $\lambda^2 < 4g/L$—λ_1, λ_2 form a complex conjugate pair (with negative real part), corresponding to a stable spiral point with (check it) clockwise rotation (see Figure 1.12).

Case (b). $\lambda^2 > 4g/L$—now we have $\lambda_1 < \lambda_2 < 0$, corresponding to a stable node.

Case (c). $\lambda^2 = 4g/L$—we have the degenerate case $\lambda_1 = \lambda_2 = -\lambda/2$, corresponding to a stable improper node.

Since energy is no longer conserved, the levels curves cannot be drawn exactly without an explicit form of the solution. However, identification of the fixed points and the local flow patterns, including actual directions, enable one to at least sketch a reasonable global phase portrait.

An interesting class of equations is the predator–prey type introduced by

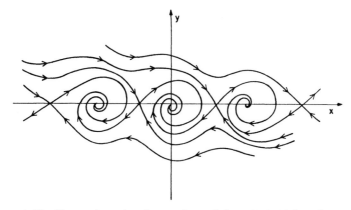

Figure 1.12 Phase plane for damped pendulum (1.4.10) for $\lambda^2 < 4g/L$.

Voltera to investigate population dynamics. A simple example is furnished by the system

$$\dot{x} = x - xy \tag{1.4.12a}$$

$$\dot{y} = -y + xy \tag{1.4.12b}$$

where x might denote a rabbit population and y might denote a fox population. The following, somewhat cynical, scenario is feasible: If $y = 0$ and $x > 0$, the rabbit population can grow without bound since there are no foxes to eat them; however, if $x = 0$ and $y > 0$, the fox population will die off because there are no rabbits to eat. With both species present, there is the possibility of a balance between these competing effects, with the hungry foxes reducing the rabbit population and enhancing their own population while at the same time running the risk of starvation if they eat too many of the rabbits. Equations (1.4.12) represent a very simple model of this process, and it is not difficult to think up various modifications corresponding to effects of overcrowding, disease, insanity, or even the appearance of killer-rabbits.

The system has the two fixed points $(x_1, y_1) = (0, 0)$ and $(x_2, y_2) = (1, 1)$ and the linearized equation

$$\frac{d}{dt}\begin{bmatrix} \delta x \\ \delta y \end{bmatrix} = \begin{bmatrix} 1 - y_i & -x_i \\ y_i & -1 + x_i \end{bmatrix}\begin{bmatrix} \delta x \\ \delta y \end{bmatrix} \tag{1.4.13}$$

where $i = 1, 2$. For the fixed point $(x_1, y_1) = (0, 0)$, the eigenvalues are easily determined to be $\lambda = \pm 1$ (i.e., a saddle point), and, on finding the respective eigenvectors $\mathbf{D}_1$ and $\mathbf{D}_2$, the general solution is found to be

$$\begin{bmatrix} \delta x \\ \delta y \end{bmatrix} = c_1 \begin{bmatrix} 1 \\ 0 \end{bmatrix} e^{+t} + c_2 \begin{bmatrix} 0 \\ 1 \end{bmatrix} e^{-t} \tag{1.4.14}$$

where c_1 and c_2 are some arbitrary constants. From this solution we observe that the incoming flow is down the y-axis and the outgoing flow is along the x-axis. For the second fixed point $(x_2, y_2) = (1, 1)$, the eigenvalues are found to be $\lambda = \pm i$, that is, an elliptic (or center) point. Setting $\delta y = 0$, $\delta x > 0$ in the equations of motion (1.4.13), we note that $\delta \dot{y} > 0$ and hence that the rotation about the elliptic fixed point is anticlockwise. Notice that this flow is consistent with the flow direction about the saddle point and that a global phase portrait can be sketched approximately but, of course, not completely (Figure 1.13).

A slightly more complicated example is the system

$$\dot{x} = x(4 - x - y) \tag{1.4.15a}$$

$$\dot{y} = y(x - 2) \tag{1.4.15b}$$

with the three fixed points $(x_1, y_1) = (0, 0)$, $(x_2, y_2) = (1, 1)$, and $(x_3, y_3) = (4, 0)$ and the linearized equations

$$\frac{d}{dt} \begin{bmatrix} \delta x \\ \delta y \end{bmatrix} = \begin{bmatrix} 4 - 2x_i - y_i & -x_i \\ y_i & x_i - 2 \end{bmatrix} \begin{bmatrix} \delta x \\ \delta y \end{bmatrix} \tag{1.4.16}$$

where $i = 1, 2, 3$. The behavior of the three fixed points is as follows:

1. $(x_1, y_1) = (0, 0)$ has eigenvalues $\lambda_{1,2} = -2, 4$ (i.e., a hyperbolic point), and the general solution is

$$\begin{bmatrix} \delta x \\ \delta y \end{bmatrix} = c_1 \begin{bmatrix} 1 \\ 0 \end{bmatrix} e^{4t} + c_2 \begin{bmatrix} 0 \\ 1 \end{bmatrix} e^{-2t} \tag{1.4.17}$$

Thus the incoming direction is down the y-axis, and the outgoing direction is along the x-axis.

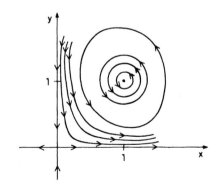

Figure 1.13 Phase plane for Eq. (1.4.12).

2. $(x_2, y_2) = (2, 2)$ has eigenvalues $\lambda_{1,2} = -1 \pm i\sqrt{3}$ (i.e., a stable spiral point). For $\delta y = 0$ and $\delta x > 0$ we obtain $\delta \dot{y} > 0$, so the spiral direction is anticlockwise.

3. $(x_3, y_3) = (4, 0)$ has eigenvalues $\lambda_{1,2} = -4, 2$ (i.e., a hyperbolic point), and the general solution is

$$\begin{bmatrix} \delta x \\ \delta y \end{bmatrix} = c_1 \begin{bmatrix} 1 \\ 0 \end{bmatrix} e^{-4t} + c_2 \begin{bmatrix} -4 \\ 6 \end{bmatrix} e^{2t} \tag{1.4.18}$$

Thus the incoming direction is along the x-axis, and the outgoing direction has a slope $\delta y / \delta x = 6/-4 = -3/2$. By taking careful note of the directions of the local flows associated with each of the fixed points, a reasonable estimate of the global phase portrait can be sketched (Figure 1.14).

It is always important to remember that the fixed-point stability analysis is based on *linearized* equations of motion. It is quite possible that the predictions thus made will not be borne out by the full nonlinear equations. Elliptic fixed points, as predicted by a linear stability analysis, are particularly susceptible to breakdown under nonlinear perturbation. A nice example is given by Bender and Orszag (1978) as follows: The system

$$\dot{x} = -y + x(x^2 + y^2) \tag{1.4.19a}$$

$$\dot{y} = x + y(x^2 + y^2) \tag{1.4.19b}$$

is, according to linear stability theory, easily determined to have an elliptic fixed point at $x = y = 0$ with anticlockwise rotation (do it!). In fact, the nonlinear equations have an unstable spiral point emanating from the origin, as may be determined from the exact solution. Multiplying (1.4.19a) by x and (1.4.19b) by y and adding, one obtains $x\dot{x} + y\dot{y} = (x^2 + y^2)^2$, which,

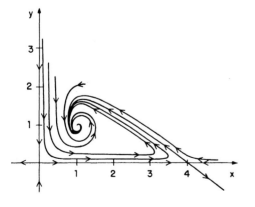

Figure 1.14 Phase plane for Eq. (1.4.15).

in the polar variable $r = \sqrt{x^2 + y^2}$, is the equation

$$\frac{1}{2} \frac{d}{dt} [r^2(t)] = r^4(t) \tag{1.4.20}$$

with the exact solution

$$r(t) = \frac{r(0)}{\sqrt{1 - 2r^2(0)t}} \tag{1.4.21}$$

Thus not only does the radius $r(t)$ grow with time, but the solution actually blows up at the critical time $t_{crit} = (2r^2(0))^{-1}$.

1.4.d Limit Cycles

Another type of behavior that will not be picked by a linear analysis is that of a limit cycle. A famous example of a system exhibiting such a feature is the Van der Pohl oscillator:

$$\ddot{x} - \lambda(1 - x^2)\dot{x} + \omega^2 x = 0 \tag{1.4.22}$$

A linear stability analysis of its first-order representation

$$\dot{x} = y \tag{1.4.23a}$$

$$\dot{y} = \lambda(1 - x^2)y - \omega^2 x \tag{1.4.23b}$$

suggests fixed points at $x = y = 0$, which will be an unstable node if $\lambda^2 > 4\omega$ and an unstable spiral point if $\lambda^2 < 4\omega^2$. Considering just the latter case, as x and y increase, (1.4.23b) will eventually be dominated by the nonlinear term (i.e., $\lim_{x,y \to \infty} \dot{y} = -\lambda x^2 y$), which would then suggest a decay back to the origin. Thus trajectories far from the origin move inwards. By continuity, there must be (at least) one solution that stays in the middle. This solution is the limit cycle, which is a closed orbit encircling the origin. Solutions starting either within or without it are attracted to it but can never cross it. An exact solution to the equation of motion (Eq. (1.4.22)) is not known, so although simple physical consideration predicts limit cycle behavior (the nonlinear term can be thought of as a position-dependent friction which changes sign at $x = 1$), the precise shape of the limit cycle must be determined numerically.

An example of a system with an analytically determinable limit cycle is

$$\dot{x} = x + y - x(x^2 + y^2) \tag{1.4.24a}$$

$$\dot{y} = -x + y - y(x^2 + y^2) \tag{1.4.24b}$$

for which linear stability analysis predicts an unstable spiral point, with clockwise rotation, about the origin. Equation (1.4.24) can be solved in the same way as (1.4.19), that is, in polar coordinates. Denoting $R(t) = r^2 = x^2 + y^2$, one finds the equation of motion R to be

$$\dot{R} + 2R^2 - 2R = 0 \tag{1.4.25}$$

This Ricatti-type equation can be exactly linearized by the substitution $\frac{1}{2}\dot{f}/f$ to the equation $\ddot{f} - 2\dot{f} = 0$, which is easily soluble. Hence

$$R(t) = x^2(t) + y^2(t) = \frac{c_1 e^{zt}}{c_1 e^{zt} + c_2} \tag{1.4.26}$$

(where c_1 and c_2 are arbitrary constants), which would suggest that the solutions to (1.4.24) tend to a limit cycle of radius 1 as $t \to \infty$. Linear stability analysis of systems higher than second order will be deferred to later sections. The situation can become very complicated, since the (local) flows along the n independent directions are no longer restricted to a plane as in the 2-D case and can become intertwined in most complicated ways.

1.5 TIME-DEPENDENT INTEGRALS*

Time-dependent integrals are rare and, accordingly, rather interesting entities. A simple example is furnished by a special case of the damped linear oscillator (1.1.6), which we again write as a first-order system (without loss of generality we set $\omega = 1$):

$$\dot{x} = y \tag{1.5.1a}$$

$$\dot{y} = -x - \lambda y \tag{1.5.1b}$$

First multiply both sides of (1.5.1a) by $2x$ and both sides of (1.5.1b) by $2y$ to obtain

$$2x\dot{x} = 2xy \tag{1.5.2a}$$

$$2y\dot{y} = -2\lambda y^2 - 2xy \tag{1.5.2b}$$

Now multiply both sides of (1.5.1a) by λy and both sides of (1.5.1b) by λx to obtain

$$\lambda \dot{x} y = \lambda y^2 \tag{1.5.3a}$$

$$\lambda x \dot{y} = -\lambda^2 xy - \lambda x^2 \tag{1.5.3b}$$

Adding together both sets of Eqs. (1.5.2) and (1.5.3), one obtains

$$2x\dot{x} + 2y\dot{y} + \lambda\dot{x}y + \lambda x\dot{y} = -\lambda y^2 - \lambda^2 xy - \lambda x^2$$

which is just

$$\frac{d}{dt}(x^2 + y^2 + \lambda xy) = -\lambda(y^2 + x^2 + \lambda xy)$$

which integrates exactly to

$$x^2 + y^2 + \lambda xy = c_1 e^{-\lambda t} \tag{1.5.4}$$

where c_1 is the constant of integration. Equation (1.5.4) can be used to express y as a function of x and t; but only for the special value of $\lambda = 2$, when the left-hand side is a perfect square, will the expression be of the simple form

$$y = \sqrt{c_1}\, e^{-t} - x \tag{1.5.5}$$

Substitution of this into (1.5.1a) leads to the exactly soluble first-order equation

$$\dot{x} + x = \sqrt{c_1}\, e^{-t} \tag{1.5.6}$$

which has the decaying solution

$$x(t) = (c_2 + t\sqrt{c_1})e^{-t} \tag{1.5.7}$$

where c_2 is the second constant of integration. This result is, of course, identical to the solution of (1.1.16) in the case of degenerate eigenvalues (which occurs when $\omega = 1$, $\lambda = 2$). For conservative systems, the presence of a time-independent first integral provides a definite geometric constraint on the phase flow. In this case, one might think of (1.5.5) as providing a constraint in the form of a straight line, given by $x + y = \text{constant}$, which shrinks exponentially toward the origin.

Another example is furnished by the system of equations

$$\dot{x} = -\tfrac{1}{2}x + y^2 \tag{1.5.8a}$$
$$\dot{y} = -\tfrac{1}{2}y - xy \tag{1.5.8b}$$

Multiplying (1.5.8a) by x and (1.5.8b) by y and adding gives

$$\frac{d}{dt}\left(\frac{1}{2}(x^2 + y^2)\right) = -\frac{1}{2}(x^2 + y^2) \tag{1.5.9}$$

which integrates to

$$x^2 + y^2 = ce^{-t} \qquad (1.5.10)$$

Substituting $y^2 = ce^{-t} - x^2$ into (1.5.8a) gives the first-order nonlinear equation

$$\dot{x} + \tfrac{1}{2} x + x^2 = ce^{-t} \qquad (1.5.11)$$

This is in the form of a Ricatti equation, which can be linearized exactly by the substitution $x = \dot{y}/y$ to yield the linear second-order equation

$$\ddot{y} + \tfrac{1}{2} \dot{y} - ce^{-t}y = 0 \qquad (1.5.12)$$

which, in principle, can be solved by standard power-series methods. For this example, one can think of the phase flow being constrained to the exponentially shrinking circle defined by (1.5.10).

1.6 NONAUTONOMOUS SYSTEMS

So far, all the systems considered have been autonomous; that is, time has only played the role of the independent variable and has not appeared explicitly in the equation of motion. A not uncommon situation is one in which a system is subjected to some form of external time-dependent force $F(t)$. Such situations could include repeated stress testing of a beam or an atom in a radiation field. Particularly interesting are those cases where $F(t)$ is a periodic function, for example, $F(t) \propto \cos(\Omega t)$. Consider as an example the damped, driven linear oscillator

$$\ddot{x} + \lambda\dot{x} + \omega^2 x = \epsilon F(t) \qquad (1.6.1)$$

where ϵ can be thought of as a "coupling parameter"—in the limit $\epsilon \to 0$, the system becomes autonomous again. In that limit, the phase space is two-dimensional, whereas for $\epsilon \neq 0$ it is important to realize that the phase space becomes *three*-dimensional with "time," thereby providing the extra dimension. To see this, consider the specific case of a periodic driving force, that is,

$$\ddot{x} + \lambda\dot{x} + \omega^2 x = \epsilon \cos(\Omega t) \qquad (1.6.2)$$

Denoting $z = \Omega t$ as a third dependent variable, (1.6.2) can be written as

$$\dot{x} = y \qquad (1.6.3a)$$

$$\dot{y} = -\lambda y - \omega^2 x + \epsilon \cos(z) \qquad (1.6.3b)$$

$$\dot{z} = \Omega \qquad (1.6.3c)$$

For $\epsilon = 0$, the solutions to (1.6.3b) (i.e., $x(t)$ and $y(t)$) are meaningfully plotted, as a continuous function of time, in the x, y phase plane. To continue this practice in the case $\epsilon \neq 0$ can be highly misleading, since now one can find phase curves intersecting each other in apparent contradiction of the uniqueness property of solutions to ordinary differential equations. Of course this is not the case, and what one is seeing is the effect of projecting the three-dimensional phase space (x, y, z) onto the two-dimensional (x, y)-plane.

The extra dimension can lead to completely new types of behavior, and for driven nonlinear systems the solutions can sometimes evolve in an apparently *chaotic*—but still completely deterministic—manner.† Little can be said about this behavior analytically, and the detailed behavior can often only be resolved on a computer. These matters will be discussed at length in later chapters. For the moment, we shall concentrate on those systems for which some analytical results can be obtained. This is mainly limited to driven linear systems; and although these cannot display (deterministic) chaos, the results obtained provide a useful background to the study of nonlinear systems.

Equations of the form (1.6.1) can be solved using the standard techniques for inhomogeneous linear equations. For the homogeneous limit of (1.6.1) (i.e., $\epsilon = 0$), we denote the two independent solutions as $x_1(t)$ and $x_2(t)$ and their (nonvanishing) Wronskian as

$$W(t) = x_1(t)x_2'(t) - x_2(t)x_1'(t)$$

The solution to (1.6.1) is written as the sum of the general homogeneous solution and a particular solution, that is, $x(t) = x_h + x_p$, where $x_h = c_1x_1 + c_2x_2$ and $x_p = ux_1 + vx_2$; here the function $u(t)$ and $v(t)$ are determined by

$$u = -\epsilon \int^t \frac{x_2(t')F(t')}{W(t')}\,dt', \quad v = \epsilon \int^t \frac{z_1(t')F(t')}{W(t')}\,dt'$$

1.6.a The Driven Oscillator

Firstly consider the undamped version of (1.6.2), that is,

$$\ddot{x} + \omega^2 x = \epsilon \cos(\Omega t) \tag{1.6.4}$$

---

†This is the first point in the main text that the word *chaos* appears. By a chaotic solution to a deterministic equation we mean a solution whose outcome is very *sensitive to initial conditions* (i.e., small changes in initial conditions lead to great differences in outcome) and whose evolution through phase space appears to be quite random.

The particular solution is easily determined to be $x_p(t) = [\epsilon/(\omega^2 - \Omega^2)]\cos(\Omega t)$ (working with the complex forms of homogeneous solution $x = e^{\pm i\omega t}$ makes the computations easier), and hence the overall solution is

$$x(t) = a\sin(\omega t + \delta) + \frac{\epsilon}{\omega^2 - \Omega^2}\cos(\Omega t) \qquad (1.6.5)$$

where the amplitude a and phase δ are given in Eq. (1.1.3). Thus the solution is a superposition of oscillations with the intrinsic frequency ω and external frequency Ω. Clearly the solution breaks down when the external and intrinsic frequencies match—this is an example of *resonance*. In order to see how the solution behaves in the limit $\Omega \to \omega$, we write the solution (1.6.5) in the form

$$x(t) = a'\sin(\omega t + \delta') + \frac{\epsilon}{\omega^2 - \Omega^2}(\cos(\Omega t) - \cos(\omega t)) \qquad (1.6.6)$$

which can always be done by suitable definition of the amplitude (a') and phase (δ'). The limit $\Omega \to \omega$ is found by L'Hôpital's rule to be

$$x(t) = a'\sin(\omega t + \delta') + \frac{\epsilon}{2\omega}t\sin(\omega) \qquad (1.6.7)$$

The second, secular, term implies unbounded linear growth of the amplitude at resonance and physically corresponds to the unrestrained absorption of energy by the system from the external field.

This behavior is in strong contrast to that of driven nonlinear oscillators. A famous example of such is the Duffing oscillator:

$$\ddot{x} + \omega^2 x + \beta x^3 = \epsilon\cos(\Omega t) \qquad (1.6.8)$$

The associated homogeneous problem is easily integrated in terms of Jacobi elliptic functions, but now the solution will have a period that is energy dependent. A resonance can still be induced by "hitting" the system with a driving frequency Ω that exactly matches the intrinsic frequency of the current motion. Now, however, as energy is absorbed and the amplitude of the oscillations increase, the intrinsic frequency also changes and takes the system out of resonance, thereby preventing secular growth. This is a case of nonlinearity stabilizing a resonance through what might be termed a "nonlinear feedback" mechanism. However, for a sufficiently strong driving force (i.e., large ϵ), there can be feedback between both the fundamental motion and its harmonics and the external field. In this regime the solution can start to exhibit chaotic behavior, oscillating in time in an apparently random manner. This will be discussed at length in Chapter 4.

1.6.b The Damped Driven Oscillator

Now let us return to the damped, linear driven problem (1.6.2). Assuming that the homogeneous problem has oscillatory (but, of course, damped) solutions, the overall solution to the inhomogeneous problem may be written as

$$x(t) = ae^{-(\lambda/2)t}\sin(\nu t + \delta) + \frac{\epsilon \cos(\Omega t + \delta')}{((\omega^2 - \Omega^2)^2 + \lambda^2\Omega^2)^{1/2}} \tag{1.6.9}$$

where the intrinsic frequency $\nu = \frac{1}{2}\sqrt{4\omega^2 - \lambda}$. The amplitude a and phase δ of the homogeneous part are determined by the initial conditions and the "external" phase $\delta' = \tan^{-1}(-\lambda\Omega/(\omega^2 - \Omega^2))$. Obviously as t increases, the homogeneous part of the solution decays and one is finally left with

$$\lim_{t \to \infty} x(t) = \frac{\epsilon \cos(\Omega t + \delta')}{((\omega^2 - \Omega^2)^2 + \lambda^2\Omega^2)^{1/2}} \tag{1.6.10}$$

and although the amplitude of the motion becomes large at resonance $\omega = \Omega$, it does not diverge. The asymptotic solution (1.6.10) corresponds to steady oscillations of constant energy, which represents a balance between the energy pumped into the system by the external field and the energy dissipated by the friction. This steady state corresponds to simple limit cycle behavior.

In the case of a *nonlinear* damped and driven system, that is, the Duffing oscillator

$$\ddot{x} + \lambda\dot{x} + \omega^2 x + \beta x^3 = \epsilon \cos(\Omega t) \tag{1.6.11}$$

much more complicated behavior can occur. For small driving forces, simple limit cycle behavior is still observed. However, as ϵ is increased, this cycle is found to bifurcate into a "double" limit cycle with twice the original period. As ϵ is further increased, further bifurcations keep on occurring (i.e., from two-cycle to four-cycle to eight-cycle, etc.) until the motion becomes quite chaotic. This highly complicated aperiodic limit cycle, which is reached at the end of the bifurcation sequence, is termed a *strange attractor*. These attractors are the mechanism by which dissipative systems (which one might imagine would always eventually decay to some boring limiting behavior) can display chaotic solutions. Much more of this will be discussed in Chapter 5.

1.7 FURTHER REMARKS ON INTEGRATION OF DIFFERENTIAL EQUATIONS

The notion of "integrating" a differential equation is clearly a delicate one. For most of the second-order examples considered so far, the existence of a constant first integral—the mechanical energy for the conservative cases—enabled us to reduce the order of the equation by one, thereby reducing the task of finding a solution to a "quadrature." Phrases such as "an exact solution" or "an analytical solution" to a differential equation have been used here (and elsewhere) rather glibly. It is certainly not a stupid question to ask what one means by such convenient phrases. After all, before the advent of elliptic functions, might not the simple nonlinear differential equations of the quartic oscillator or pendulum-type have appeared to be insoluble? Indeed one can ask what happens "beyond" the elliptic functions, that is, for differential equations like (1.2.4) with powers of x higher than x^3. Here at least there is still an integral of the motion (the energy) that enables one to get down to a quadrature—these integrals are termed *hyperelliptic integrals*. However, now the inversion becomes much more complicated, and many deep results from algebraic geometry come into play.

For second-order systems that are not conservative, the question of "integration" becomes more tricky. In some cases, as discussed in Section 1.5, we were able to find a time-dependent integral that allowed a reduction of order. Clearly though, these are rather special cases. Other awkward equations are systems such as

$$\ddot{y} = 6y^2 + t \tag{1.7.1}$$

$$\ddot{y} = 2y^3 + ty + \alpha \tag{1.7.2}$$

known as the first and second Painlevé transcendents, respectively. These do not have known first integrals and cannot be "integrated by quadratures." Clearly though, equations like these obviously have a "solution," but the extent to which this solution can be represented "analytically" as opposed to a piece of computer output is a nontrivial issue.

One factor that is now believed to be important in discussing this issue is that of analytic structure—namely, the behavior of the solution in the complex plane. Fortunately, a number of important properties, such as the type of singularity a solution can exhibit, can be determined directly from the equations of motion without recourse to an actual solution (which, of course, we may not "know"). It may be that if the analytic structure is sufficiently "simple," the associated equation may in some sense be "integrable." This will be discussed further in Chapter 8.

To some extent, questions of integrability for autonomous second-order systems may be rather academic. For these cases, the motion is always confined to the two-dimensional phase plane and can never exhibit particularly complicated or chaotic motion. As soon as the phase space acquires a third dimension (or greater), as might arise for a nonautonomous second-order system or a third order autonomous system, the question of integrability becomes much more critical, since now chaotic behavior is possible. Now the existence of "integrals of the motion" becomes very important since, as we have seen, they provide a geometric constraint to the phase flow. Consider the following autonomous third-order system

$$\dot{x} = y \tag{1.7.3a}$$

$$\dot{y} = -xz + x \tag{1.7.3b}$$

$$\dot{z} = xy \tag{1.7.3c}$$

We first observe from (1.7.3a and 1.7.3c) that $x\dot{x} = \dot{z}$. This identifies the (constant) integral of motion

$$\tfrac{1}{2}x^2 - z = I_1 \tag{1.7.4}$$

so now the orbits in the 3-D phase space (x, y, z) are seen to be restricted to a surface defined by Eq. (1.7.4). This integral can now be used to reduce the third-order system (1.7.3) to a second-order system

$$\ddot{x} = x(1 - I_1) - \tfrac{1}{2}x^3 \tag{1.7.5}$$

This second-order system can, in turn, be reduced to a first-order system by using the second integral

$$\tfrac{1}{2}(\dot{x}^2 - (1 - I_1)x^2 + \tfrac{1}{4}x^4) = I_2 \tag{1.7.6}$$

which can again be thought of as providing another geometrical constraint on the phase flow. Equation (1.7.6) can be integrated in terms of Jacobi elliptic functions. Thus the identification of the two integrals I_1 and I_2 has enabled one to reduce the solution of the original third-order system down to a single "quadrature"—in this sense, one can think of having "completely integrated" the system.

Now consider the system

$$\dot{x} = y + \beta \tag{1.7.7a}$$

$$\dot{y} = -xz + x \tag{1.7.7b}$$

$$\dot{z} = xy + \alpha x \tag{1.7.7c}$$

where α and β are constants. Here one finds that unless $\alpha = \beta$, integrals like I_1 (and then I_2) cannot apparently be found—so now the system would appear not to be "integratable."† So without the assistance of numerical

---

†The word "integratable" is abominable English but may serve a useful purpose. For Hamiltonian systems (Chapter 2), the term "integrable" has a very special geometrical implication that is distinct from non-Hamiltonian systems (such as (1.7.3), which can nonetheless be "integrated").

studies, it is not clear where in the 3-D phase space the solution might wander.

The simple (and carefully designed!) examples (1.7.3) and (1.7.7) raise many fundamental questions. Firstly, how does one find integrals of the motion (if they exist)? As the order of the equations gets higher and their functional form becomes more complicated, this is a very difficult task. In fact, there is no systematic procedure for doing this—one has to rely on experience, luck, and, in desperation, prayer! It may be, as some brilliant work of the Russian mathematician Sofya Kovalevskaya a hundred years ago implied, that the existence of integrals may be tied up with the analytic structure of the differential equations. Another fundamental question concerns the actual *number* of integrals one needs to find in order to effect a complete "integration." This is where it is important to distinguish between different classes of dynamical system. For those that are Hamiltonian (which (1.7.3) and (1.7.7) are not), some very strong results are available. As will be discussed in the next chapter, if the system has as many integrals (N) as "degrees of freedom" (i.e., half the dimension of the phase space), then the system can be "integrated by quadratures". Briefly, this stems from very special geometrical properties of the Hamiltonian—N integrals are enough to restrict the flow in a particular way which can then be parameterized in terms of a special set of coordinates that are trivially integrated. For non-Hamiltonian systems, the situation is less happy. For an nth-order system, $\dot{x} = f_i(x_1, \ldots, x_n)$, $i = 1, \ldots, n$, one needs $n - 1$ integrals in order to effect a complete integration.

APPENDIX 1.1 ELLIPTIC FUNCTIONS

To reduce Eq. (1.2.7), namely,

$$\left(\frac{dx}{dt}\right)^2 = e(x - \alpha)(x - \beta)(x - \gamma)(x - \delta) \equiv eG^2(x) \tag{1.A.1}$$

to the standard (Legendre) form

$$\left(\frac{dy}{dt}\right)^2 = (1 - y^2)(1 - k^2 y^2) \equiv H^2(y) \tag{1.A.2}$$

we follow Davis (1962). From (1.A.1) we can write

$$\frac{1}{e^{1/2} G(x)} \frac{dx}{dt} = 1 \tag{1.A.3}$$

Introducing the variables

$$y^2 = \frac{(\beta - \delta)(x - \alpha)}{(\alpha - \delta)(x - \beta)}, \quad k^2 = \frac{(\beta - \gamma)(\alpha - \delta)}{(\alpha - \gamma)(\beta - \delta)}, \quad M^2 = \frac{(\beta - \delta)(\alpha - \gamma)}{4}$$

$$\tag{1.A.4}$$

we obtain

$$\frac{M}{G(x)} \frac{dx}{dt} = \frac{1}{H(y)} \frac{dy}{dt} \qquad (1.A.5)$$

and hence

$$\frac{1}{e^{1/2} G(x)} \frac{dx}{dt} = \frac{1}{H(y)} \frac{dy}{d(e^{1/2} Mt)} = 1 \qquad (1.A.6)$$

Thus

$$y = \mathrm{sn}(e^{1/2} Mt, k) \qquad (1.A.7)$$

and hence the original variable x can be solved for by using the first equation in (1.A.4).

Elliptic integrals of the *first kind* are the integrals

$$F(x, k) = \int_0^x \frac{dx'}{\sqrt{(1 - x'^2)(1 - k^2 x'^2)}}, \qquad k^2 < 1 \qquad (1.A.8)$$

or its equivalent

$$F(\theta, k) = \int_0^\theta \frac{d\theta'}{\sqrt{1 - k^2 \sin^2 \theta'}} \qquad (1.A.9)$$

where $x = \sin \theta$. Elliptic integrals of the *second kind* are

$$E(x, k) = \int_0^x \sqrt{\frac{1 - k^2 x'^2}{1 - x'^2}} \, dx', \qquad k^2 < 1 \qquad (1.A.10)$$

or its equivalent

$$E(\theta, k) = \int_0^\theta \sqrt{1 - k^2 \sin^2 \theta'} \, d\theta' \qquad (1.A.11)$$

where again $x = \sin \theta$. Elliptic integrals of the *third kind* are

$$\Pi(x, n, k) = \int_0^x \frac{dx'}{(1 + nx'^2)\sqrt{(1 - x'^2)(1 - k^2 x'^2)}} \qquad (1.A.12)$$

or its equivalent

$$\Pi(\theta, n, k) = \int_0^\theta \frac{d\theta'}{(1 + n \sin^2 \theta')\sqrt{1 - k^2 \sin^2 \theta'}} \qquad (1.A.13)$$

under the same change of variables as before.

There are an enormous number of relationships between elliptic functions starting with the basic elliptic integral

$$u = \int_0^\theta \frac{d\theta'}{\sqrt{1 - k^2 \sin^2 \theta'}} \qquad (1.A.14)$$

The following functions are defined:

$$\text{sn}(u, k) = \sin \theta \qquad (1.A.15a)$$

$$\text{cn}(u, k) = \cos \theta \qquad (1.A.15b)$$

$$\text{dn}(u, k) = \sqrt{1 - k^2 \sin^2 \theta} \qquad (1.A.15c)$$

$$\text{am}(u, k) = \theta \qquad (1.A.15d)$$

$$\text{tn}(u, k) = \frac{\text{sn}(u, k)}{\text{cn}(u, k)} = \tan \theta \qquad (1.A.15e)$$

Some of the special values and relationships are

$$\text{sn}(0) = 1, \quad \text{cn}(0) = 1, \quad \text{dn}(0) = 1, \quad \text{am}(0) = 0 \qquad (1.A.16)$$

$$\text{sn}^2(u) + \text{cn}^2(u) = 1 \qquad (1.A.17)$$

$$\text{sn}(-u) = -\text{sn}(u), \quad \text{cn}(-u) = \text{cn}(u) \qquad (1.A.18)$$

$$\text{sn}(u, 0) = \sin(u), \quad \text{cn}(u, 0) = \cos(u) \qquad (1.A.19)$$

$$\text{sn}(u, 1) = \tanh(u) = \frac{e^u - e^{-u}}{e^u + e^{-u}} \qquad (1.A.20)$$

$$\text{cn}(u, 1) = \text{dn}(u, 1) = \text{sech}(u) = \frac{1}{e^u + e^{-u}} \qquad (1.A.21)$$

SOURCES AND REFERENCES

Arnold, V. I., *Ordinary Differential Equations*, MIT Press, Cambridge, Massachusetts, 1978.

Bender, C. M. and S. A. Orszag, *Advanced Mathematical Methods for Scientists and Engineers*, McGraw-Hill, New York, 1978.

Davis, H. T., *Introduction to Nonlinear Differential and Integral Equations*, Dover, New York, 1962.

Ince, E. L., *Ordinary Differential Equations*, Dover, New York, 1956.

2

HAMILTONIAN DYNAMICS

2.1 LAGRANGIAN FORMULATION OF MECHANICS

In this section we discuss some of the basic principles of the Lagrangian
formulation of classical mechanics. This provides the necessary background
to learn about the Hamiltonian formulation, which, in turn, provides the
natural framework in which to investigate the ideas of integrability and
nonintegrability in a wide class of mechanical systems. Many of the
differential equations discussed in Chapter 1 describe the motion of a
particle moving in some force field (represented as the gradient of a
potential energy function) and, as such, they are examples of *Newtonian
equations* of motion.† Since Newton's pioneering work, the "laws" of
mechanics have been the subject of ever more general and elegant formulations. General equations of motion can be seductively derived by
invoking such fundamental principles as the homogeneity of space and time
and the use of an almost magical variational principle ("Hamilton's
principle") to the extent that the resulting "laws" would appear to have
been determined from purely deductive ("absolute") principles. Nonetheless, it should always be remembered that all these results are at some point
based on experimental facts and human experience—they have just stood
the test of time remarkably well. (For example, there was no reason or
evidence to doubt that Newton's laws could describe microscopic systems
until spectroscopic evidence came along to indicate the need for quantum
mechanics.)

†The reader is assumed to be familiar with the basics of Newtonian mechanics and such
concepts as constraints and generalized coordinates (see, for example, Chapter I of Goldstein
(1980)).

2.1.a The Lagrangian Function and Hamilton's Principle

If one considers a mechanical system consisting of a collection of particles—interacting amongst each other according to well-defined force laws—then experience has shown that the "state of the system" is completely described by the set of all the positions and velocities of the particles. The coordinate frame need not be cartesian, as was the case in Newton's work, and the description can be effected by means of some set of "generalized coordinates" q_i ($i = 1, \ldots, n$) and "generalized velocities" $\dot{q}_i$ ($i = 1, \ldots, n$). (The use of generalized coordinates relieves a system from the explicit presence of holonomic constraints.)

If a system moves from a position at some time t_1, labelled by the coordinate set $\mathbf{q}^{(1)} = q_1(t_1), \ldots, q_n(t_1)$, to a position $\mathbf{q}^{(2)} = q_1(t_2), \ldots, q_n(t_2)$ at another time t_2, then the actual motion can be determined from Hamilton's principle of least action. This requires that the integral of the so-called *Lagrangian* function takes the minimum possible value between the initial and final times. For the moment, we treat the Lagrangian as a "black box", merely stating that it can only be some function of those variables on which the state of a system can depend (i.e., the generalized coordinates, velocities, and time), namely,

$$L = L(q_1, \ldots, q_n, \dot{q}_1, \ldots, \dot{q}_n, t) \tag{2.1.1}$$

The famous "principle of least action," or "Hamilton's principle," requires that the *action integral*

$$W = \int_{t_1}^{t_2} L(\mathbf{q}, \dot{\mathbf{q}}, t)\, dt \tag{2.1.2}$$

be a minimum. For the moment, we drop the subscript on the q_i's and $\dot{q}_i$'s and assume a single degree of freedom. The positions $q^{(1)}$ and $q^{(2)}$ at the initial and final times t_1 and t_2 are assumed fixed. (Allowing the end points to also vary with time has other important consequences.) There can be many different paths $q(t)$ connecting $q^{(1)}$ and $q^{(2)}$, and the aim is to find those that *extremize* (which usually means minimize but can, in fact, also result in a maximum) the action (2.1.2). This is done by looking at the effect of a "first variation," that is, adding small excursions along the path which vanish at either end (i.e., $\delta q(t_1) = \delta q(t_2) = 0$). A remarkable feature of this procedure is that we are considering the effect of these variations about a path which we do not yet know. The first variation of the action W is then determined by

$$\delta W = \int_{t_1}^{t_2} L(q + \delta q, \dot{q} + \delta \dot{q}, t)\, dt - \int_{t_1}^{t_2} L(q, \dot{q}, t)\, dt \tag{2.1.3}$$

By expanding the first integrand to first order, one obtains the variation

$$\delta W = \int_{t_1}^{t_2} \left(\frac{\partial L}{\partial q} \delta q + \frac{\partial L}{\partial \dot{q}} \delta \dot{q} \right) dt \qquad (2.1.4)$$

Using $\delta \dot{q} = d\delta q/dt$ and integrating the second term by parts yields

$$\delta W = \left| \frac{\partial L}{\partial \dot{q}} \delta q \right|_{t_1}^{t_2} + \int_{t_1}^{t_2} \left(\frac{\partial L}{\partial q} - \frac{d}{dt} \left(\frac{\partial L}{\partial \dot{q}} \right) \right) \delta q \, dt \qquad (2.1.5)$$

By the end-point condition, the first term on the right-hand side vanishes. If the variation is to be an extremum, then $\delta W = 0$; this can only occur if the integrand vanishes, that is,

$$\frac{\partial L}{\partial q} - \frac{d}{dt} \left(\frac{\partial L}{\partial \dot{q}} \right) = 0 \qquad (2.1.6)$$

For n degrees of freedom $q_1, \ldots, q_n$, the variation must be effected for each variable independently, that is, $q_i(t) + \delta q_i(t)$. The net result is the set of equations

$$\frac{d}{dt} \left(\frac{\partial L}{\partial \dot{q}_i} \right) - \frac{\partial L}{\partial q_i} = 0 \qquad (2.1.7)$$

which are the celebrated *Lagrange's equations*. If the (correct!) form of the Lagrangian is known for the given mechanical system, then the set of second-order equations (2.1.7) are the equations of motion for the system and, given the initial data $q_i(0)$, $\dot{q}_i(0)$ $(i = 1, \ldots, n)$ will determine the entire history of the system.†

In determining the correct form for the Lagrangian function it is interesting to see how far one can go in making this choice by invoking only the most basic principles. In their brilliant text on mechanics, Landau and Lifshitz (1960) persuasively argue that, for a free particle at least, the principles of homogeneity of time and isotropy of space‡ determine that the Lagrangian can only be proportional to the square of the (generalized) velocities. If the constant of proportionality is taken to be half the particle mass, then the Lagrangian for a system of noninteracting particles is just their total kinetic energy in rectilinear coordinates, that is,

---

†So powerful did this deterministic framework appear that Laplace was led to say "We ought then to regard the present state of the universe as the effect of its preceeding state and as the cause of its succeeding state."

‡These two properties ensure that the motion can be considered in the context of an "inertial frame," that is, independent of its "absolute" position in space and time.

$$L = T = \sum_{i=1}^{n} \frac{1}{2} m_i \dot{q}_i^2$$

Beyond this, "experimental" facts have to be invoked in that if the particles interact amongst each other according to some force law contained in a "potential energy function" $V(q_1, \ldots, q_n)$, then, to quote Landau and Lifshitz (1960), "experience has shown that" the correct form of the Lagrangian is

$$L = T - V = \sum_{i=1}^{n} \tfrac{1}{2} m_i \dot{q}_i^2 - V(q_1, \ldots, q_n) \qquad (2.1.8)$$

The potential energy function is such that the force acting on each particle is determined by

$$F_i = -\frac{\partial}{\partial q_i} V(q_1, \ldots, q_n) \qquad (2.1.9)$$

(This provides a definition of the potential energy, since it ensures that the net work done by a system in traversing a closed path in the configuration (i.e., coordinate) space is zero.) For velocity-independent potentials, Lagrange's equations (Eqs. (2.1.7)) become

$$m\ddot{q}_i = -\frac{\partial V}{\partial q_i} \qquad (2.1.10)$$

which, in the case of cartesian coordinates, are just Newton's equations.

2.1.b Properties of the Lagrangian

Given the Lagrangian for a system of (interacting) particles, a number of interesting properties of that system can be deduced. First we consider the total time derivative of the Lagrangian, that is,

$$\frac{d}{dt} L(q, \dot{q}, t) = \sum_i \frac{\partial L}{\partial q_i} \dot{q}_i + \sum_i \frac{\partial L}{\partial \dot{q}_i} \ddot{q}_i + \frac{\partial L}{\partial t}$$

$$= \sum_i \frac{d}{dt}\left(\dot{q}_i \frac{\partial L}{\partial \dot{q}_i} \right) + \frac{\partial L}{\partial t}$$

Thus

$$\frac{d}{dt}\left(\sum_i \dot{q}_i \frac{\partial L}{\partial \dot{q}_i} - L \right) = -\frac{\partial L}{\partial t} \qquad (2.1.11)$$

For a closed system (i.e., one which does not interact with any external forces), the homogeneity of time ensures that L does not depend explicitly on time, and thus the "energy"

$$E = \sum_i \dot{q}_i \frac{\partial L}{\partial \dot{q}_i} - L \qquad (2.1.12)$$

is a constant of the motion; that is, the system considered is conservative (see Chapter 1). Furthermore, using the definition of L in (2.1.8), it is easily deduced that

$$E = T + V \qquad (2.1.13)$$

Using L, one may define the generalized forces

$$F_i = \frac{\partial L}{\partial q_i} \qquad (2.1.14)$$

and, most importantly, the *generalized momenta*

$$p_i = \frac{\partial L}{\partial \dot{q}_i} \qquad (2.1.15)$$

where only for cartesian coordinates do we obtain $p_i = m\dot{q}_i$. With these two definitions, Lagrange's equations can then be cast in the form

$$\dot{p}_i = F_i, \qquad i = 1, \ldots, n \qquad (2.1.16)$$

Clearly, if any one of the generalized coordinates, say q_k, is missing from the Lagrangian, we obtain the associated generalized force $F_k = 0$ and hence (from (2.1.6)) the corresponding generalized momentum $p_k = $ constant. Missing coordinates are sometimes referred to as *cyclic*—clearly, cyclic coordinates simplify the integration of the equations of motion. With the above definition of the p_i, the energy of the system (2.1.12) can then be written as

$$E = \sum_{i=1}^{n} p_i \dot{q}_i - L \qquad (2.1.17)$$

If a system is closed and space is homogeneous, then the net effect of all the particle forces must be zero, that is, $\sum_{i=1}^{n} F_i = 0$ (in Newton's third law this is stated, for two bodies, as "action and reaction are equal and opposite"). In this case we may deduce from (2.1.16) that $\sum_{i=1}^{n} p_i = $ constant; that is, the overall translation of a system of particles is constant.

Thus we see that by invoking such fundamental principles as homogeneity of time and space (for closed systems), one may deduce basic conservation laws such as the conservation of energy and total momentum. These results are examples of a deep and general result known as *Noether's theorem*, which states that for every group of transformations that leaves the Lagrangian function invariant, there is an associated conserved quantity; for example, invariance under translation in time and space leads to conservation of energy and (linear) momentum, respectively. Another simple example is a system invariant under rotation—here angular momentum will be conserved.

2.1.c Properties of the Generalized Momenta

In the Lagrangian picture of mechanics, the generalized coordinates and velocities are considered as independent variables. Now we have also defined another set of variables, the generalized momenta p_i (Eq. (2.1.15)). The simple connection between the generalized momenta and velocities in the cartesian coordinate system is deceptive in that the p_i are truly a completely independent (from the q_i and $\dot{q}_i$) set of variables. The differences are deep and are most clearly seen when viewed from a more geometric point of view.† So deep and elegant is this structure that Arnold (1978) has described Hamiltonian mechanics (i.e., the description of mechanics in terms of the p_i and q_i) as "geometry in phase space."

In a more traditional framework, an important property of the p_i that distinguishes them from the q_i is that they are expressible as the gradient of a scalar field—this is a simple way of demonstrating their covariant properties. To see this, we return to the action integral (2.1.2). For a given extremal path, that is, a path satisfying Eqs. (2.1.7), this is just a definite integral, that is, the value of the action along the path connecting $q(t_1)$ and $q(t_2)$. A variational principle can still be applied to (2.1.2)—but this time to see how the action varies between neighboring extremal paths with the same initial point but different final points. So now one looks at the variation of W for $q = q(t) + \delta q$, where $q(t)$ is an actual extremal path with $q^{(1)} = q(t_1)$ fixed and $q^{(2)} = q(t_2) + \delta q(t_2)$ as the varying end point. The

---

†The velocities are examples of tangent vectors and transform as *contravariant* variables. The combined set of variables q_i, $\dot{q}_i$ $(i = 1, \ldots, n)$ forms a $2n$-dimensional manifold known as a *tangent bundle* (TM), and the Lagrangian then becomes the mapping of the tangent-bundle space to a scalar field, namely, $L: TM \rightarrow R$ (i.e., it converts the set q_i, $\dot{q}_i$ to a real number). Momenta have completely different geometric properties. They transform as *covariant* variables, and the *phase space* made up of the set q_i, p_i $(i = 1, \ldots, n)$ is a $2n$-dimensional *symplectic* manifold with very different geometric properties from the tangent bundle space of Lagrangian mechanics. For a full account, the reader is referred to V. I. Arnold's (1978) wonderful book *Mathematical Methods of Classical Mechanics*. An introductory account of these concepts is given in Appendix 2.2.

variation leads to the same result as (2.1.5), that is,

$$\delta W = \left| \frac{\partial L}{\partial \dot{q}} \, \delta q \right|_{t_1}^{t_2} + \int_{t_1}^{t_2} \left(\frac{\partial L}{\partial q} - \frac{d}{dt} \left(\frac{\partial L}{\partial \dot{q}} \right) \right) \delta q \, dt \qquad (2.1.18)$$

but since the path is assumed to be extremal, the integral vanishes and one is left with just the end-point contributions. Since the initial point is fixed, we have $\delta q(t_1) = 0$; and denoting $\delta q(t_2)$ as just δq and using $p = \partial L / \partial \dot{q}$, one has $\delta W = p \, \delta q$ or

$$\delta W = \sum_{i=1}^{n} p_i \, \delta q_i \qquad (2.1.19)$$

for a system of n degrees of freedom. From this result follows the property that the p_i are *gradients of the action*, that is,

$$p_i = \frac{\partial W}{\partial q_i} \qquad (2.1.20)$$

at a given time along a given extremal path. By contrast, note that it is not possible to express $\dot{q}$ as the gradient of a scalar field.

2.2 HAMILTONIAN FORMULATION OF MECHANICS

Hamiltonian mechanics is the description of a mechanical system in terms of generalized coordinates q_i and generalized momenta p_i. Although the Hamiltonian formulation of classical mechanics contains the same physical information as the Lagrangian picture, it is far better suited for the formulation of quantum mechanics, statistical mechanics, and perturbation theory. In particular, the use of Hamiltonian phase space provides the ideal framework for a discussion of the concepts of integrability and nonintegrability and the description of the chaotic phenomena that can be exhibited by nonintegrable systems.

2.2.a Transformation to the Hamiltonian Picture

To effect the transformation from the Lagrangian description involving the q_i and the $\dot{q}_i$ to the Hamiltonian description, one uses the standard technique of Legendre's transformation (see Appendix 2.1). The Legendre transform of $L = L(\mathbf{q}, \dot{\mathbf{q}}, t)$ with respect to $\dot{\mathbf{q}}$ to a new function in which $\dot{\mathbf{q}}$ is expressed in terms of $\mathbf{p}$ is

$$H(\mathbf{p}, \mathbf{q}, t) = \sum_{i=1}^{n} p_i \dot{q}_i - L(\mathbf{q}, \dot{\mathbf{q}}, t) \qquad (2.2.1)$$

where $\mathbf{p}$, $\mathbf{q}$, and $\dot{\mathbf{q}}$ are the n-component vectors $\mathbf{p} = (p_1, \ldots, p_n)$, $\mathbf{q} = (q_1, \ldots, q_n)$, $\dot{\mathbf{q}} = (\dot{q}_1, \ldots, \dot{q}_n)$ and the new function H is the *Hamiltonian* of the system. The crucial relationship is that which defines p_i in terms of the q_i and $\dot{q}_i$, namely,

$$p_i(\mathbf{q}, \dot{\mathbf{q}}, t) = \frac{\partial L}{\partial \dot{q}_i} (\mathbf{q}, \dot{\mathbf{q}}, t) \qquad (2.2.2)$$

which, providing

$$\det \left| \frac{\partial^2 L}{\partial \dot{q}_i \, \partial \dot{q}_j} \right| \neq 0 \qquad (2.2.3)$$

can be inverted to express the $\dot{q}_i$ in terms of the p_i with (at this stage) q_i and t treated as parameters. As a simple example of this transformation, consider the Lagrangian

$$L = \sum_{i=1}^{n} \frac{1}{2} m_i \dot{q}_i^2 - V(q_1, \ldots, q_n) \qquad (2.2.4)$$

From this we determine

$$p_i = \frac{\partial L}{\partial \dot{q}_i} = m\dot{q}_1 \qquad (2.2.5)$$

and since condition (2.2.3) is satisfied, the inverse is (trivially in this case)

$$\dot{q}_i = \frac{p_i}{m_i} \qquad (2.2.6)$$

and thus the Hamiltonian is

$$
\begin{aligned}
H(\mathbf{p}, \mathbf{q}) &= \sum_{i=1}^{n} p_i \left(\frac{p_i}{m_i} \right) - \left\{ \sum_{i=1}^{n} \frac{1}{2} m_i \left(\frac{p_i}{m_i} \right)^2 - V(q_1, \ldots, q_n) \right\} \\
&= \sum_{i=1}^{n} \frac{1}{2 m_i} p_i^2 + V(q_1, \ldots, q_n) \qquad (2.2.7)
\end{aligned}
$$

Although the Lagrangian (2.2.4) describes an important class of mechanical system, the simplicity of the relationship between the p_i and $\dot{q}_i$ is such that some of the subtlety is lost. A standard example of a less trivial transformation is provided by the case of a particle moving under gravity but constrained to a smooth wire frame of specified shape.†

†This is an example of an holonomic constraint.

Consider a bead of mass m sliding smoothly on a wire of shape $z = f(x)$ in the vertical (z, x)-plane. To begin with, think of unconstrained motion in this plane. This will require, in the Lagrangian description, two co-ordinates and velocities (i.e., x, z and $\dot{x}$, $\dot{z}$); thus the kinetic energy is $T = \frac{1}{2} m(\dot{x}^2 + \dot{z}^2)$. Now introduce the relationship between x and z, that is, $z = f(x)$, from which one obtains $\dot{z} = \dot{x}(df/dx) \equiv \dot{x} f'(x)$. Thus one finds

$$T = \tfrac{1}{2}m(\dot{x}^2 + \dot{z}^2) = \tfrac{1}{2}m\dot{x}^2(1 + (f'(x))^2) \qquad (2.2.8)$$

The potential energy is just that due to gravity (i.e., $V = mgz = mgf(x)$), and so the Lagrangian is

$$L = \tfrac{1}{2}m\dot{x}^2[1 + (f'(x))^2] - mgf(x) \qquad (2.2.9)$$

From Eq. (2.1.5) the generalized momentum is

$$p = \frac{\partial L}{\partial \dot{x}} = m\dot{x}(1 + (f'(x))^2) \qquad (2.2.10)$$

and hence

$$\dot{x} = \frac{p}{m(1 + (f'(x))^2)} \qquad (2.2.11)$$

Thus from (2.2.1) the Hamiltonian is

$$H(p, x) = p\frac{p}{m(1 + (f'(x))^2)} - \left\{ \left(\frac{m}{2}\right) \frac{p^2}{m^2(1 + (f'(x))^2)^2} (1 + (f'(x))^2) - mgf(x) \right\}$$

$$= \frac{p^2}{2m(1 + (f'(x))^2)} + mgf(x) \qquad (2.2.12)$$

2.2.b Hamilton's Equations

Given the Lagrangian for a system, Lagrange's equations of motion were derived from Hamilton's principle. Clearly, we now want to derive equations of motion for the Hamiltonian formulations of the problem. These can also be derived on the basis of a variational principle (see Section 2.3.c for a discussion of this) but are more directly seen as follows. The differential of H, defined by (2.2.1), is

$$dH = \sum_i p_i\, d\dot{q}_i + \dot{q}_i\, dp_i - \frac{\partial L}{\partial \dot{q}_i}\, d\dot{q}_i - \frac{\partial L}{\partial q_i}\, dq_i - \frac{\partial L}{\partial t}\, dt \qquad (2.2.13)$$

The first and third terms on the right-hand side cancel by the definition $p_i = \partial L/\partial \dot{q}_1$; and using the relation $\dot{p}_i = \partial L/\partial q_i$ (Eq. (2.1.6)), one obtains

$$dH = \sum \dot{q}_1 \, dp_i - \dot{p}_i \, dq_i - \frac{\partial L}{\partial t} \, dt \qquad (2.2.14)$$

Thus one obtains the famous "canonical" or *Hamiltonian equations* of motion, namely,

$$\dot{q}_i = \frac{\partial H}{\partial p_i}, \qquad \dot{p}_i = -\frac{\partial H}{\partial q_i} \qquad (2.2.15)$$

plus the additional relationship† (for explicitly time-dependent systems)

$$\frac{\partial H}{\partial t} = -\frac{\partial L}{\partial t} \qquad (2.2.16)$$

The system of equations (2.2.15) form a set of $2n$ first-order equations in contrast to the set of n second-order equations obtained in the Lagrangian description. Although in Chapter 1 we saw that a second-order equation, say of the form $\ddot{x} = f(x)$, could be written as a pair of first-order equations by introducing a new variable $y = \dot{x}$, it should be clear that they are not necessarily in Hamiltonian form. Consider the sliding-bead problem. Hamilton's equations are

$$\dot{x} = \frac{\partial H}{\partial p} = \frac{p}{m(1 + (f'(x))^2)} \qquad (2.2.17a)$$

$$\dot{p} = -\frac{\partial H}{\partial x} = \frac{p^2 f'(x) f''(x)}{m(1 + (f'(x))^2)^2} - mgf'(x) \qquad (2.2.17b)$$

On the other hand, Lagrange's equation is

$$\ddot{x} = -\frac{f'(x)(\dot{x}^2 f''(x) + g)}{1 + (f'(x))^2} \qquad (2.2.18)$$

Introducing the variable $y = \dot{x}$, one obtains the pair

$$\dot{x} = y \qquad (2.2.19a)$$

$$\dot{y} = -\frac{f'(y^2 f'' + g)}{1 + (f')^2} \qquad (2.2.19b)$$

which are clearly quite different from the pair (2.2.17a) and (2.2.17b).

---

†In addition, one may easily demonstrate that $dH/dt = \partial H/\partial t$. Comparison with the canonical equations (2.2.15) suggests that, formally, one could also consider $-H$ and t to be a pair of canonical variables. This concept is especially valuable for time-dependent Hamiltonians where one often considers the $(2n + 2)$-dimensional "extended" phase space of $p_1, \ldots, p_n, q_1, \ldots, q_n; -H, t$. (See the later discussion of the Poincaré–Cartan invariant.)

Hamilton's equations (2.2.15) have a number of important properties; for the moment, we discuss them in the context of a time-independent Hamiltonian. First of all, the set of $2n$ variables $q_1, \ldots, q_n, p_1, \ldots, p_n$—often called the "canonical" or "canonically conjugate" variables (e.g., "p_i is the momentum conjugate to q_i")—define a $2n$-dimensional *phase space* (cf. the discussion of phase space in Chapter 1). The solution to Hamilton's equations

$$q_i(t) = q_i(\mathbf{q}_0, \mathbf{p}_0, t), \qquad p_i(t) = p_i(\mathbf{q}_0, \mathbf{p}_0, t) \qquad (2.2.20)$$

where $\mathbf{q}_0 = (q_1(0), \ldots, q_n(0))$, $\mathbf{p}_0 = (p_1(0), \ldots, p_n(0))$ are the set of initial conditions, define the mechanical state of the system at time t. As time evolves, $\mathbf{q}(t)$, $\mathbf{p}(t)$ map out a phase-space trajectory which explores certain regions of the phase space. Precisely what regions these are is *the* fundamental issue—which we shall soon discuss.

It is easy to see that Eqs. (2.2.15) satisfy the "incompressibility" condition

$$\sum_{i=1}^{n} \left(\frac{\partial \dot{q}_i}{\partial q_i} + \frac{\partial \dot{p}_i}{\partial p_i} \right) = 0 \qquad (2.2.21)$$

Imagine a blob of phase-space "fluid"—Eq. (2.2.21) is just a statement that this blob has zero divergence. Thus a volume element in phase space is preserved under the Hamiltonian flow—this is *Liouville's theorem* and is one of the most fundamental properties of Hamiltonian systems.† In the sliding-bead problem, for example, it is easily seen from Eqs. (2.2.17) that the phase-space flow is indeed divergenceless. Notice on the other hand that the pair (2.2.19) deduced from the Lagrangian does not preserve volume (area, to be more precise, in this case) in the (x, y) "phase space."

So symmetric are Hamilton's equation in p and q that it seems natural to consider the variables p_i and q_i on very much of an equal footing. Often it is convenient to introduce a single "set" of $2n$ coordinates z_i, where $\mathbf{z} = (q_i, \ldots, q_n, p_i, \ldots, p_n)$. Thus for a given Hamiltonian $H = H(\mathbf{q}, \mathbf{p}) = H(\mathbf{z})$, Hamilton's equations can be written in the concise form

$$\dot{\mathbf{z}} = \mathbf{J} \cdot \nabla H(\mathbf{z}) \qquad (2.2.22)$$

where $\nabla = (\partial z_1, \ldots, \partial z_{2n})$ and the $2n \times 2n$ matrix $\mathbf{J}$ is termed the *symplectic* matrix

$$\mathbf{J} = \begin{pmatrix} 0 & \mathbb{1} \\ -\mathbb{1} & 0 \end{pmatrix} \qquad (2.2.23)$$

where $\mathbb{1}$ is the $n \times n$ unit matrix.

---

†A more geometric account of Liouville's theorem is given in Appendix 2.2.

2.2.c Poisson Brackets*

Of particular importance is our ability to integrate Hamilton's equations. For systems with just one degree of freedom (i.e., just a single pair of canonical variables (p, q)), we can integrate the pair of first-order equations in the ways discussed in Chapter 1. However, whether it be one or many degrees of freedom, the crucial step is to identify the integrals of motion. In the Hamiltonian picture, the time dependence of dynamical quantities can be formulated very elegantly. Consider some function $f = f(p, q, t)$; then

$$
\begin{aligned}
\frac{df}{dt} &= \sum_{i=1}^{n} \left(\frac{dq_i}{dt} \frac{\partial f}{\partial q_i} + \frac{dp_i}{dt} \frac{\partial f}{\partial p_i} \right) + \frac{\partial f}{\partial t} \\
&= \sum_{i=1}^{n} \left(\frac{\partial H}{\partial p_i} \frac{\partial f}{\partial q_i} - \frac{\partial H}{\partial q_i} \frac{\partial f}{\partial p_i} \right) + \frac{\partial f}{\partial t} \\
&= [H, f] + \frac{\partial f}{\partial t}
\end{aligned}
\tag{2.2.24}
$$

where $[H, f]$ is the *Poisson bracket* of f with H. There is a close analogy between the Poisson brackets of classical mechanics and the commutator brackets of quantum mechanics. In fact, one can write the Poisson bracket for any pair of dynamical quantities, for example,

$$
[g, f] = \sum_{i=1}^{n} \left(\frac{\partial g}{\partial p_i} \frac{\partial f}{\partial q_i} - \frac{\partial g}{\partial q_i} \frac{\partial f}{\partial p_i} \right)
\tag{2.2.25}
$$

If a quantity is explicitly time independent (i.e., $f = f(p, q)$ and its Poisson bracket with H vanishes), then it is clear from (2.2.24) that f is a constant of motion. Obviously, since the Poisson bracket of H with itself is zero, the energy of a time-independent system (i.e., $H = E$) is a constant of motion.

Following from its definition (Eq. (2.2.25)), the Poisson bracket may be shown to have a variety of properties. For the three given functions f, g, h, one finds

$$
[f, g] = -[g, f]
\tag{2.2.26a}
$$

$$
[f + g, h] = [f, h] + [g, h]
\tag{2.2.26b}
$$

$$
[fg, h] = f[g, h] + g[f, h]
\tag{2.2.26c}
$$

$$
[f, [g, h]] + [g, [h, f]] + [h, [f, g]] = 0
\tag{2.2.26d}
$$

the last of which, with its characteristic cyclic structure, is known as *Jacobi's identity*. The set of properties (2.2.26) shows that the Poisson brackets satisfy what is known as a *Lie algebra*. There is nothing to stop one from choosing the various functions f, g, h to be just individual canonical variables, in which case one obtains relations of the form

$$[q_i, q_j] = 0, \qquad [p_i, p_j] = 0, \qquad [p_i, q_j] = \delta_{ij} \qquad (2.2.27)$$

which are closely analogous to those obtained in quantum mechanics (e.g., the third relation becomes $[\hat{p}_i, \hat{q}_j] = -i\hbar\delta_{ij}$). If f and g are both constants of motion (i.e., $[H, f] = [H, g] = 0$), then it follows by Poisson's theorem that the bracket between f and g is also a constant of motion, that is, $[f, g] =$ constant. This is easily seen from the Jacobi identity (2.2.26d), that is, $[f, [g, H]] + [g, [H, f]] + [H, [f, g]] = 0$. Since the first two brackets vanish (because f and g are constants), we are immediately left with the desired result $[H, [f, g]] = 0$, which indicates that $[f, g]$ is also a constant. However, Poisson's theorem may not always be very useful in practice (i.e., for constructing new integrals of motion), since the bracket $[f, g]$ may just be a simple constant (e.g., zero) or just a function of the original integrals f and g.

In Chapter 1 we saw that, in general, a system of n first-order equations requires $n - 1$ integrals (these include both the nontrivial "integrals of motion" and the trivial "constants of integration") in order to effect a complete "integration." Does this mean, then, that for the system of $2n$ equations of a Hamiltonian system we require $2n - 1$ integrals of motion to solve the problem? Fortunately, as already mentioned, it turns out that, owing to the special symplectic structure of Hamilton's equations, one only requires n integrals of motion. However, to see how this miracle occurs, it is useful to first of all learn about what are termed *canonical transformations*. These are the transformations of variables for which the Hamiltonian structure of the system is still preserved.

2.3 CANONICAL TRANSFORMATIONS

In the Lagrangian description of a system (i.e., the description in terms of generalized coordinates and velocities q_i, $\dot{q}_i$), it is sometimes convenient to transform to some new set of generalized coordinates, that is,

$$Q_i = Q_i(q_1, \ldots, q_n) \qquad (2.3.1)$$

to simplify the integration of the equations of motion (e.g., a transformation from cartesian to polar coordinates). In the Hamiltonian description, there are now two sets of independent variables, the p_i and q_i $(i = 1, \ldots, n)$, which, as we have discussed, are very much on an equal footing. Thus we now have to consider the possibility of transformations from one set of phase-space variables (p_i, q_i) to some new set (P_i, Q_i), that is,

$$P_i = P_i(q_1, \ldots, q_n, p_1, \ldots, p_n) \qquad (2.3.2)$$

$$Q_i = Q_i(q_1, \ldots, q_n, p_1, \ldots, p_n)$$

Notice that the new P_i and Q_i can be, in general, functions of both the old p_i and q_i. Those cases in which the transformations just involve making the new P_i and Q_i functions of only the old p_i and q_i, respectively (i.e., as in Eq. (2.3.1)), are referred to as *point transformations*. Transformations of the form (2.3.2) are referred to as *canonical transformations* if the symplectic structure of the system is still preserved. Loosely speaking (a more precise, geometric definition will be given later), this means that the canonical form of Hamilton's equations are still preserved, that is,

$$\dot{Q}_i = \frac{\partial}{\partial P_i} H'(\mathbf{Q}, \mathbf{P}), \qquad \dot{P}_i = -\frac{\partial}{\partial Q_i} H'(\mathbf{Q}, \mathbf{P}) \qquad (2.3.3)$$

where $H' = H'(\mathbf{Q}(\mathbf{q}, \mathbf{p}), \mathbf{P}(\mathbf{q}, \mathbf{p}))$ is the transformed Hamiltonian. (The transformation of $H(\mathbf{p}, \mathbf{q})$ to $H'(\mathbf{P}, \mathbf{Q})$ is not always just a simple substitution of variables—see later.)

2.3.a The Preservation of Phase Volume

A fundamental property of canonical transformations is that phase volume is preserved.† If $\prod_{i=1}^{n} dp_i\, dq_i$ represents a volume element in the "old" phase space and $\prod_{i=1}^{n} dP_i\, dQ_i$ represents a volume element in the "new" phase space, then we require that

$$\int \prod_{i=1}^{n} dp_i\, dq_i = \int \prod_{i=1}^{n} dP_i\, dQ_i \qquad (2.3.4)$$

where the integral sign represents a $2n$-dimensional integration over a

---

†In fact, the preservation of phase volume is just one of a hierarchy of quantities, preserved under canonical transformation, known as the Poincaré invariants. The first of these is the invariant

$$\int\int \sum_{i=1}^{n} dp_i dq_i = \int\int \sum_{i=1}^{n} dP_i dQ_i$$

which represents the sum of areas (of a phase-space element) projected onto the set of (p_i, q_i) planes. In geometric language this is expressed in terms of the "*differential 2-form*," that is,

$$\sum_{i=1}^{n} dp_i \wedge dq_i = \sum_{i=1}^{n} dP_i \wedge dQ_i$$

where $\wedge$ denotes the so called wedge product. This result provides a rigorous, geometric definition of canonical transformations. All other invariances, including (2.3.4), follow from this (see Appendix 2.2).

prescribed volume in phase space. The two integrals are related by the Jacobian of transformation, that is,

$$\int \prod_{i=1}^{n} dP_i \, dQ_i = \int \frac{\partial(P_1, \ldots, P_n, Q_1, \ldots, Q_n)}{\partial(p_1, \ldots, p_n, q_1, \ldots, q_n)} \prod_{i=1}^{n} dp_i \, dq_i \qquad (2.3.5)$$

Thus a volume preserving transformation must have unit Jacobian, namely,

$$\frac{\partial(P_1, \ldots, P_n, Q_1, \ldots, Q_n)}{\partial(p_1, \ldots, p_n, q_1, \ldots, q_n)} = \frac{\partial(p_1, \ldots, p_n, q_1, \ldots, q_n)}{\partial(P_1, \ldots, P_n, Q_1, \ldots, Q_n)} = 1 \qquad (2.3.6)$$

Consider the very simple example

$$Q = -p, \qquad P = q \qquad (2.3.7)$$

Then

$$\frac{\partial(P, Q)}{\partial(p, q)} = \begin{vmatrix} \dfrac{\partial P}{\partial p} & \dfrac{\partial Q}{\partial p} \\ \dfrac{\partial P}{\partial q} & \dfrac{\partial Q}{\partial q} \end{vmatrix} = \begin{vmatrix} 0 & -1 \\ 1 & 0 \end{vmatrix} = 1 \qquad (2.3.8)$$

which therefore shows that (2.3.7) is a volume-preserving (canonical) transformation. The transformation (2.3.7) demonstrates on just how equal a footing the p and q are; that is, they can be interchanged—but with a sign change. Notice that if we did not make that sign change (e.g., $Q = p$, $P = q$), the Jacobian would be -1. In fact, this need for the sign change should not be surprising since it is required to preserve the form of Hamilton's equations (2.3.3) under interchange of P and Q.

An example of a noncanonical transformation is that from polar to cartesian coordinates, that is,

$$q = P \cos Q, \qquad p = P \sin Q \qquad (2.3.9)$$

since

$$\frac{\partial(q, p)}{\partial(Q, P)} = \begin{vmatrix} -P \sin Q & P \cos Q \\ \cos Q & \sin Q \end{vmatrix} = -P \qquad (2.3.10)$$

which indicates that phase volume is not preserved.

Liouville's theorem is the statement that phase volume is preserved under the Hamiltonian flow—we saw this in Section 2.2.b as an almost obvious "incompressibility condition" that follows from the form of Hamilton's equations. In fact, we can couch Liouville's theorem in the language of canonical transformations as follows. Consider some phase-space trajectory along which some initial q_0, p_0 at time t_0 evolve to some q_1, p_1 at a (short) time t later, that is,

$$q_1 = q(t + \delta t) = q_0 + \delta t \left.\frac{dq}{dt}\right|_{t=t_0} + O(\delta t^2)$$

$$= q_0 + \delta t \frac{\partial}{\partial p_0} H(q_0, p_0, t) + O(\delta t^2)$$

$$p_1 = p(t + \delta t) = p_0 + \delta t \left.\frac{dp}{dt}\right|_{t=t_0} + O(\delta t^2)$$

$$= p_0 - \delta t \frac{\partial}{\partial q_0} H(q_0 \, p_0, t) + O(\delta t^2)$$

If the transformations from q_0, p_0 to q_1, p_1 is, in fact, a canonical transformation from one set of variables to the other, then the Jacobian $\partial(q_1, p_1)/\partial(q_0, p_0)$ must be unity. We find

$$\frac{\partial(q_1, p_1)}{\partial(q_0, p_0)} = \begin{vmatrix} \dfrac{\partial q_1}{\partial q_0} & \dfrac{\partial p_1}{\partial q_0} \\ \dfrac{\partial q_1}{\partial p_0} & \dfrac{\partial p_1}{\partial p_0} \end{vmatrix} = \begin{vmatrix} 1 + \delta t \dfrac{\partial^2 H}{\partial q_0 \, \partial p_0} & -\delta t \dfrac{\partial^2 H}{\partial q_0^2} \\ \delta t \dfrac{\partial^2 H}{\partial p_0^2} & 1 - \delta t \dfrac{\partial^2 H}{\partial q_0 \, \partial p_0} \end{vmatrix}$$

$$= 1 + O(\delta t^2)$$

$$= 1 \quad \text{in} \quad \lim \delta t \to 0$$

Note that the vanishing term is $O(\delta t^2)$ rather than $O(\delta t)$. Because this change is proportional to $O(\delta t^2)$, it follows that over any finite period of time (i.e., any multiple of δt), the total change of area goes as $O(\delta t)$—which vanishes in the limit $\delta t \to 0$. Thus the "infinitesimal transformation" generated by the Hamiltonian itself is a canonical transformation. The phase volume in the variables q_0, p_0 is preserved under transformation (due to the Hamiltonian flow) to the "new" variables q_1, p_1—which is, of course, just a statement of Liouville's Theorem.

2.3.b The Optimal Transformation

The practical use of canonical transformations (although they certainly have an elegant structure in their own right) is to find those transformations that make the integration of Hamilton's equations as simple as possible. The optimal case is the one in which all the Q_i are cyclic; that is, the transformed Hamiltonian depends only on the new momenta P_i:

$$H(p_1, \ldots, p_n, q_1, \ldots, q_n) \longrightarrow H'(P_1, \ldots, P_n) \qquad (2.3.11)$$

Hamilton's equation then becomes very simple since

$$\dot{P}_i = -\frac{\partial H'}{\partial Q_i} = 0, \qquad \text{i.e., } P_i = \text{const.}, \quad i = 1, \ldots, n \qquad (2.3.12a)$$

$$\dot{Q}_i = \frac{\partial H'}{\partial P_i} = f_i(P_i, \ldots, P_n) \qquad (2.3.12b)$$

where the f_i are some time-independent function of the P_i. The equation for Q_i can then be immediately integrated, that is,

$$Q_i = f_i t + \delta_i, \qquad i = 1, \ldots, n \qquad (2.3.12c)$$

where $\delta_i = Q_i(0)$ are a set of arbitrary constants determined by the initial conditions. Clearly, the new set of "momenta" P_i are constants of the motion. Thus if we can find them we are able to effect a complete integration of the equations of motion. The P_i and δ_i constitute a set of $2n$ integrals. The n P_i are the set of nontrivial constants of motion (or first integrals) that enable one to "perform" the integration, and the n δ_i are the set of trivial constants of integration that enable one to "complete" the integration. (If need be, these solutions can then be transformed back, in principle at least, to the original representation in terms of "old" p_i's and q_i's.) Of course we have to be able to do two things: (1) find these magical new variables and (2) know how to correctly transform the Hamiltonian into its new representation.

2.3.c Generating Functions

Canonical transformations are effected by means of so-called *generating functions*.† One way to introduce them is through a variational principle. Although formally elegant, it is perhaps easier, at least for time-independent problems, to proceed via a simpler route that only involves the principle of phase-volume preservation. (Here we follow the presentation of

†It is important to emphasize that generating functions are more than formalism—as sometimes appears on a first reading. They are extremely useful. They enable one to find, directly, both the "new" canonical P and Q and their relationship to the "old" p and q. This is, in effect, "two for the price of one." If one did not use a generating function but just started off with some $Q = Q(q, p)$, one would probably have to work quite hard to find the corresponding canonical $P = P(q, p)$.

Percival and Richards (1982)). For the moment, consider just one degree of freedom and the canonical sets of variables (p, q) and (P, Q). Then by phase-volume preservation, we obtain

$$\iint_R dp\, dq = \iint_R dP\, dQ \tag{2.3.13}$$

over some closed region R. By Stokes' theorem, the double integrals may be replaced by a line integral about the closed path $\mathscr{C}$ enclosing R, that is,

$$\oint_{\mathscr{C}} p\, dq = \oint_{\mathscr{C}} P\, dQ \tag{2.3.14}$$

Now we are assuming that the P, Q are some function of the p, q (i.e., $P = P(p, q)$, $Q = Q(p, q)$), but there is nothing to stop us from writing this as a mixed dependence between variables (i.e., $P = P(Q, q)$, $p = p(Q, q)$, where now Q and q are considered as the independent variables). Thus (2.3.14) can be written as

$$\oint_{\mathscr{C}} [p(Q, q)\, dq - P(Q, q)\, dQ] = 0 \tag{2.3.15}$$

This implies that the integral must be the differential of some function—call it $F_1(Q, q)$. One can now write

$$\oint_{\mathscr{C}} [p\, dq - P\, dQ] = \oint_{\mathscr{C}} dF_1(Q, q) = \oint_{\mathscr{C}} \left(\frac{\partial F_1}{\partial Q}\, dQ + \frac{\partial F_1}{\partial q}\, dq \right) \tag{2.3.16}$$

and equating coefficients of dq and dQ, one observes that

$$p = \frac{\partial}{\partial q} F_1(Q, q) \tag{2.3.17a}$$

$$P = -\frac{\partial}{\partial Q} F_1(Q, q) \tag{2.3.17b}$$

Equation (2.3.17a) provides a relationship between p and (q, Q), which then has to be inverted (which will be possible if $\partial^2 F_1/\partial q\, \partial Q \neq 0$) to give the dependence $Q = Q(q, p)$. This is then substituted in (2.3.17b) to obtain the other desired relation $P = P(q, p)$.

In the above example, (q, Q) were chosen as the independent variables—obviously, other combinations (e.g., (P, q), (Q, p), (P, p)) that mix the two sets together are also possible. Consider the first of these, that is, the (P, q) combinations. To proceed, we note the identity

$$\oint d(pQ) = \oint p \, dQ + \oint Q \, dp \qquad (2.3.18)$$

substituting this in (2.3.15) and introducing a new differential $dF_2(P, q)$, we obtain, analogous to (2.3.16),

$$\oint_{\mathscr{C}} [p \, dq + Q \, dp] = \oint_{\mathscr{C}} dF_2(P, q) = \oint_{\mathscr{C}} \left[\frac{\partial F_2}{\partial P} \, dP + \frac{\partial F_2}{\partial q} \, dq \right] \quad (2.3.19)$$

from which follows the pair of relationships

$$p = \frac{\partial}{\partial q} F_2(P, q) \qquad (2.3.20a)$$

$$Q = \frac{\partial}{\partial P} F_2(P, q) \qquad (2.3.20b)$$

A very simple example of the F_2 generating functions is $F_2(P, q) = Pq$, which, from (2.3.20), is just seen to be the identity transformations:

$$p = \frac{\partial F_2}{\partial q} = P, \qquad Q = \frac{\partial F_2}{\partial P} = q \qquad (2.3.21)$$

From the above discussions it should be clear that there are two other generating functions $F_3(Q, p)$ and $F_4(P, p)$. For our purposes, F_2 is the most important. Discussion of the general properties and interrelationships between the F_i $(i = 1, \ldots, 4)$ can be found in any standard mechanics text.

If the canonical transformations is time independent, the transformation of the "old" Hamiltonian $H(p, q)$ to the "new" Hamiltonian $H'(P, Q)$ is a straightforward change of variables, that is,

$$H' = H'(P, Q) = H(p(P, Q), q(P, Q)) \qquad (2.3.22)$$

It is not difficult to verify, using the chain rule and the Jacobian property $\partial(P, Q)/\partial(p, q) = 1$, that

$$\dot{Q} = \frac{\partial H'}{\partial P} = \frac{\partial H}{\partial P} \qquad (2.3.23)$$

$$\dot{P} = -\frac{\partial H'}{\partial Q} = -\frac{\partial H}{\partial Q} \qquad (2.3.24)$$

Explicitly, the first of these follows from

$$\frac{\partial H'}{\partial P} = \dot{Q} = \frac{\partial Q}{\partial q}\dot{q} + \frac{\partial Q}{\partial p}\dot{p} = \frac{\partial Q}{\partial q}\frac{\partial H}{\partial p} - \frac{\partial Q}{\partial p}\frac{\partial H}{\partial q}$$

$$= \frac{\partial Q}{\partial q}\left(\frac{\partial H}{\partial Q}\frac{\partial Q}{\partial p} + \frac{\partial H}{\partial P}\frac{\partial P}{\partial p}\right) - \frac{\partial Q}{\partial p}\left(\frac{\partial H}{\partial Q}\frac{\partial Q}{\partial q} + \frac{\partial H}{\partial P}\frac{\partial P}{\partial q}\right)$$

$$= \frac{\partial H}{\partial P}\left(\frac{\partial Q}{\partial q}\frac{\partial P}{\partial p} - \frac{\partial Q}{\partial p}\frac{\partial P}{\partial q}\right)$$

$$= \frac{\partial H}{\partial P}$$

where we have used the chain rule in the second line and the Jacobian property in the third. Equation (2.3.24) is derived in an analgous fashion.

The transformation of the Hamiltonian in the case of an explicitly time-dependent transformation (e.g., $F_2 = F_2(P, q, t)$) is less straightforward than for the time-independent case. Although the required transformation can be derived using the same types of argument given above, a more elegant route is via a variational principle. The idea is to (formally) consider Hamilton's principle in phase space, that is, to write the action integral in terms of the Hamiltonian and require (cf. (2.1.2))

$$\delta \int_{t_1}^{t_2} (p\dot{q} - H(p, q, t))\, dt = 0 \qquad (2.3.25)$$

(There are some subtle considerations here. For example, the variation of (2.1.2) only requires that the end-point contributions in configuration space (i.e., $\delta q(t_1)$ and $\delta q(t_2)$), must vanish. For (2.3.25), one also has to consider what to do with the variations of p at the end points: Interestingly enough (2.3.25) can be used to derive Hamilton's equations without having to specify these conditions. These delicate issues are further discussed by Goldstein (1980).) For the present purposes, we use $\dot{q}\, dt = dq$ and write (2.3.25) as

$$\delta \int_{t_1}^{t_2} (p\, dq - H(p, q, t))\, dt = 0 \qquad (2.3.26)$$

The same principle must hold for any other canonical pair of variables (P, Q), so we can also write†

$$\delta \int_{t_1}^{t_2} (P\, dQ - H'(P, Q, t))\, dt = 0 \qquad (2.3.27)$$

†The quantity $p\, dq - H\, dt$ is known as the *Poincaré–Cartan invariant*. In geometric language, it is an example of a "differential 1-form." Such a 1-form is invariant under canonical transformations.

These two integrals can differ, at most, by the total differential of some function of the canonical variables and t (provided that end-point contributions in the (p, q) and (P, Q) spaces, respectively, vanish). Calling this total derivative dF, we can therefore write

$$p\, dq - H\, dt = P\, dQ - H'\, dt + dF \tag{2.3.28}$$

From this we immediately see that

$$H'(P, Q, t) = H(p, q, t) + \frac{\partial F}{\partial t} \tag{2.3.29}$$

If we identify F as the generating function, this gives us the desired rule for transforming time-dependent Hamiltonians. To complete the story, consider the case of F being a function of q and Q, that is, a type-1 generating function $F = F_1(Q, q, t)$. In this case, (2.3.28) immediately gives, in addition to (2.3.29),

$$p = \frac{\partial}{\partial q} F_1(Q, q, t) \tag{2.3.30a}$$

and

$$P = -\frac{\partial}{\partial Q} F_1(Q, q, t) \tag{2.3.30b}$$

If, on the other hand, F is of type 2 (i.e., a function of P, q, and t), then we rewrite (2.3.28) as

$$d(F + PQ) = p\, dq + Q\, dP + (H' - H)\, dt \tag{2.3.31}$$

from which we obtain (setting $F + PQ \equiv F_2(P, q, t)$, where Q is written as a function of P, q, and t)

$$p = \frac{\partial}{\partial q} F_2(P, q, t) \tag{2.3.32a}$$

and

$$Q = \frac{\partial}{\partial P} F_2(P, q, t) \tag{2.3.32b}$$

as well as (2.3.29). Similar arguments can be carried through for all four types of generating function, and we have the general result

$$H'(P, Q, t) = H(p, q, t) + \frac{\partial F_i}{\partial t}, \qquad i = 1, 2, 3, 4 \tag{2.3.33}$$

In the next section we will show how all this canonical formalism can be used to explicitly integrate an important class of Hamiltonian systems.

2.4 HAMILTON–JACOBI EQUATION AND ACTION-ANGLE VARIABLES

Our goal is to find the canonical transformation to a set of constant conjugate momenta (as in Eq. (2.3.11)). To achieve this, we actually have to *find* a suitable generating function. This will be of the F_2 type, that is, a function of the "old" coordinates $q_1, \ldots, q_n$ and "new" momenta, which, for conventions sake, we denote as $\alpha_1, \ldots, \alpha_n$. We denote the generating function as $S = S(q_1, \ldots, q_n, \alpha_1, \ldots, \alpha_n)$. From Eqs. (2.3.25) we have the relations

$$p_i = \frac{\partial}{\partial q_i} S(q_1, \ldots, q_n, \alpha_1, \ldots, \alpha_n) \qquad (2.4.1a)$$

$$\beta_i = \frac{\partial}{\partial \alpha_i} S(q_1, \ldots, q_n, \alpha_1, \ldots, \alpha_n) \qquad (2.4.1b)$$

where the β_i are the "new" coordinates conjugate to the α_i. From (2.3.22) the relationship between the "new" Hamiltonian $H'(\alpha_1, \ldots, \alpha_n)$ and the "old" $H(q_1, \ldots, q_n, p_1, \ldots, p_n)$ is then

$$H\left(q_1, \ldots, q_n, \frac{\partial S}{\partial q_1}, \ldots, \frac{\partial S}{\partial q_n}\right) = H'(\alpha_1, \ldots, \alpha_n) \qquad (2.4.2)$$

Here the right-hand side of (2.4.2) is to be viewed as a *constant* quantity, that is, the value of the Hamiltonian. Equation (2.4.2) is a first-order partial differential equation for S in n independent variables $(q_i, i = 1, \ldots, n)$ known as the *time-independent Hamilton–Jacobi equation*.† Equations of this type have in their "complete solution" n independent constants of integration—these can be taken as the set of α_i, $i = 1, \ldots, n$. Solving the Hamilton–Jacobi equation is equivalent to solving, in effect, the canonical equations of motion. Except for a class of mechanical systems known as

---

†For time-dependent Hamiltonians the *time-dependent Hamilton–Jacobi equation* takes the form

$$\frac{\partial S}{\partial t} + H\left(q_1, \ldots, q_n, \frac{\partial S}{\partial q_1}, \ldots, \frac{\partial S}{\partial q_n}, t\right) = 0$$

where now the generating function S depends on t. For conservative systems the time dependence separates (i.e., $S = S(q_1, \ldots, q_n, \alpha_1, \ldots, \alpha_n) - Et$) and the time-independent equation (2.4.2) is reclaimed with $E = H'(\alpha_1, \ldots, \alpha_n)$.

separable—to be discussed later—this is, in general, a very difficult task (as indeed it should be in conformity with the principle of "conservation of difficulty"—an everyday experience for physicists!). We can, at least, see the form of the solution S from relations (2.4.1). For a fixed set of α_i, one has

$$dS = \sum_{i=1}^{n} \frac{\partial S}{\partial q_i} dq_i = \sum_{i=1}^{n} p_i \, dq_i \qquad (2.4.3)$$

Thus S is the line integral

$$S = \int_{q_0}^{q} \sum_{i=1}^{n} p_i \, dq_i \qquad (2.4.4)$$

where $q_0 = q_1(0), \ldots, q_n(0)$ is the initial point on the classical trajectory, with the given set of α_i, and $q = q_1(t), \ldots, q_n(t)$ is a moving (i.e., time-dependent) point on that path. Obviously, to know the explicit form of S we need to know the actual classical path from q_0 to q, that is, the solution to the problem.

2.4.a Hamilton–Jacobi Equation for One Degree of Freedom

For systems of one degree of freedom, the Hamilton–Jacobi equation is fairly easy to solve. The old Hamiltonian is just a function of one pair of canonical variables (i.e., $H = H(p, q)$), and the "new" Hamiltonian is just a function of one (constant) canonical momentum (i.e., $H' = H'(\alpha)$). For time-independent problems, the trick is to set the constant of integration α equal to the Hamiltonian itself, that is, $H' = \alpha$ (α is, of course, just the energy of the system). The Hamilton–Jacobi equation is now

$$H\left(q, \frac{\partial}{\partial q} S(q, \alpha)\right) = \alpha \qquad (2.4.5)$$

with the generating function relations

$$p = \frac{\partial}{\partial q} S(q, \alpha) \qquad (2.4.6a)$$

$$\beta = \frac{\partial}{\partial \alpha} S(q, \alpha) \qquad (2.4.6b)$$

Since the transformation is canonical, the equations of motion for the transformed Hamiltonian are

$$\dot{\alpha} = -\frac{\partial H'}{\partial \beta} = 0$$

$$\dot{\beta} = \frac{\partial H'}{\partial \alpha} = 1$$

which are trivially integrated to give $\alpha = $ constant $= H'$ (by definition) and

$$\beta = t - t_0 \qquad (2.4.7)$$

Note that for our one-degree-of-freedom system, we now have two integrals, namely, the nontrivial constant of motion α and the trivial constant of integration t_0. Combining (2.4.7) with (2.4.6b) and (2.4.4) gives

$$t - t_0 = \int_{q_0}^{q} \frac{\partial}{\partial \alpha} p(q, \alpha) \, dq \qquad (2.4.8)$$

For simple particle-in-potential-field motion, the Hamiltonian takes the form (cf. Eq. (2.2.7))

$$H(p, q) = \frac{p^2}{2m} + V(q) \qquad (2.4.9)$$

in which case, since $H(p, q) = H'(\alpha) = \alpha$, we obtain

$$p(q, \alpha) = \pm \sqrt{2m(\alpha - V(q))} \qquad (2.4.10)$$

and thus (2.4.8) becomes the quadrature

$$t - t_0 = \sqrt{\frac{m}{2}} \int_{q_0}^{q} \frac{dq}{\sqrt{\alpha - V(q)}} \qquad (2.4.11)$$

Since for conservative systems the constant of integration α is nothing more than the mechanical energy $E = H$, we have ended up with the quadrature (1.1.9) of Chapter 1. It now might seem as though we have expended a vast amount of energy on fancy formalism in order to obtain a result that we already knew on page 2! However, as we shall now see, the above exercise is useful in that it provides the background for finding a particularly useful type of canonical variables—the *action-angle variables*—which play an important role in describing the properties of systems of many degrees of freedom.

2.4.b Action-Angle Variables for One Degree of Freedom

From the investigations in Chapter 1 for bounded Hamiltonian systems, we saw that the phase-space trajectories were confined to *closed* invariant curves in the phase plane. The motion is thus, of course, periodic, returning to the same point in (p, q)-space after a characteristic period $2\pi/\omega$, where ω is the frequency of motion to complete a full cycle. The idea of action-angle variables is to find the pair of conjugate variables such that the conjugate "coordinate" increases by 2π after each complete period of motion. Denoting these variables as I, θ, where I is the constant conjugate momentum, the pair of generating function relations are

$$p = \frac{\partial}{\partial q} S(q, I) \tag{2.4.12a}$$

$$\theta = \frac{\partial}{\partial I} S(q, I) \tag{2.4.12b}$$

and the Hamilton–Jacobi equation is (cf. (2.4.5))

$$H\left(q, \frac{\partial S}{\partial q}\right) = \alpha = H'(I) \tag{2.4.13}$$

For a path with a fixed value of α (and hence fixed value of I, since $\alpha = H'(I)$), one obtains from (2.4.12b)

$$\frac{d\theta}{dq} = \frac{\partial}{\partial I}\left(\frac{\partial S}{\partial q}\right) \tag{2.4.14}$$

Our requirement is that on one cycle around the invariant curve (denoted by $\mathscr{C}$), with that fixed value of α, the change in θ is 2π, that is,

$$2\pi = \oint_{\mathscr{C}} d\theta = \frac{\partial}{\partial I} \oint_{\mathscr{C}} \frac{\partial S}{\partial q} dq = \frac{\partial}{\partial I} \oint_{\mathscr{C}} p \, dq \tag{2.4.15}$$

This condition can be satisfied if

$$I = \frac{1}{2\pi} \oint_{\mathscr{C}} p \, dq \tag{2.4.16}$$

which is taken as the definition of the action variable. A little care is required in the treatment of the constant of integration α and the constant conjugate momentum I. The Hamilton–Jacobi equation (2.4.13) is still solved for S as a function of q and α, that is,

$$S = \int_{q_0}^{q} p(q, \alpha) \, dq \tag{2.4.17}$$

and then the action variable is computed from (2.4.16) as

$$I = \frac{1}{2\pi} \oint_{\mathscr{C}} p(q, \alpha)\, dq \qquad (2.4.18)$$

around the invariant curve with the fixed value of $\alpha = H'(I) = E$. (By integrating around the closed loop $\mathscr{C}$ the multivaluedness of p in (2.4.10) is resolved in that both branches are involved in the integration.) The canonical equations for the transformed Hamiltonian are

$$\dot{I} = -\frac{\partial}{\partial \theta} H'(I) = 0 \qquad (2.4.19a)$$

$$\dot{\theta} = \frac{\partial}{\partial I} H'(I) = \omega(I) \qquad (2.4.19b)$$

which are integrated immediately to give

$$I = \text{const.} \qquad (2.4.20a)$$

$$\theta = \omega(I)t + \delta \qquad (2.4.20b)$$

where $\omega(I)$ is the characteristic frequency of motion (cf. Eq. (1.1.12)) and $\delta = \theta(0)$. Having obtained the action variable from (2.4.18) and hence the precise connection between it and α, one can then write S in (2.4.17) in terms of its dependence on q and I. Now the generating relation (2.4.12) can be used to obtain the explicit dependence of p and q in terms of I and θ, that is, $p = p(I, \theta)$, $q = (I, \theta)$.

A straightforward example is provided by the simple harmonic oscillator with Hamiltonian

$$H = \tfrac{1}{2}(p^2 + \omega^2 q^2) \qquad (2.4.21)$$

The Hamiltonian–Jacobi equation is

$$\frac{1}{2}\left(\frac{\partial S}{\partial q}\right)^2 + \frac{1}{2}\omega^2 q^2 = \alpha \qquad (2.4.22)$$

where α is the constant of integration which we take to be the total mechanical energy $E = H$. The action variable is thus

$$I = \frac{1}{2\pi} \oint_{\mathscr{C}} \sqrt{2(E - \tfrac{1}{2}\omega^2 q^2)}\, dq \qquad (2.4.23)$$

where the closed path $\mathscr{C}$ is the round trip between the turning points

$\pm\sqrt{2E}/\omega$. The integral (2.4.23) is easily evaluated to give

$$I = \frac{E}{\omega} \tag{2.4.24}$$

This now gives the relation between the constant of integration $\alpha = E$ and the new constant "momentum" I, that is, $\alpha = E = H'(I) = I\omega$. The generating function can now be written as

$$S(q, I) = \int_{q_0}^{q} \sqrt{2(I\omega - \tfrac{1}{2}\omega^2 q^2)} \, dq \tag{2.4.25}$$

From the generating function relation (2.4.12b), one can explicitly obtain $q = q(I, \theta)$, which is easily found to be

$$q = \sqrt{\frac{2I}{\omega}} \sin(\theta + \delta) \tag{2.4.26}$$

where δ is some phase shift ($\delta = \sin^{-1}(q_0\omega/\sqrt{2E})$). Since Hamilton's equations are so easily integrated (i.e., $\theta = \omega t + \delta$), the solution (2.4.26) is immediately identified with that obtained in Eq. (1.1.11).

A less trivial example is provided by the system (1.1.13), which in Hamiltonian form is expressed as

$$H = \tfrac{1}{2}p^2 + \tfrac{1}{4}\beta q^4 \tag{2.4.27}$$

The action is found to be

$$I = \frac{2^{1/2}}{3\,\pi^{3/2}\beta^{1/4}} \Gamma^2(\tfrac{1}{4}) E^{3/4} \tag{2.4.28}$$

and hence the "new" Hamiltonian as a function of action is expressed as

$$H(I) = I^{4/3}\left(\frac{3^4\,\pi^6\beta}{2^2\Gamma^8(\tfrac{1}{4})}\right)^{1/3} \tag{2.4.29}$$

This is now a nonlinear problem because from Hamilton's equation we see that frequency is action dependent, that is, $\omega(I) = \text{const.} \times I^{1/3}$.

2.5 INTEGRABLE HAMILTONIANS

For systems of one degree of freedom, a solution always seems possible—whether it is by the methods discussed in Chapter 1 or by the

Hamilton–Jacobi equation technique just described. However, we must now face up to the problem of systems with many degrees of freedom.

2.5.a Separable Systems

The Hamilton–Jacobi equation (2.4.2) for n degrees of freedom cannot, in general, be solved unless it is *separable*, that is, the generating function can be written as a sum of n-terms—each depending on only one coordinate, namely,

$$S(q_1, \ldots, q_n, \alpha_1, \ldots, \alpha_n) = \sum_{k=1}^{n} S_k(q_k, \alpha_1, \ldots, \alpha_n) \qquad (2.5.1)$$

A rather simple class of systems for which this separation is obviously valid is one in which the Hamiltonian itself is the sum of n-independent parts of the form

$$H(p_1, \ldots, p_n, q_1, \ldots, q_n) = \sum_{k=1}^{n} H_k(p_k, q_k) \qquad (2.5.2)$$

(e.g., a system of n-uncoupled oscillators). In this case the Hamilton–Jacobi equation is simply

$$H_k\left(\frac{\partial S}{\partial q_k}, q_k\right) = \alpha_k, \qquad k = 1, \ldots, n \qquad (2.5.3)$$

where the α_k are related by $\alpha = \alpha_1 + \alpha_2 + \cdots + a_n = H'$ and α is the value of the transformed Hamiltonian H'. Often the Hamiltonian will not be in such a convenient form as (2.5.2), but there is still the hope that the Hamilton–Jacobi equation can be separated in some other (orthogonal) coordinate system. Some examples of separable coordinate systems are given in Landau and Lifshitz (1960).

Assuming, though, that a separation of the form (2.5.1) is still possible, the relation

$$p_k = \frac{\partial}{\partial q_k} S_k(q_k, \alpha_1, \ldots, \alpha_n) \qquad (2.5.4)$$

shows that each p_k is a function of only one q_k. Furthermore, if the motion is periodic in each of the q_k (i.e., some form of libration (or rotation) in each degree of freedom), then a set of action variables can be introduced:

$$I_k = \frac{1}{2\pi} \oint_{\mathscr{C}_k} p_k(q_k, \alpha_1, \ldots, \alpha_n) \, dq_k \qquad (2.5.5)$$

where $\mathcal{C}_k$ is the closed path corresponding to a complete librational motion. Having obtained the I_k (and their relation with the α_k), they can be substituted into S and the second generating function relation can be invoked, that is,

$$\theta_k = \frac{\partial S}{\partial I_k} = \sum_{m=1}^{n} \frac{\partial}{\partial I_k} S_m(q_m, I_1, \ldots, I_n) \qquad (2.5.6)$$

where θ_k is the angle-variable conjugate to I_k. The transformed Hamiltonian $H'(I_1, \ldots, I_k)$ provides the canonical equations

$$\dot{I}_k = -\frac{\partial}{\partial \theta_k} H'(I_1, \ldots, I_n) = 0 \qquad (2.5.7a)$$

$$\dot{\theta}_k = \frac{\partial}{\partial I_k} H'(I_1, \ldots, I_n) = \omega_k(I_1, \ldots, I_n) \qquad (2.5.7b)$$

where ω_k is the frequency associated with each degree of freedom, which are trivially integrated to give

$$I_k = \text{const.} \qquad (2.5.8a)$$

$$\theta_k = \omega_k(I_1, \ldots, I_n)t + \delta_k \qquad (2.5.8b)$$

where δ_k are some arbitrary constants ("phase shifts"). Note that the set of $I_1, \ldots, I_n, \delta_1, \ldots, \delta_n$ constitute the set of $2n$ constants of integration that are required in the integration of a system of $2n$ first-order equations. Clearly, though, the I_k are a rather special set of constants—since once we have them, the symplectic structure of the Hamiltonian immediately gives us the other set.

2.5.b Properties of Integrable Systems

In general, however, it is not obvious that a given Hamiltonian will be separable and that the action variables can be defined as in (2.5.5). From all our preceeding discussions it should be clear that the key to integrating a Hamiltonian system of n degrees of freedom is to find n *independent integrals (constants) of motion*. If these can be put in the form of n constant conjugate momenta, Hamilton's equations are then trivially integrated (even if the transformation back to the original p and q is not trivial). Furthermore, if it is possible to find n independent "librational" paths $\mathcal{C}_k$, a set of action variables can be defined explicitly. As we shall see, this does not, in fact, require separability of the form (2.5.1).

If $F_i(\mathbf{p}, \mathbf{q})$ is an integral of motion, then along any phase-space trajectory its value is (of course) constant, that is,

$$F_i(\mathbf{p}(t), \mathbf{q}(t)) = f_i \qquad (2.5.9)$$

In terms of the Poisson bracket notation of Section 2.2.c, we can write

$$[H, F_i] = 0 \qquad (2.5.10)$$

A Hamiltonian system is said to be *completely integrable* if there exist n integrals of the motion $F_1, \ldots, F_n$, with say $F_1 = H$, which are in *involution*. This latter term means that the F_i all commute with each other, that is,

$$[F_i, F_j] = 0, \qquad i = 1, \ldots, n, \qquad j = 1, \ldots, n \qquad (2.5.11)$$

The significance of this property will become clear shortly.

The existence of the n integrals F_i means that the phase-space trajectories will be confined to some n-dimensional manifold, M, in the $2n$-dimensional phase space (recall (Chapter 1) that for a system of one degree of freedom, the existence of one integral confined the motion to an invariant (one-dimensional) curve). What we can now show is that the manifold M has the topology of an n-*dimensional torus*. Recalling the symplectic matrix defined in Eq. (2.2.23), we define the "velocity" fields

$$\boldsymbol{\xi}_i = \mathbf{J} \cdot \boldsymbol{\nabla} F_i, \qquad i = 1, \ldots, n \qquad (2.5.12)$$

where $\boldsymbol{\nabla} = (\partial q_1, \ldots, \partial q_n, \partial p_1, \ldots, \partial p_n)$. If $F_i = H$, then (2.5.12) defines the actual Hamiltonian flow, which, by virtue of the existence of the set of integrals $F_1, \ldots, F_n$, must wholly lie on M. Thus the vector field $\boldsymbol{\xi}_i$ is *tangent* to M. In fact, all n "velocity" fields $\boldsymbol{\xi}_i$ are tangent to M and, by virtue of the involutive property (2.5.11), are all linearly independent. A standard theorem in topology (the Poincaré–Hopf theorem—sometimes called the "hairy ball" theorem) tells us that an n-dimensional manifold (M) for which one can construct n independent vector fields tangent to M has the topology of an n-dimensional torus. (This notion is easily seen by comparing the act of combing the hairs on a 2-D torus or a 2-D sphere—in the later case, a hair will always stick up at the pole—see Figure 2.1.)

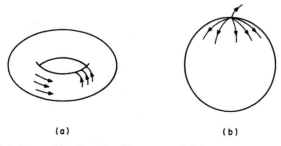

(a)   (b)

Figure 2.1 (a) Smoothly "combed" vector fields on a 2-torus. (b) Singular point of vector field on a 2-sphere.

The existence of these tori in phase space then provides the means of defining action variables in an invariant (i.e., representation-independent) way. The n-torus is a naturally periodic object and can be considered as the direct product of n independent 2π periodicities. In other words, one can define n topologically independent closed paths $\mathscr{C}_k$ on the torus where none of the $\mathscr{C}_k$ can be deformed continuously into each other or shrunk to zero (see Figure 2.2). The set of action variables can thus be defined as

$$I_k = \frac{1}{2\pi} \oint_{\mathscr{C}_k} \sum_{m=1}^{n} p_m \, dq_m \qquad (2.5.13)$$

and from the generating function $S = S(q_1, \ldots, q_n, I_1, \ldots, I_n)$, one can obtain the conjugate-angle variables

$$\theta_k = \frac{\partial}{\partial I_k} S(q_1, \ldots, q_n, I_1, \ldots, I_n) \qquad (2.5.14)$$

Hamilton's equations in the action-angle representation are

$$\dot{I}_k = -\frac{\partial}{\partial \theta_k} H'(I_1, \ldots, I_n) = 0 \qquad (2.5.15a)$$

$$\dot{\theta}_k = \frac{\partial}{\partial I_k} H'(I_1, \ldots, I_n) = \omega_k(I_1, \ldots, I_n) \qquad (2.5.15b)$$

It is important to note that if the system is completely integrable, the transformation to action-angle variables is global. That is, the whole phase

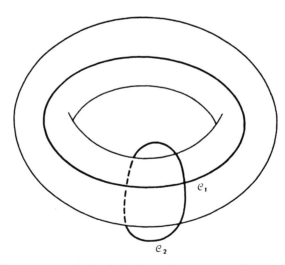

Figure 2.2 The two topologically independent curves $\mathscr{C}_1$ and $\mathscr{C}_2$ on a 2-torus.

space is filled with tori (although there can also be some multidimensional separatrices—see later), and a given trajectory will lie, forever, on one torus or another. A given set of initial conditions $(q_1(0), \ldots, q_n(0), p_1(0), \ldots, p_n(0))$ fixes particular values of the integrals, that is, $f_i = F_i(\mathbf{p}(0), \mathbf{q}(0))$. This set of F_i determines on which torus the trajectory lies (i.e., the actual values of the set of actions I_i, $i = 1, \ldots, n$); and the value of the angle variables θ_i, at a given time, determines the position of the trajectory on that torus.

For a conservative system of n degrees of freedom, that is,

$$E = H(p_1, \ldots, p_n, q_1, \ldots, q_n) \qquad (2.5.16)$$

which is completely integrable, the following important dimensions can be identified:

   (i) Phase space: $2n$-dimensional
   (ii) Energy shell: $2n - 1$-dimensional
   (iii) Tori: n-dimensional

An illustrative table is easily drawn up:

Degrees of Freedom n	1	2	3	4
Dimension of phase space:	2	4	6	8
Dimension of energy shell:	1	3	5	7
Dimension of tori:	1	2	3	4

A number of significant points can be deduced from these numbers. Firstly, for one-degree-of-freedom systems, the energy shell and tori are the same one-dimensional manifold. Formally this means that these systems are *ergodic*.† Secondly, for $n = 2$, the two-dimensional tori are embedded in the three-dimensional energy shell. This means that they divide the energy shell into an inside and outside. Thus if there was somehow a "gap" between tori (which occurs for nonintegrable systems), then a trajectory in that gap cannot escape from it. Thus, thirdly, for $n \geq 3$, trajectories in "gaps" between higher-dimensional tori can escape to other regions of the energy

---

†In statistical mechanics the term *ergodic* means that any system trajectory "uniformly" explores the energy shell. *Uniformly* means that the "time average" of a given quantity, say $f(\mathbf{p}, \mathbf{q})$, must equal its "phase average," that is,

$$\lim_{T \to \infty} \int_{-T}^{T} f(\mathbf{p}(t), \mathbf{q}(t)) \, dt = \int d\mathbf{p} \int d\mathbf{q} \, f(\mathbf{p}, \mathbf{q}) \, \delta(E - H(\mathbf{p}, \mathbf{q}))$$

where the $\delta(E - H(\mathbf{p}, \mathbf{q}))$ represents the "microcanonical ensemble," that is, the measure on the energy shell $E = H$.

shell. This gives rise to a phenomenon known as *Arnold diffusion*. (See Lichtenberg and Lieberman (1983) for a discussion of this topic.)

2.5.c Examples of Integrable Systems

There are a number of simple examples of multidimensional integrable systems. The first is the two-dimensional harmonic oscillator with Hamiltonian

$$H = \tfrac{1}{2}(p_1^2 + p_2^2 + \omega_1^2 q_1^2 + \omega_2^2 q_2^2) \tag{2.5.17}$$

The two integrals of motion are just the Hamiltonians (energies) associated with each mode, that is,

$$F_1 = E_1 = \tfrac{1}{2}(p_1^2 + \omega_1^2 q_1^2), \qquad F_2 = E_2 = \tfrac{1}{2}(p_2^2 + \omega_2^2 q_2^2) \tag{2.5.18}$$

This system is of the type defined in Eq. (2.5.2), for which the Hamilton–Jacobi equation is obviously separable. The motion consists of independent librations in the (p_1, q_1) and (p_2, q_2) planes, and the actions are easily determined to be

$$I_1 = \frac{1}{2\pi} \oint_{\mathscr{C}_1} p_1(q_1, E_1, E_2)\, dq_1, \qquad I_2 = \frac{1}{2\pi} \oint_{\mathscr{C}_2} p_2(q_2, E_1, E_2)\, dq_2 \tag{2.5.19}$$

where the circuits $\mathscr{C}_1$ and $\mathscr{C}_2$ are the round trips between the librational turning points $\pm\sqrt{2E_1}/\omega_1$ and $\pm\sqrt{2E_2}/\omega_2$, respectively. The Hamiltonian (2.5.17) in action variables is simply

$$H(I_1, I_2) = I_1\omega_1 + I_2\omega_2 \tag{2.5.20}$$

For conservative two-degree-of-freedom systems such as this, it is instructive to plot contours of constant energy $E = H(I_1, I_2)$ in the I_1, I_2 action plane (see Figure 2.3). Each point on one of these surfaces cor-

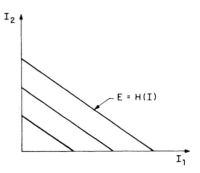

Figure 2.3 Constant energy surfaces in the (I_1, I_2)-plane for a two-dimensional simple harmonic oscillator.

responds to a particular value of I_1, I_2 and, hence, to a particular torus in the phase space.

Another simple two-dimensional system is a free particle, of mass m, in a box, that is,

$$H = \frac{1}{2m} (p_x^2 + p_y^2), \qquad 0 < x < a, 0 < y < b \qquad (2.5.21)$$

Since the coordinates x and y are cyclic, the momenta (or, to be more precise, the moduli of the momenta, since there is a sign change, e.g., $p_x \rightarrow - p_x$, on collision with a wall) are the constants of motion. The actions are just

$$I_1 = \frac{1}{2\pi} \oint_{\mathscr{C}_x} p_x \, dx = \frac{a}{\pi} |p_x|, \qquad I_2 = \frac{1}{2\pi} \oint_{\mathscr{C}_y} p_y \, dy = \frac{b}{\pi} |p_y| \qquad (2.5.22)$$

and hence the transformed Hamiltonian is

$$H(I_1, I_2) = \frac{\pi^2}{2m} \left(\frac{I_1^2}{a^2} + \frac{I_2^2}{b^2} \right) \qquad (2.5.23)$$

Notice that this is a nonlinear system since from Hamilton's equations the frequencies are seen to be action-dependent, that is,

$$\omega_1 = \frac{\partial H}{\partial I_1} = \frac{\pi^2}{ma^2} I_1, \qquad \omega_2 = \frac{\partial H}{\partial I_2} = \frac{\pi^2}{mb^2} I_2 \qquad (2.5.24)$$

The "action surfaces" of constant energy are shown in Figure 2.4. Notice that the normal $\nabla_{\mathbf{I}} H(I_1, I_2)$ (where $\nabla_{\mathbf{I}} = (\partial I_1, \partial I_2)$) direction changes smoothly along a given surface—this corresponds to the fact that now the frequencies vary from torus to torus.† (Obviously, $\nabla_{\mathbf{I}} H(\mathbf{I}) = (\omega_1(\mathbf{I}), \omega_2(\mathbf{I})$).)

A final simple example is a motion on the plane under a central force potential, that is,

$$H = \frac{p_r^2}{2m} + \frac{p_\varphi^2}{2mr^2} + V(r) \qquad (2.5.25)$$

where r and φ are the plane–polar coordinates. Notice that the angular coordinate is cyclic, so the angular momentum p_φ is (obviously) a conserved

---

†Such a system is termed *nondegenerate* and satisfies the condition

$$\det \left| \frac{\partial^2 H}{\partial I_i \, \partial I_j} \right| \neq 0$$

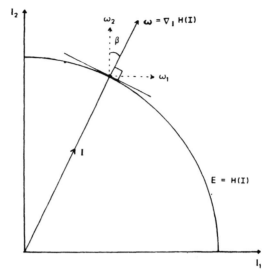

Figure 2.4 Constant energy surface for the particle-in-a-box. The vector **I** identifies a particular torus on the surface, and the normal **ω** at that point determines the associated frequency vector. (Here $\beta = \tan^{-1}(\omega_1/\omega_2)$.)

quantity. Thus the two integrals are

$$F_1 = p_\varphi, \qquad F_2 = H(p_r, p_\varphi, r) = E \qquad (2.5.26)$$

For rotational motion the closed path needed to define the corresponding action integral is just rotation of φ from 0 to 2π. Thus the two action integrals are

$$I_1 = \frac{1}{2\pi} \oint p_\varphi \, d\varphi = p_\varphi \qquad (2.5.27a)$$

$$I_2 = \frac{1}{2\pi} \oint p_r \, dr = \frac{1}{\pi} \int_{r_1}^{r_2} \sqrt{2m(E - I_1^2/2mr^2 - V(r))} \, dr \qquad (2.5.27b)$$

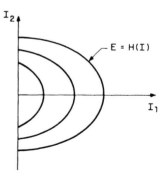

Figure 2.5 Constant energy surfaces for a typical central force Hamiltonian.

where r_1 and r_2 are the librational turning points. The precise form of I_2 will depend on $V(r)$. For a typical repulsive-core–attractive-tail-type potential, one would expect to see the "action surfaces" sketched in Figure 2.5.

2.5.d Motion on the Tori

Owing to the periodic nature of the (Hamiltonian) flow on tori, one can express a given dynamical quantity $f(\mathbf{p}, \mathbf{q})$ as a multiple Fourier series in the angle variables $\theta_1, \ldots, \theta_n$. Thus, for example, the q_i can be expressed as

$$q_i(t) = \sum_{k_i=-\infty}^{\infty} \cdots \sum_{k_n=-\infty}^{\infty} a^{(i)}_{k_1 k_2, \ldots, k_n} e^{i(k_1\theta_1 + k_2\theta_2 + \cdots + k_n\theta_n)}$$

$$= \sum_{k_1} \cdots \sum_{k_n} a^{(i)}_{k_1, \ldots, k_n} e^{i(k_1\omega_1 + \cdots + k_n\omega_n)t + i(k_1\delta_1, \ldots, k_n\delta_n)} \qquad (2.5.28)$$

where the Fourier coefficients $a^{(i)}_{k_1, \ldots, k_n}$, which depend on the actions, are determined by

$$a^{(i)}_{\mathbf{k}}(\mathbf{I}) = \int_0^{2\pi} d\theta_1 \cdots \int_0^{2\pi} d\theta_n q_i(\mathbf{I}, \boldsymbol{\theta}) \, e^{i(k_1\theta_1 + \cdots + k_n\theta_n)} \qquad (2.5.29)$$

where $q_i(\mathbf{I}, \boldsymbol{\theta})$ denotes the $q_i(t)$ expressed as functions of the action-angle variables. Variables that can be expressed in such multiple series are usually referred to as *multiply periodic*. Their precise behavior will be determined by the values of the frequencies ω_i. If the frequencies are not rationally related, the motion on a given torus (i.e., the torus with that set of ω_i) will never exactly repeat itself. Such orbits are usually termed *quasi-periodic*. Thus a single orbit will eventually cover the torus uniformly; that is, flow is ergodic on the torus.† This is easily seen for the 2-D case. Here the 2-torus is topologically equivalent to the unit square with identified edges (see Figure 2.6). If the ratio ω_1/ω_2 is irrational, one can easily verify the ergodicity of the motion. This was apparently first proved by Jacobi in 1835 but had been considered nonrigorously by the fourteenth-century scholar Nicholas Oresme (ca. 1325–ca. 1382). If, on the other hand, the frequencies are all rationally related, the motion will eventually repeat itself—these are usually referred to as *closed* orbits. For the two-dimensional case, this

---

†Ergodicity on the torus means, again, that time average equals phase average. Using the action-angle variables for the given torus, this means that

$$\lim_{T \to \infty} \int_{-T}^{T} f(\mathbf{p}(t), \mathbf{q}(t)) \, dt = \frac{1}{(2\pi)^n} \int_0^{2\pi} f(\mathbf{p}(\mathbf{I}, \boldsymbol{\theta}), \mathbf{q}(\mathbf{I}, \boldsymbol{\theta})) \, d\boldsymbol{\theta}$$

More on ergodicity and related concepts will be found in Chapter 4.

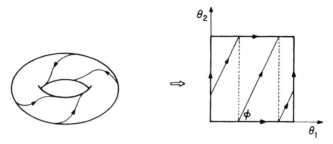

Figure 2.6 Equivalence between a 2-torus and (unit) square with identified edges.

requires just a single relation of the form

$$\frac{\omega_1}{\omega_2} = \frac{n}{m} \tag{2.5.30}$$

For flow on a torus with this frequency ratio, the orbit will close after m cycles of θ_1 and n cycles of θ_2. For the n-dimensional case, closure requires $n - 1$ relations of the form

$$\sum_{i=1}^{n} k_i \omega_i = 0 \tag{2.5.31}$$

A completely integrable system is termed *nondegenerate* if the condition

$$\det\left|\frac{\partial \omega_i(\mathbf{I})}{\partial I_j}\right| = \det\left|\frac{\partial^2 H(\mathbf{I})}{\partial I_i\,\partial I_j}\right| \neq 0 \tag{2.5.32}$$

is satisfied. This ensures that the frequencies vary from torus to torus (i.e., the system is nonlinear). Thus on a given energy shell, some tori will be covered by closed orbits, whereas other tori will be covered by quasi-periodic orbits. Although there are an infinity of rational numbers, they form what is known as a set of zero measure in the space of real numbers; that is, they are infinitely outnumbered by the irrationals. Thus for a nondegenerate system the set of tori with closed orbits, although dense, will be infinitely outnumbered by the tori covered with quasi-periodic orbits. Nonetheless, we shall see that the closed orbits play a significant role in determining the properties of integrable systems under perturbation.

2.5.e Fundamental Issues

This brings us to some fundamental issues. So far we have only discussed completely integrable systems, that is, ones which have as many first integrals as degrees of freedom. What happens if the system does not have a set of n first integrals? Another way of putting this is to ask what happens to an integrable system under nontrivial perturbation. Are the tori preserved, albeit in some "distorted form," or are they destroyed? Resolution of this fundamental issue became one of the main tasks of nineteenth- and twentieth-century mechanics and was only resolved in the early 1960s. This will be discussed in detail in the following chapters.

Another fundamental question is: Given some Hamiltonian system, how does one know if it is integrable in the first place? One way of answering this question is to try and identify explicitly the n integrals of motion—if they exist. Except for special cases (e.g., separable systems), this is usually an impossible task—one has to be a genius or lucky, or both. The resolution of this issue, that is, the identification of integrable systems, is still being researched. Some of the current ideas on this topic will be examined in Chapter 8.

APPENDIX 2.1   LEGENDRE TRANSFORMATIONS

A curve (or surface) can be represented by either a set of points or by a set of tangent planes as shown in Figure 2.7a. The Legendre transformation provides the transformation between these two representations. Here we follow the geometrical construction given by Arnold (1978). Consider the curve $y = f(x)$, which is a convex function, that is,

$$\frac{\partial^2 f}{\partial x^2} > 0 \qquad\qquad (2.\text{A}.1)$$

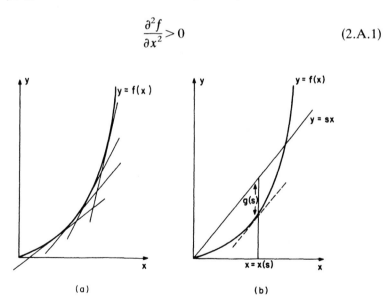

(a)   (b)

Figure 2.7 (a) Curve $y = f(x)$ and associated set of tangent planes. (b) Construction of Legendre transform variable $g(s)$.

The Legendre transformation of f is a new function g of a new variable s constructed as shown in Figure 2.7b. We see that $g(s)$ is the maximum vertical distance between the straight line $y = sx$ and $f(x)$, that is,

$$g(s) = sx - f(x) = G(s, x(s)) \tag{2.A.2}$$

Since the point $x(s)$ is defined by the maximum condition

$$\frac{\partial G}{\partial x} = s - f'(x) = 0 \tag{2.A.3}$$

we see that the new variable s is just the slope of $f(x)$, that is,

$$s = f'(x) \tag{2.A.4}$$

and since f is convex, the point $x = x(s)$ is unique (if it exists).

As a simple illustration, consider the curve defined by the Hamiltonian function $H(p)$, that is, $y = H(p)$. Following the above construction (now in the (y, p)-plane), the new function, let us call it $L(s)$, is therefore

$$L(s) = sp - H(p) \tag{2.A.5}$$

Now in this case the new variable is $s = \partial H / \partial p = \dot{q}$, so the Legendre transform is

$$L(\dot{q}) = \dot{q}p - H(p) \tag{2.A.6}$$

where p is expressed as a function of $\dot{q}$ through the relationship $\dot{q} = \partial H(p)/\partial p$. This example illustrates the general classical result

$$H(\mathbf{p}, \mathbf{q}) = \sum_{i=1}^{n} p_i \dot{q}_i - L \tag{2.A.7}$$

and

$$L(\mathbf{q}, \dot{\mathbf{q}}) = \sum_{i=1}^{n} p_i \dot{q}_i - H \tag{2.A.8}$$

where in (2.A.7) the $\dot{q}_i$ are expressed as functions of the p_i through

$$p_i = \frac{\partial}{\partial \dot{q}_i} L(\mathbf{q}, \dot{\mathbf{q}}) \tag{2.A.9}$$

and in (2.A.8) the p_i are expressed as functions of the $\dot{q}_i$ through

$$\dot{q}_i = \frac{\partial}{\partial p} H(\mathbf{p}, \mathbf{q}) \tag{2.A.10}$$

In order that both transformations be unique, we require that L and H be convex functions (cf. (2.2.3)) of $\dot{q}_i$ and p_i, respectively.

APPENDIX 2.2   GEOMETRIC CONCEPTS IN CLASSICAL MECHANICS

We first distinguish between *covariant* and *contravariant* vectors. Consider a vector $\mathbf{a} = (a_1, \ldots, a_n)$, where each of the a_i is a functiuon of the coordinates $\mathbf{x} = (x_1, \ldots, x_n)$, that is, $a_j = a_j(\mathbf{x})$. Now consider changing to a new coordinate set $\mathbf{y} = (y_1, \ldots, y_n)$. The vector $\mathbf{a}$ is termed contravariant if it transforms according to the rule

$$\bar{a}_i = \sum_{j=1}^{n} \frac{\partial y_i}{\partial x_j} a_j \tag{2.A.11}$$

where $\bar{a}_i$ are the a_j expressed in terms of the new variables y_i. If $\mathbf{a}$ is contravariant, it is common to label its components with a superscript, that is, $\mathbf{a} = (a^1, \ldots, a^n)$. However, here we will use subscript labeling throughout. A useful example of a contravariant vector is the infinitesimal separation between adjacent points, that is, $dx = (dx_1, \ldots, dx_n)$. These infinitesimals transform as (using the chain rule)

$$dy_i = \sum_{j=1}^{n} \frac{\partial y_i}{\partial x_j} dx_j \tag{2.A.12}$$

As another example consider the following operator (whose significance will become clear shortly):

$$\xi = \sum_{i=1}^{n} \xi_i \frac{\partial}{\partial x_i} \tag{2.A.13}$$

where $\xi_i = \xi_i(x_1, \ldots, x_n)$. Transforming (2.A.13) to the y_i coordinates gives

$$\xi = \sum_{i=1}^{n} \sum_{j=1}^{n} \xi_i \frac{\partial y_j}{\partial x_i} \frac{\partial}{\partial y_j} \equiv \sum_{j=1}^{n} \bar{\xi}_j \frac{\partial}{\partial y_j} \tag{2.A.14}$$

where

$$\bar{\xi}_j = \sum_{i=1}^{n} \xi_i \frac{\partial y_j}{\partial x_i} \tag{2.A.15}$$

that is, the quantities ξ_i transform as contravariant vectors.

By contrast, the vector $\mathbf{b} = (b_1, \ldots, b_n)$ is termed covariant if it transforms under the coordinate change $\mathbf{x} \to \mathbf{y}$ as

$$\bar{b}_i = \sum_{j=1}^{n} \frac{\partial x_j}{\partial y_i} b_j \tag{2.A.16}$$

Vectors that are gradients of a scalar field are examples of covariant vectors. Thus if $b_i = \partial F(\mathbf{x})/\partial x_i$ and $\bar{b}_i = \partial F(\mathbf{y})/\partial y_i$, it is easily seen that

$$\bar{b}_i = \sum_{j=1}^{n} \frac{\partial x_j}{\partial y_i} \frac{\partial F}{\partial x_j} \equiv \sum_{j=1}^{n} \frac{\partial x_j}{\partial y_i} b_j \tag{2.A.17}$$

By demonstrating that the generalized momenta p_i can be represented as the gradient of the action, that is, Eq. (2.1.20), we see that they are covariant vectors. (A common notation is to label the components of covariant vectors with subscripts in contrast to the superscripts used for contravariant quantities.)

The operator introduced in (2.A.13) is an example of a *tangent vector*. When it acts on some scalar function $h = h(x_1, \ldots, x_n)$, at a specified point $\mathbf{x} = \mathbf{X}$, it gives the directional derivative of h at that point, that is,

$$\xi h = \sum_{i=1}^{n} \xi_i \frac{\partial h}{\partial x_i}\bigg|_{\mathbf{x}=\mathbf{X}} \tag{2.A.18}$$

Consider now some curve $\boldsymbol{\varphi} = \boldsymbol{\varphi}(s)$, parameterized by the variable which passes through the point $\mathbf{X}$ at the s value $s = 0$. If $\boldsymbol{\varphi} = (\varphi_1, \ldots, \varphi_n)$ has the coordinate representation $\varphi_i(s) = x_i$, then the directional derivative of any quantity along $\boldsymbol{\varphi}(s)$ at the point $\mathbf{X}$ is determined by the tangent vector

$$\xi = \frac{d}{ds}\bigg|_{s=0} = \sum_{i=1}^{n} \xi_i \frac{\partial}{\partial x_i}\bigg|_{\mathbf{x}=\mathbf{X}} \tag{2.A.19}$$

where

$$\xi_i = \frac{d\varphi_i}{ds}\bigg|_{s=0} \tag{2.A.20}$$

Obviously, if $\boldsymbol{\varphi}$ is just a system trajectory $\mathbf{q}(t) = (q_1(t), \ldots, q_n(t))$, where we identify s as time, we see that the ξ_i are just the velocity components $\dot{q}_i(t)$. Thus the velocities can be identified as contravariant vectors.

Through a given point $\mathbf{X}$ many different trajectories can pass and this gives rise to a whole set of possible tangent vectors. This collection of vectors at $\mathbf{X}$ gives rise to a vector space termed the *tangent space* at the point $\mathbf{X}$. It is denoted by $TM_{\mathbf{X}}$ where M is the *manifold*, that is, the n-dimensional "space" which the system inhabits. The *tangent bundle* is simply the collection of all possible tangent spaces at all possible points $\mathbf{X}$ in M. It is denoted by TM.

In the Lagrangian description of a dynamical system the state of the system is described by its positions q_i and velocities $\dot{q}_i$. Thus at any point $\mathbf{Q}$, at any instant t we have the tangent vector

$$\xi = \sum_i \dot{q}_i(t) \left. \frac{\partial}{\partial q_i} \right|_{\mathbf{q}=\mathbf{Q}} \tag{2.A.21}$$

that is, the state of the system can be identified with a point in the tangent bundle. The value of the Lagrangian can thus be regarded as the mapping of TM to a scalar field, that is, $L:\text{TM} \to \mathcal{R}$ (where $\mathcal{R}$ is the space of real numbers).

The Hamiltonian description of mechanics is in terms of coordinates q_i and conjugate momenta p_i, the latter quantities transforming as covariant vectors. The associated $2n$-dimensional phase space is a symplectic manifold which has many special properties. The most important property of Hamiltonian systems is the preservation of phase volume under the (Hamiltonian) flow. To appreciate this at a geometric level requires the language of *differential forms*.†

A differential *one-form*—here denoted by the symbol ω^1—is a quantity (here defined for two dimensions) of the form

$$\omega^1 = b_1 dx_1 + b_2 dx_2 \tag{2.A.22}$$

where $b_1 = b_1(x_1, x_2)$ and $b_2 = b_2(x_1, x_2)$ are the components of a covariant vector. Now consider what happens to ω^1 under a change of variables $(\mathbf{x} \to \mathbf{y})$. Using (2.A.12) for the transformation of the contravariant quantities dx_i, we have

$$\omega^1 = \sum_{i=1}^{2} b_i \, dx_i = \sum_{i=1}^{2} \sum_{j=1}^{2} b_i \frac{\partial x_i}{\partial y_j} dy_j = \sum_{j=1}^{2} \bar{b}_j \, dy_j \tag{2.A.23}$$

where

$$\bar{b}_j = \sum_{i=1}^{2} b_i \frac{\partial x_i}{\partial y_j} \tag{2.A.24}$$

which is the usual rule for covariant transformations. Thus from (2.A.23) we see that the 1-form ω^1 is invariant under transformation of variables.

In Hamiltonian mechanics the frequently occurring quantity

$$\omega^1 = \sum_{i=1}^{n} p_i \, dq_i \tag{2.A.25}$$

---

†According to Arnold (1978), "Hamiltonian mechanics cannot be understood without the use of differential forms."

is thus seen to be an example of a 1-form. If the phase space is "extended" to include time as an independent variable, with conjugate variable $-H$, then we can also construct the 1-form

$$\omega^1 = \sum_{i=1}^{n} p_i \, dq_i - H \, dt \qquad (2.A.26)$$

known as the Poincaré–Cartan 1-form. As described in Section (2.3.c), such a 1-form is invariant under canonical transformation, that is,

$$\sum_{i=1}^{n} p_i \, dq_i - H \, dt = \sum_{i=1}^{n} P_i \, dQ_i - H' \, dt + dF \qquad (2.A.27)$$

where $H = H(p_1, \ldots, p_n, q_1, \ldots, q_n, t)$, $H' = H'(P_1, \ldots, P_n, Q_1, \ldots, Q_n, t)$, and dF is a total differential which turns out to be the generating function effecting the canonical transformation between the p_i, q_i and P_i, Q_i variables. A fundamental property of the 1-form (2.A.26) is that its integral around a closed curve, $\mathscr{C}$, encompassing a tube of trajectories, that is,

$$\oint_{\mathscr{C}} \omega^1 = \oint_{\mathscr{C}} \sum_{i=1}^{n} p_i \, dq_i - H \, dt \qquad (2.A.28)$$

is the same for any other closed curve $\mathscr{C}'$ encircling the same tube of trajectories, that is,

$$\oint_{\mathscr{C}} \omega^1 = \oint_{\mathscr{C}'} \omega^1 \qquad (2.A.29)$$

As a special choice of $\mathscr{C}$ and $\mathscr{C}'$, we can imagine $\mathscr{C}$ identifying a particular set of initial conditions $p_i(0)$, $q_i(0)$ at $t = 0$ and following the "evolution" of $\mathscr{C}$ under the Hamiltonian flow. At time $t = T$, $\mathscr{C}$ will have evolved into the (still closed) curve $\mathscr{C}_T$. In the extended phase space $(p_n, \ldots, p_n, q_1, \ldots, q_n, t)$ the two curves $\mathscr{C}$ (at $t = 0$) and $\mathscr{C}_T$ (at $t = T$) correspond to closed curves lying in different $t = $ constant "planes." We can write (2.A.29) as

$$\oint_{\mathscr{C}} \sum_{i=1}^{n} p_i \, dq_i - H \, dt = \oint_{\mathscr{C}_T} \sum_{i=1}^{n} p_i \, dq_i - H \, dt \qquad (2.A.30)$$

but, since the integrations are being carried out at fixed time slices in the extended phase space, $dt = 0$ and hence (2.A.30) reduces to

$$\oint_{\mathscr{C}} \sum_{i=1}^{n} p_i \, dq_i = \oint_{\mathscr{C}_T} \sum_{i=1}^{n} p_i \, dq_i \qquad (2.A.31)$$

The 1-form $\sum_{i=1}^{n} p_i \, dq_i$ is sometimes called Poincaré's relative integral invariant.

Of particular importance in Hamiltonian mechanics are differential 2-*forms*. These 2-forms can be constructed from the "product" of 1-forms but now the "product" is the *exterior* or *wedge* product. It is different from the usual product in that it obeys the "antisymmetry" rules

$$dx_i \wedge dx_j = -dx_j \wedge dx_i \tag{2.A.32}$$

and

$$dx_i \wedge dx_i = 0 \tag{2.A.33}$$

where the wedge symbol $\wedge$ denotes the exterior product. The exterior product of functions and differentials obeys the rules

$$b_i \wedge dx_j = dx_j \wedge b_i = b_i \, dx_j \tag{2.A.34}$$

and hence

$$dx_i \wedge (b_i \, dx_j) = b_i \, dx_i \wedge dx_j \tag{2.A.35}$$

Consider now the exterior product between the 1-forms $\theta_1 = b_1 \, dx_1 + b_2 \, dx_2$ and $\theta_2 = c_1 \, dx_1 + c_2 \, dx_2$. Following the above rules we find

$$\begin{aligned}
\theta_1 \wedge \theta_2 &= (b_1 \, dx_1 + b_2 \, dx_2) \wedge (c_1 \, dx_1 + c_2 \, dx_2) \\
&= (b_1 c_2 - b_2 c_1) \, dx_1 \wedge dx_2 \\
&= -\theta_2 \wedge \theta_1
\end{aligned} \tag{2.A.36}$$

The form $\theta_1 \wedge \theta_2$ is an example of a differential 2-form and we denote it by the symbol ω^2.

Under the change of variables $x_1 = x_1(y_1, y_2)$, $x_2 = x_2(y_1, y_2)$ we have

$$dx_1 = \frac{\partial x_1}{\partial y_1} \, dy_1 + \frac{\partial x_1}{\partial y_2} \, dy_2 \tag{2.A.37}$$

and

$$dx_2 = \frac{\partial x_2}{\partial y_1} \, dy_1 + \frac{\partial x_2}{\partial y_2} \, dy_2 \tag{2.A.38}$$

hence the 2-form $dx_1 \wedge dx_2$ transforms as

$$\begin{aligned}
dx_1 \wedge dx_2 &= \left(\frac{\partial x_1}{\partial y_1} \frac{\partial x_2}{\partial y_2} - \frac{\partial x_1}{\partial y_2} \frac{\partial x_2}{\partial y_1} \right) dy_1 \wedge dy_2 \\
&= \frac{\partial(x_1, x_2)}{\partial(y_1, y_2)} \, dy_1 \wedge dy_2
\end{aligned} \tag{2.A.39}$$

when $\partial(x_1, x_2)/\partial(y_1, y_2)$ is, of course, just the Jacobian of transformation. Thus in dynamical terms, if the transformation is the canonical transformation between one set of conjugate variables (p, q) and another set (P, Q), we have

$$dp \wedge dq = dP \wedge dQ \qquad (2.A.40)$$

where for a canonical transformation we are assuming a unit Jacobian. This preservation of 2-form under canonical transformation is a fundamental property of Hamiltonian systems and can be generalized to any number of degrees of freedom, that is,

$$\sum_{i=1}^{n} dp_i \wedge dq_i = \sum_{i=1}^{n} dP_i \wedge dQ_i \qquad (2.A.41)$$

Just as 2-forms can be constructed from the wedge product of 1-forms, it is also possible to construct wedge products between 2-forms. The wedge product of $\omega^2 = \sum_{i=1}^{n} dp_i \wedge dq_i$ with itself gives (using the rules described above) the 4-form

$$\omega^4 = \omega^2 \wedge \omega^2 = \sum_{i \leq j} dp_i \wedge dp_j \wedge dq_i \wedge dq_j \qquad (2.A.42)$$

This process can be repeated to construct ever "higher" forms (for example, $\omega^8 = \omega^4 \wedge \omega^4 = \omega^2 \wedge \omega^2 \wedge \omega^2 \wedge \omega^2$) all the way up to the n-fold wedge product $\omega^{2n} = \omega^2 \wedge \omega^2 \wedge \cdots \wedge \omega^2$, at which point the form becomes the single term

$$\omega^{2n} = \prod_{i=1}^{n} dp_i \wedge dq_i \qquad (2.A.43)$$

which is just the volume element for n-dimensional phase space. All the differential forms from ω^2 to ω^{2n} are invariant under canonical transformation, with the invariance of ω^{2n} just being a statement of Liouville's theorem.

Two-forms may also be obtained from 1-forms by means of *exterior differentiation*. If ω^1 is the 1-form $\sum_{i=1}^{n} b_i \, dx_i$, then its exterior derivative $d\omega^1$ is given by

$$d\omega^1 = d\left(\sum_{i=1}^{n} b_i \, dx_i \right) = \sum_{i=1}^{n} db_i \wedge dx_i \qquad (2.A.44)$$

which we recognize as a 2-form. As an explicit example consider the two-dimensional 1-form $\omega^1 = b_1(x_1, x_2) \, dx_1 + b_2(x_1, x_2) \, dx_2$; then

$$d\omega^1 = db_1 \wedge dx_1 + db_2 \wedge dx_2$$

$$= \left(\frac{\partial b_1}{\partial x_1} dx_1 + \frac{\partial b_1}{\partial x_2} dx_2 \right) \wedge dx_1 + \left(\frac{\partial b_2}{\partial x_1} dx_1 + \frac{\partial b_2}{\partial x_2} dx_2 \right) \wedge dx_2$$

$$= \left(\frac{\partial b_2}{\partial x_1} - \frac{\partial b_1}{\partial x_2} \right) dx_1 \wedge dx_2 \tag{2.A.45}$$

This last result is very suggestive of Green's theorem for the transformation of a line integral around a closed curve $\mathscr{C}$ to an area integral over the domain D (enclosed by $\mathscr{C}$). In the x, y plane this takes the form (Stokes' theorem)

$$\oint_{\mathscr{C}} f(x, y) \, dx + g(x, y) \, dy = \int \int_D \left(\frac{\partial g}{\partial x} - \frac{\partial f}{\partial y} \right) dx \, dy \tag{2.A.46}$$

In the language of differential forms this is elegantly expressed as

$$\oint_{\mathscr{C}} \omega^1 = \int \int_D \omega^2 \tag{2.A.47}$$

where ω^1 is the 1-form $f \, dx + g \, dy$ and $\omega^2 = d\omega^1$ is the 2-form $df \wedge dx + dg \wedge dy$.

For a one-degree-of-freedom Hamiltonian we can write (2.A.47) as

$$\oint_{\mathscr{C}} p \, dq = \int \int_A dp \wedge dq \tag{2.A.48}$$

where $\mathscr{C}$ is a closed curve taken on a $t = $ constant plane in the extended phase space (p, q, t). There is no difficulty in interpreting (2.A.48)—it just states that the integral around the closed curve $\mathscr{C}$ in the p, q *plane* equals the area integral over the domain A enclosed by $\mathscr{C}$. The result (2.A.47) can be written down for systems with any number of degrees of freedom. By exterior differentiation we have

$$d \left(\sum_{i=1}^{n} p_i \, dq_i \right) = \sum_{i=1}^{n} dp_i \wedge dq_i \tag{2.A.49}$$

but now the interpretation of the integration is more subtle. What we must write is

$$\oint_{\mathscr{C}} \sum_{i=1}^{n} p_i \, dq_i = \sum_{i=1}^{n} \int \int_{A_i} dp_i \wedge dq_i \tag{2.A.50}$$

where now the A_i are the set of areas found by projecting the closed curve $\mathscr{C}$ (on a $t = $ constant slice of the extended phase space) onto each of the (p_i, q_i) planes. (This is sketched in Fig. 4.5, Chapter 4.) Using the result (2.A.31) we see that this sum of projected areas is preserved under the Hamiltonian flow, that is,

$$\sum_{i=1}^{n} \iint_{A_i} dp_i \wedge dq_i = \sum_{i=1}^{n} \iint_{A_{i,T}} dp_i \wedge dq_i \qquad (2.A.51)$$

where the $A_{i,T}$ are the projected areas associated with the curve $\mathscr{C}_T$. These ideas can be extended to the higher forms $\omega^4, \omega^8, \ldots, \omega^{2n}$. In fact, all n of these forms are conserved under the phase flow (or other canonical transformations). This set of "conserved" quantities is known as the Poincaré invariants. All these "conservation laws" follow from the result (2.A.51) for the 2-form and it is thus this property (of the 2-form) that provides the fundamental definition of canonical transformations. As already mentioned, Liouville's theorem is seen to be a statement about the conservation of the form ω^{2n}, namely

$$\int \cdots \int_V \prod_{i=1}^{n} dp_i \wedge dq_i = \int \cdots \int_{V_T} \prod_{i=1}^{n} dp_i \wedge dq_i \qquad (2.A.52)$$

that is, phase volume preservation under the Hamiltonian flow, or phase volume preservation under canonical transformation, namely

$$\int \cdots \int_{V_{pq}} \prod_{i=1}^{n} dp_i \wedge dq_i = \int \cdots \int_{V_{PQ}} \prod_{i=1}^{n} dP_i \wedge dQ_i \qquad (2.A.53)$$

(In (2.A.52) V_T is just the "evolved" initial volume V and in (2.A.53) V_{PQ} is the volume V_{pq} transformed from (p, q) space to (P, Q) space.)

SOURCES AND REFERENCES

Arnold, V. I., *Mathematical Methods of Classical Mechanics*, Springer-Verlag, New York, 1978.

Born, M., *The Classical Mechanics of the Atom*, Ungar, New York, 1960.

Goldstein, H., *Classical Mechanics*, 2nd ed., Addison-Wesley, Reading, Massachusetts, 1980.

Landau, L. D. and E. M. Lifshitz, *Mechanics*, 2nd ed., Pergamon Press, Oxford, 1960.

Lichtenberg, A. J., and M. A. Lieberman, *Regular and Stochastic Motion*, Springer-Verlag, New York, 1983.

Percival, I. C. and D. Richards, *Introduction to Dynamics*, Cambridge University Press, Cambridge, 1982.

3

CLASSICAL PERTURBATION THEORY

3.1 ELEMENTARY PERTURBATION THEORY

Completely integrable Hamiltonian systems are exceptional.† Nonetheless, despite their rarity, integrable systems play an important role in our understanding of nonintegrable systems. This is because it is often convenient to represent a given Hamiltonian system in the form of an integrable part H_0 plus some (small) nonintegrable *perturbation H_1*, that is,

$$H(\mathbf{p}, \mathbf{q}) = H_0(\mathbf{p}, \mathbf{q}) + \epsilon H_1(\mathbf{p}, \mathbf{q}) \qquad (3.1.1)$$

where the perturbation parameter ϵ is assumed $\ll 1$. The idea is to find approximate solutions to $H(\mathbf{p}, \mathbf{q})$ in the form of the "exact" solutions to H_0 plus small corrections due to $H_1(\mathbf{p}, \mathbf{q})$. This is the subject of perturbation theory, and the most famous example is in its application to planetary motion in our solar system. The motion of the earth about the sun—treated as a strictly "two body" problem—is an exactly soluble (i.e., integrable) problem. (The earth orbits the sun in Keplerian orbits). However, other planets, particularly Jupiter, provide a non-negligible influence that can be considered as a small perturbation to the two-body problem. This "three-body" problem, even when stripped down to its

†There are some significant results to this effect. For example, Siegel has shown that in certain classes of Hamiltonian system, the nonintegrable ones form a dense set (Siegel and Moser, 1971) and Markus and Meyer (1974) have proved that generic Hamiltonian dynamical systems are neither integrable nor ergodic.

simplest form, has proven to be a formidable task in that all efforts—including those of some of the most famous mathematicians—failed to produce *convergent* perturbation expansions for the approximate orbits. This famous/infamous problem will be discussed further in due course, but for the moment we will describe some simpler examples of perturbation theory which can indicate the profound problems that lie ahead.

The central idea of perturbation theory is to expand the solution $x(t)$ in a power series in ϵ, that is,

$$x(t) = x_0(t) + \epsilon x_1(t) + \epsilon^2 x_2(t) + \cdots \qquad (3.1.2)$$

where $x_0(t)$ is the exact solution to the integrable part of the problem, and the corrections $x_1(t)$, $x_2(t)$, ... are calculated recursively. Clearly, the expansion is such that in the limit $\epsilon \rightarrow 0$ we retain only the "integrable" part of the problem. Provided that ϵ is sufficiently small, it might be hoped that only a few terms in the series (3.1.2) will provide an accurate representation of the "true" solution—although even for a small ϵ, there is no guarantee that this will be so for long times—in other words, for any expansion of the form (3.1.2) we are faced with the major issue of its convergence. Some simple, yet instructive, examples of series expansions can be found in the problem of finding approximate roots to polynomial equations.

3.1.a Regular Perturbation Series

Consider the simple quadratic equation

$$x^2 + x - 6\epsilon = 0 \qquad (3.1.3)$$

where ϵ is assumed small. The "zero-order" problem equivalent to the "integrable" part of a dynamical problem, $x^2 + x = 0$—has the two roots $x = 0, -1$. We now attempt to construct the roots of the "perturbed" problem (3.1.3) as a power-series expansion of the form

$$x = \sum_{n=0}^{\infty} a_n \epsilon^n = a_0 + \sum_{n=1}^{\infty} a_n \epsilon^n \qquad (3.1.4)$$

where a_0 is one of the two zero-order roots about which we are expanding. Substituting the expansion (3.1.4) into (3.1.3) and equating each power of ϵ to zero, we obtain up to $O(\epsilon^2)$:

$$O(\epsilon^0): \quad a_0^2 + a_0 = 0 \qquad (3.1.5a)$$

$$O(\epsilon^1): \quad 2a_0 a_1 + a_1 - 6 = 0 \qquad (3.1.5b)$$

$$O(\epsilon^2): \quad a_1^2 + 2a_0 a_2 + a_2 = 0 \qquad (3.1.5c)$$

Equation (3.1.5a) just gives us the expected zero-order roots. For each choice of a_0, we can successively solve (3.1.5b), (3.1.5c), and so on, to higher orders of ϵ. Thus the two roots of (3.1.3) are easily found to have the expansion

$$x_1 = 6\epsilon - 36\epsilon^2 + O(\epsilon^3)$$
$$x_2 = -1 - 6\epsilon + 36\epsilon^2 - O(\epsilon^3)$$

So, for example, for the choice $\epsilon = 0.01$, these two roots are (to order ϵ^2) $x_1 = 0.0564$ and -1.0564, which agree to the third decimal place with the exact solutions. Expansions of the form (3.1.4) are examples of a *regular perturbation* series in that they have a finite radius of convergence and the solutions tend to the zero-order results in the limit $\epsilon \to 0$.

3.1.b Singular Perturbation Series*

Now consider the polynomial

$$\epsilon x^2 + x - 1 = 0 \qquad (3.1.6)$$

This is clearly not a regular problem since in the limit $\epsilon \to 0$ the zero-order system has only one root, whereas the perturbed problem has two. Situations like this, that is, where the $\epsilon = 0$ limit is fundamentally different from the neighboring (i.e., small ϵ) behavior, are termed *singular perturbation problems*. In this case the expansion may not take the form of a power series, but if it does it will have a vanishing radius of convergence. Although we will not be considering such cases for our mechanical problems, it is instructive to complete this exercise.† The unperturbed part of (3.1.6) (i.e., $x - 1 = 0$) has the obvious zero-order root $x = 1$. The singular part of the problem is associated with the other root, which goes to ∞ in the limit $\epsilon \to 0$, as may easily be verified, in this case, from the exact solution. However, the root $x = 1$ clearly behaves "regularly" as $\epsilon \to 0$, so we can construct regular perturbation expansion about it (i.e., $x_1 = 1 + a_1\epsilon + a_2\epsilon^2 + \cdots$). On substitution into (3.1.6), the expansion coefficients $a_1, a_2, \ldots$ are easily determined and we obtain the expansion

$$x_1 = 1 - \epsilon + 2\epsilon^2 + O(\epsilon^3) \qquad (3.1.7)$$

The second root goes to ∞ as $\epsilon \to 0$, implying that it behaves as some inverse power of ϵ (i.e., $x = O(1/\epsilon^n)$). This suggests a rescaling of Eq. (3.1.6) by means of the substitution $x = y/\epsilon^n$. By balancing the terms ϵx^2 and x in (3.1.6) (this is the method of "detailed balance" described, for

---

†Singular perturbation problems will appear again in Chapter 6 in the context of semiclassical mechanics.

example, in Bender and Orszag (1978)), one sees that the only consistent scaling is

$$x = \epsilon^{-1} y \tag{3.1.8}$$

which leads to the equation

$$y^2 + y - \epsilon = 0 \tag{3.1.9}$$

This is now in the form of a regular problem with zero-order roots $y_1 = 0$, $y_2 = 1$. Regular expansions of the form $y = \sum_{n=0}^{\infty} b_n \epsilon^n$ are easily determined and one finds that

$$y_1 = \epsilon - \epsilon^2 + 2\epsilon^3 + O(\epsilon^4)$$
$$y_2 = -1 - \epsilon + \epsilon^2 - 2\epsilon^3 + O(\epsilon^4)$$

Transforming back to the original variable x by means of (3.1.8), we find that the two roots are

$$x = 1 - \epsilon + 2\epsilon^2 + O(\epsilon^3) \tag{3.1.10a}$$

$$x = -\frac{1}{\epsilon} - 1 + \epsilon - 2\epsilon^2 + O(\epsilon^3) \tag{3.1.10b}$$

Equation (3.1.10a) is nothing more than the regular root (3.1.7), and (3.1.10b) is the singular root. For this rather trivial problem, one may confirm that (3.1.10a) and (3.1.10b) are indeed correct from the exact solution $x = (1/2\epsilon)(-1 \pm \sqrt{1 + 4\epsilon})$ by making the standard binomial expansion of the square root.

3.1.c   Regular Perturbation Series for Differential Equations

A simple example of a regular perturbation expansion solution to a differential equation is provided by the first-order system

$$\frac{dx}{dt} = x + \epsilon x^2 \tag{3.1.11}$$

with initial condition $x(0) = A$ and the assumption that $0 < \epsilon \ll 1$. (Here we follow the presentation of Percival and Richards (1982)). Using the expansion (3.1.2) and equating powers of ϵ, we find

$$O(\epsilon^0): \quad \dot{x}_0 = x_0 \tag{3.1.12a}$$

$$O(\epsilon^1): \quad \dot{x}_1 = x_1 + x_0^2 \tag{3.1.12b}$$

$$O(\epsilon^2): \quad \dot{x}_2 = x_2 + 2x_1 x_0 \tag{3.1.12c}$$

The zero-order (integrable) part of the problem (Eq. (3.1.12a)) is easily solved to give

$$x_0(t) = Ae^t \tag{3.1.13}$$

Substitution of this into (3.1.12b) gives

$$\dot{x}_1 = x_1 + A^2 e^{2t} \tag{3.1.14}$$

with the initial condition $x_1(0) = 0$. (By setting all $x_n(0) = 0$ for $n \geq 1$, we ensure that the solution to (3.1.11) satisfies $x(0) = A$ for all ϵ.) The linear inhomogeneous equation (Eq. (3.1.14)) is easily solved to give

$$x_1(t) = A^2 e^t (e^t - 1) \tag{3.1.15}$$

which may, in turn, be substituted into (3.1.12c), yielding the result

$$x_2(t) = A^3 e^t (e^t - 1)^2 \tag{3.1.16}$$

Thus to $O(\epsilon^2)$ the solution of (3.1.11) is

$$x(t) = Ae^t[1 + \epsilon A(e^t - 1) + \epsilon^2 A^2 (e^t - 1)^2] + O(\epsilon^3) \tag{3.1.17}$$

Equation (3.1.11) can also be solved exactly by forming the linearized equation by means of the substitution $x = 1/y$ (an alternative linearization is provided by $x = \dot{y}/y$). Either way, one obtains the result

$$x(t) = \frac{Ae^t}{(1 - \epsilon A(e^t - 1))} \tag{3.1.18}$$

Expanding the denominator gives the result $x(t) = Ae^t \sum_{n=0}^{\infty} [\epsilon A(e^t - 1)]^n$, which obviously matches (3.1.17) up to $O(\epsilon^2)$. This expansion clearly has a radius of convergence determined by $\epsilon A(e^t - 1) < 1$. Thus there is a critical time t_c beyond which the series becomes invalid. This time is easily determined from the condition $A(e^{t_c} - 1) = 1$, that is,

$$t_c = \ln\left(\frac{1 + \epsilon A}{\epsilon A}\right) \tag{3.1.19}$$

For completeness we briefly consider the generalization of (3.1.11), namely

$$\frac{dx}{dt} = x + \epsilon x^\alpha \tag{3.1.20}$$

where α is any power. This equation (often referred to as a *Bernoulli equation*) can be linearized by the substitution $y = x^{1-\alpha}$, leading to the result

$$x(t) = e^t(x(0)^{1-\alpha} + \epsilon(1 - e^{-(1-a)t}))^{1/(1-a)} \tag{3.1.21}$$

This may again be represented as a power series with the critical time determined by

$$t_c = \frac{1}{\alpha - 1} \ln\left(\frac{x(0)^{1-\alpha} + \epsilon}{\epsilon}\right) \tag{3.1.22}$$

from which we deduce that the power-series expansion will converge for all (positive) $t < t_c$ provided that $\alpha > 1$.

We now turn to the series expansion solution of a "real" mechanical problem, namely, the standard example of a perturbed harmonic oscillator of the form

$$\ddot{x} + \omega_0^2 x - \epsilon x^3 = 0 \tag{3.1.23}$$

In fact, for the choice $\epsilon = \omega_0^2/6$ this system corresponds to the pendulum differential equation expanded to third order in the angular displacement x. Although (3.1.23) can be easily integrated in terms of elliptic functions, we want to investigate the validity of the small ϵ expansion about the simple harmonic oscillator solution. Following the presentation of Lictenberg and Lieberman (1983) we substitute (3.1.2) into (3.1.23), and obtain

$$O(\epsilon^0): \quad \ddot{x}_0 + \omega_0^2 x_0 = 0 \tag{3.1.24a}$$

$$O(\epsilon^1): \quad \ddot{x}_1 + \omega_0^2 x_1 = x_0^3 \tag{3.1.24b}$$

$$O(\epsilon^2): \quad \ddot{x}_2 + \omega_0^2 x_2 = 3x_0^2 x_1 \tag{3.1.24c}$$

and so on. The zero-order equation (Eq. (3.1.24a)) has the solution $x_0(t) = A \cos(\omega_0 t)$ for the initial condition $\dot{x}_0(t) = 0$, $x_0(t) = A$. Substitution of x_0 into (3.1.24b) gives

$$\ddot{x}_1 + \omega_0^2 x_1 = \tfrac{1}{4} A^3 (\cos(3\omega_0 t) + 3 \cos(\omega_0 t)) \tag{3.1.25}$$

where we have used the trigonometric identity $\cos^3 x = \frac{1}{4}(\cos(3x) + 3\cos(x))$. Equation (3.1.25) is a linear inhomogeneous equation which can be solved by the standard methods described in Chapter 1. The general solution takes the form

$$x_1(t) = A_1 \cos(\omega_0 t) + B_1 \sin(\omega_0 t) - \frac{A^3}{32\omega_0^2} \cos(3\omega_0 t) + \frac{3A^3}{8\omega_0}(\omega_0 t) \sin(\omega_0 t) \tag{3.1.26}$$

which has the secular term $t \sin(\omega_0 t)$ because of the fact that in (3.1.25) the second right-hand "driving" term is in resonance with intrinsic oscillator frequency. Clearly, the small ϵ expansion has failed—even at first order—since we know that the nonlinearity of the system stabilizes the resonance. The way out of this difficulty was found by Poincaré and others (Poincaré, 1892), who showed that *both* the amplitude x and frequency ω must be expanded in powers of ϵ. (In anticipation of what we shall soon learn about "canonical perturbation" theory of Hamiltonian systems, we comment that both sets of canonical variables (i.e., the p and q) must be expanded in powers of ϵ—so in hindsight, this approach should not be too surprising.) In fact, we now assume that $x = x(\omega t)$ is a periodic function of the variable ωt with period 2π (hints of the angle variable) and expand both x and ω as

$$x = x_0 + \epsilon x_1 + \epsilon^2 x_2 + \cdots, \qquad \omega = \omega_0 + \epsilon \omega_1 + \epsilon \omega_2 + \cdots$$

Now using the relation

$$\frac{d^2}{dt^2} = \omega^2 \frac{d^2}{d(\omega t)^2} = (\omega_0 + \epsilon \omega_1 + \epsilon^2 \omega_2 + \cdots)^2 \frac{d^2}{d(\omega t)^2}$$

in Eq. (3.1.23) and equating powers of ϵ as before, one obtains

$$O(\epsilon^0): \quad \omega_0^2(x_0'' + x_0^2) = 0 \tag{3.1.27a}$$

$$O(\epsilon^1): \quad \omega_0^2 x_1'' + 2\omega_0 \omega_1 x_0'' + \omega_0^2 x_1 = x_0^3 \tag{3.1.27b}$$

$$O(\epsilon^2): \quad \omega_0^2 x_2'' + 2\omega_0 \omega_1 x_1 x_1'' + (\omega_1^2 + 2\omega_0 \omega_2)x_0'' + \omega_0^2 x_2 = 3x_0^2 x_1 \tag{3.1.27c}$$

and so on, where prime denotes differentiation with respect to the new variable ωt. Equation (3.1.27a) again has the standard linear solution $x_0(t) = A \cos(\omega t)$, which, on substitution into (3.1.27b), gives

$$\omega_0^2 x_1'' + \omega_0^2 x_1 = 2A\omega_0 \omega_1 \cos(\omega_0 t) + \tfrac{1}{4} A^3 \cos(3\omega_0 t) + \tfrac{3}{4} A^3 \cos(\omega_0 t) \tag{3.1.28}$$

Since we are requiring that $x = x(\omega t)$ be a periodic function, the terms causing secular behavior (i.e., those proportional to $\cos(\omega_0 t)$) in (3.1.28) must be eliminated. This can be achieved by choosing

$$\omega_1 = -\tfrac{3}{8} A^2 \omega_0 \tag{3.1.29}$$

which leads to the well-behaved periodic solution of (3.1.24b)

$$x_1(t) = A_1 \cos(\omega_0 t) + B_1 \sin(\omega_0 t) - \frac{A^3}{32\omega_0^2} \cos(3\omega_0 t) \tag{3.1.30}$$

This procedure can be continued to higher orders in ϵ with the correct choices of higher-order frequency corrections eliminating secular terms at every order. Notice that this approach requires not only x and ω to be expanded in powers of ϵ but also the assumption that x be a periodic function of the variable ωt. This idea can be given a much more general (and elegant) formulation known as *canonical perturbation theory*.

3.2 CANONICAL PERTURBATION THEORY

Canonical perturbation theory exploits the special properties of action-angle variables. By and large, it is very successful for autonomous systems of one degree of freedom. For systems of two or more degrees of freedom, it succinctly shows up the profound difficulties associated with solving the "many-body" problem. As we shall see, it is these difficulties that are, in a sense, the "seeds" of chaotic behavior. These difficulties were considered, until recently, to be virtually insuperable. For now, we just concentrate on one-degree-of-freedom systems. A careful discussion of these will facilitate our investigation of the many-body problem described in Section 3.3.

The starting point is to express the system as a function of the action-angle variable (I, θ) of the zero-order system, that is,

$$H(I, \theta) = H_0(I) + \epsilon H_1(I, \theta) \tag{3.2.1}$$

These zero-order action–action variables are still perfectly good canonical variables; and by expressing the perturbation term H_1 in terms of them, the canonical equations of motion for (3.2.1) are simply

$$\dot{I} = -\frac{\partial}{\partial \theta} H(I, \theta) \tag{3.2.2a}$$

$$\dot{\theta} = \frac{\partial}{\partial I} H(I, \theta) \tag{3.2.2b}$$

Although in (3.2.1) we have put the perturbation as an order ϵ ($\epsilon \ll 1$) correction, it may be that the perturbing forces can themselves be ordered in a power series in ϵ, that is,

$$H(I, \theta) = H_0(I) + \epsilon H_1(I, \theta) + \epsilon^2 H_2(I, \theta) + O(\epsilon^3) \tag{3.2.3}$$

and so on. The basic idea of canonical perturbation theory is to find a new set of action-angle variables (J, φ) for the perturbed system $H(I, \theta)$ such that there is a canonical transformation to a new Hamiltonian which is a function only of the J, that is, $H(I, \theta) \rightarrow K(J)$.† If this is achieved, (3.2.3)

---

†My colleague Professor H. Goldstein has suggested that if H is called the Hamiltonian, K should be called the Kamiltonian.

becomes a completely integrable system whose equations of motion are trivially integrated.

3.2.a Perturbation Series for the Hamilton–Jacobi Equation

For the zero-order system, Hamilton's equations are just

$$\dot{I} = -\frac{\partial}{\partial \theta} H_0(I) = 0 \qquad (3.2.4a)$$

$$\dot{\theta} = \frac{\partial}{\partial I} H_0(I) = \omega_0(I) \qquad (3.2.4b)$$

and for the perturbed system they become

$$\dot{I} = -\frac{\partial}{\partial \theta} H(I, \theta) = -\epsilon \frac{\partial H_1}{\partial \theta} - \epsilon^2 \frac{\partial H_2}{\partial \theta} + O(\epsilon^3) \qquad (3.2.5a)$$

$$\dot{\theta} = \frac{\partial}{\partial I} H(I, \theta) = \omega_0(I) + \epsilon \frac{\partial H_1}{\partial I} + \epsilon^2 \frac{\partial H_2}{\partial I} + O(\epsilon^3) \qquad (3.2.5b)$$

The aim is to find the generating function $S = S(\theta, J)$, where θ is the old angle (coordinate) and J is the new action (momentum) (cf. the type-F_2 generating function in Section 2.3), which effects the desired transformation through the standard relations

$$I = \frac{\partial}{\partial \theta} S(\theta, J), \qquad \varphi = \frac{\partial}{\partial J} S(\theta, J) \qquad (3.2.6)$$

S is also expanded in power series in ϵ, that is,

$$S = S_0 + \epsilon S_1 + \epsilon^2 S_2 + \cdots \qquad (3.2.7)$$

where $S_0 = J\theta$ is the identity generator (see Eq. (2.3.21)). Using the relations (3.2.6), one obtains the time-independent Hamilton–Jacobi equation

$$H_0\left(\frac{\partial S}{\partial \theta}\right) + \epsilon H_1\left(\frac{\partial S}{\partial \theta}, \theta\right) + \epsilon^2 H_2\left(\frac{\partial S}{\partial \theta}, \theta\right) + \cdots = K(J) \qquad (3.2.8)$$

Furthermore, one now expands the "new" Hamiltonian $K(J)$ in an ϵ series, that is,

$$K(J) = K_0(J) + \epsilon K_1(J) + \epsilon^2 K_2(J) + \cdots$$

The next stage is to use the series expansion (3.2.7) for S in each of the terms on the right-hand side of (3.2.8) and expand these terms out in a Taylor series. Thus for example, the term involving H_0 is expanded as

$$H_0\left(\frac{\partial S}{\partial \theta}\right) = H_0\left(J + \epsilon \frac{\partial S_1}{\partial \theta} + \epsilon^2 \frac{\partial S_2}{\partial \theta} + \cdots\right)$$

$$= H_0(J) + \epsilon \frac{\partial S_1}{\partial \theta} \frac{\partial H_0}{\partial J} + \epsilon^2 \left\{\frac{1}{2}\left(\frac{\partial S_1}{\partial \theta}\right)^2 \frac{\partial^2 H_0}{\partial J^2} + \frac{\partial S_2}{\partial \theta} \frac{\partial H_0}{\partial J}\right\} + O(\epsilon^3)$$

Expanding up to terms of $O(\epsilon^2)$, the Hamilton–Jacobi equation (Eq. (3.2.8)) thus becomes

$$H_0(J) + \epsilon \left\{\frac{\partial S_1}{\partial \theta} \frac{\partial H_0}{\partial J} + H_1\right\} + \epsilon^2 \left\{\frac{1}{2}\left(\frac{\partial S_1}{\partial \theta}\right)^2 \frac{\partial^2 H_0}{\partial J^2} + \frac{\partial S_2}{\partial \theta} \frac{\partial H_0}{\partial J} + \frac{\partial S_1}{\partial \theta} \frac{\partial H_1}{\partial J} + H_2\right\}$$

$$= K_0(J) + \epsilon K_1(J) + \epsilon^2 K_2(J)$$

Equating powers of ϵ leads to the system of equations

$$O(\epsilon^0): \quad H_0(J) = K_0(J) \tag{3.2.9a}$$

$$O(\epsilon^1): \quad \frac{\partial S_1}{\partial \theta} \frac{\partial H_0}{\partial J} + H_1(J, \theta) = K_1(J) \tag{3.2.9b}$$

$$O(\epsilon^2): \quad \frac{1}{2}\left(\frac{\partial S_1}{\partial \theta}\right)^2 \frac{\partial^2 H_0}{\partial J^2} + \frac{\partial S_2}{\partial \theta} \frac{\partial H_0}{\partial J} + \frac{\partial S_1}{\partial \theta} \frac{\partial H_1}{\partial J} + H_2(J, \theta) = K_2(J) \tag{3.2.9c}$$

3.2.b Solutions to First Order in ϵ

From (3.2.9a) we see that we can write $K_0(J) = H_0(J)$; that is, $K_0(J)$ is found by replacing I by J in the zero-order Hamiltonian H_0. At $O(\epsilon)$ (i.e., Eq. (3.2.9b)), we have (again) replaced I by J in the terms involving H_0 and H_1 (their functional form remaining unaltered) and write

$$K_1(J) = \omega_0(J)\frac{\partial}{\partial \theta} S_1(\theta, J) + H_1(J, \theta) \tag{3.2.10}$$

At this point, one exploits the periodicity of the motion in the (old) angle variable θ and averages Eq. (3.2.10) over this variable. Since S_1 and subsequent S_i, $i = 2, 3, \ldots$, are assumed periodic in θ and we require that the mean of their derivative vanishes, one obtains the first-order energy correction

$$K_1(J) = \overline{H_1}(J, \theta) \tag{3.2.11}$$

where

$$\overline{H_1}(J, \theta) = \frac{1}{2\pi} \int_0^{2\pi} H_1(J, \theta) \, d\theta \qquad (3.2.12)$$

In some cases it may be that $\overline{H_1}$ itself is zero. Having obtained $K_1(J)$, we can now solve (3.2.10) for S_1, that is,

$$\frac{\partial}{\partial \theta} S_1(\theta, J) = \frac{1}{\omega_0(J)} [K_1(J) - H_1(J, \theta)] \qquad (3.2.13)$$

Since $K_1(J)$ is just the "mean" part of $H_1(J, \theta)$ (see Eq. (3.2.11)), it is convenient to denote the right-hand side of (3.2.13) as $\tilde{H}_1 = H_1(J, \theta) - K_1(J)$; that is, $\tilde{H}_1$ represents the "periodic" part of H_1. So now we write (3.2.13) as

$$\frac{\partial}{\partial \theta} S_1(\theta, J) = -\frac{1}{\omega_0(J)} \tilde{H}_1(J, \theta) \qquad (3.2.14)$$

Both sides of (3.2.14) are periodic in θ, and we represent them as Fourier series in the angle variable θ, that is,

$$\tilde{H}_1(J, \theta) = \sum_{k=1}^{\infty} A_k(J) e^{ik\theta} \qquad (3.2.15a)$$

$$S_1(J, \theta) = \sum_{k=0}^{\infty} B_k(J) e^{ik\theta} \qquad (3.2.15b)$$

The summation in (3.2.15a) excludes $k = 0$ since, by definition, $\tilde{H}_1$ is taken to be purely periodic (i.e., there is no constant term). Although S_1 can include an arbitrary function of J (corresponding to the $k = 0$ term in (3.2.15b)), it is convenient to set it to zero and from now on all such summation will exclude the zero term.

Equating coefficients in (3.2.14) we find that

$$B_k(J) = \frac{i}{k\omega_0(J)} A_k(J) \qquad (3.2.16)$$

where the $A_k(J)$ are assumed to be known. Thus the first-order generating function takes the form

$$S_1(J, \theta) = \sum_k \frac{iA_k(J)}{k\omega_0(J)} e^{ik\theta} \qquad (3.2.17)$$

The generating function relations (3.2.6) can now be used to find the "new" action-angle variable correct to first order in ϵ,† that is,

$$\varphi = \theta + \epsilon \frac{\partial}{\partial J} S_1(J, \theta) \qquad (3.2.18a)$$

$$J = I - \epsilon \frac{\partial}{\partial \theta} S_1(J, \theta) \qquad (3.2.18b)$$

and the corrected frequency is given by

$$\omega(J) = \omega_0(J) + \epsilon \frac{\partial}{\partial J} K_1(J) \qquad (3.2.18c)$$

Thus the perturbed motion consists of the unperturbed motion plus $O(\epsilon)$ oscillatory corrections. Secular terms do not appear. (A potential source of difficulty is if the zero order frequency $\omega_0(I)$ is, or is close to, zero—this can occur for motion near a separatrix.)

3.2.c Solutions to Higher Order in ϵ

So far, we have only taken the calculation to first order in ϵ. At second order in ϵ (3.2.9c) we have (again replacing I by J in the terms involving H_2 as well as H_1 and H_0)

$$\frac{\partial H_0}{\partial J} \frac{\partial S_2}{\partial \theta} = K_2(J) - H_2(J, \theta) - \frac{1}{2} \left(\frac{\partial S_1}{\partial \theta} \right)^2 \frac{\partial^2}{\partial J^2} H_0(J) - \frac{\partial S_1}{\partial \theta} \frac{\partial}{\partial J} H_1(J, \theta) \quad (3.2.19)$$

which, on averaging over θ, gives the second-order energy corrections

$$K_2(J) = \overline{H_2(J, \theta)} - \frac{1}{2} \overline{\left(\frac{\partial S_1}{\partial \theta} \right)^2 \frac{\partial^2}{\partial J^2} H_0(J)} - \overline{\frac{\partial H_1}{\partial \theta} \frac{\partial}{\partial J} H_1(J, \theta)} \quad (3.2.20)$$

One can now solve for S_2 from

$$\frac{\partial}{\partial \theta} S_2(\theta, J) = -\frac{1}{\omega_0(J)} \left[\tilde{H}_2(J, \theta) - \frac{1}{2} \left(\frac{\partial S_1}{\partial \theta} \right)^2 \frac{\partial^2}{\partial J^2} H_0(J) - \frac{\partial S_1}{\partial \theta} \frac{\partial}{\partial J} H_1(J, \theta) \right]$$

$$(3.2.21)$$

where $\tilde{H}_2(J, \theta)$ represents the periodic part of H_2 (which is defined exactly like $\tilde{H}_1(J, \theta)$). This procedure can be carried to all orders in ϵ (see, for

---

†In order to evaluate the J to first order in ϵ in (3.2.18b), the argument J in $S_1(J, \theta)$ is simply replaced by the symbol I.

example, Lichtenberg and Lieberman (1983) or Born (1960)). Thus at order ϵ^n, one obtains an equation of the form

$$\frac{\partial H_0}{\partial J}\frac{\partial S_n}{\partial \theta} = K_n(J) - V_n(\theta, J) \tag{3.2.22}$$

where V_n is the collection of terms computed at lower orders of ϵ. Using the averaging procedure, one obtains

$$K_n(J) = \overline{V_n}(\theta, J) \tag{3.2.23}$$

and hence

$$\frac{\partial}{\partial \theta} S_n(J, \theta) = -\frac{1}{\omega_0(J)}\tilde{V}_n(\theta, J) \tag{3.2.24}$$

where $\tilde{V}_n$ represents the periodic part of V_n. Equation (3.2.24) is then solved by expanding both sides in Fourier series and equating coefficients.

3.2.d The Perturbed Oscillator

As a simple example of the method, we return to (3.1.23) and solve it by canonical perturbation theory. This equation is expressed in the form

$$H(p, q) = H_0(p, q) + \epsilon H_1(p, q) \tag{3.2.25}$$

where

$$H_0(p, q) = \tfrac{1}{2}(p^2 + \omega_0^2 q^2) \tag{3.2.26a}$$

and

$$H_1(p, q) = q^3 \tag{3.2.26b}$$

The action-angle solution to (3.2.25) was already determined in Chapter 2, that is,

$$H_0(I) = I\omega_0 \tag{3.2.27a}$$

and

$$q = \sqrt{\frac{2I}{\omega_0}}\sin\theta \tag{3.2.27b}$$

where $\theta = \omega_0 t + \delta$. The result (3.2.27b) can be used to express H_1 in terms of the zero-order action-angle variables, that is,

$$H_1(I, \theta) = \left(\frac{2I}{\omega_0}\right)^{3/2}\sin^3\theta \tag{3.2.28}$$

Following the procedures just described, we find at $O(\epsilon)$

$$\omega_0 \frac{\partial}{\partial \theta} S_1(\theta, J) = K_1(J) - H_1(J, \theta) \tag{3.2.29}$$

and, on averaging over θ, it is found that $\overline{H_1(J, \theta)} = 0$, that is, there is no first-order correction to the energy. Equation (3.2.29) is easily solved for S_1 to give

$$S_1 = \frac{1}{\omega_0} \left(\frac{2J}{\omega_0}\right)^{3/2} \left[\frac{1}{3} \sin^2 \theta \cos \theta + \frac{2}{3} \cos \theta\right] \tag{3.2.30}$$

At $O(\epsilon^2)$ one obtains

$$\omega_0 \frac{\partial}{\partial \theta} S_2(\theta, J) = K_2(J) - \frac{1}{2} \left(\frac{\partial S_1}{\partial \theta}\right) \frac{\partial}{\partial J} H_1(J, \theta) \tag{3.2.31}$$

where the term involving $(\partial S_1/\partial \theta)^2$ (cf. (Eq. (3.2.9c)) is absent, since for this linear zero-order problem we have $\partial^2 H_0/\partial I^2 = 0$. Using (3.2.30) and (3.2.29) and averaging over θ, the second-order energy correction is easily found to be

$$K_2(J) = -\frac{15 \pi J^2}{2\omega_0^4} \tag{3.2.32}$$

Thus to second order in ϵ, the "new" Hamiltonian takes the form

$$K(J) = \omega_0 J - \epsilon^2 \frac{15 \pi J^2}{4\omega_0^4} \tag{3.2.33}$$

and hence the new frequency is

$$\omega(J) = \frac{\partial}{\partial J} K(J) = \omega_0 - \epsilon^2 \frac{15 \pi J}{2\omega_0^4} \tag{3.2.34}$$

which displays the (here weak) amplitude dependence typical of nonlinear systems.

3.3 MANY DEGREES OF FREEDOM AND THE PROBLEM OF SMALL DIVISORS

For systems of two or more degrees of freedom, canonical perturbation theory is formulated in exactly the same way as before—but now profound difficulties arise, even at first order in ϵ. Consider the system

$$H(\mathbf{I}, \boldsymbol{\theta}) = H_0(\mathbf{I}) + \epsilon H_1(\mathbf{I}, \boldsymbol{\theta}) \tag{3.3.1}$$

where $\mathbf{I} = (I_1, \ldots, I_n)$ and $\boldsymbol{\theta} = (\theta_1, \ldots, \theta_n)$ are the n-dimensional vectors of actions and angles, respectively. Hamilton's equations are expressed as

$$\dot{\mathbf{I}} = -\boldsymbol{\nabla}_\theta H_0(\mathbf{I}) = 0 \tag{3.3.2a}$$

and

$$\dot{\boldsymbol{\theta}} = \boldsymbol{\nabla}_I H_0(\boldsymbol{I}) = \boldsymbol{\omega}_0(\mathbf{I}) \tag{3.3.2b}$$

where $\boldsymbol{\nabla}_\theta = (\partial\theta_1, \ldots, \partial\theta_n)$, $\boldsymbol{\nabla}_I = (\partial I_1, \ldots, \partial I_n)$, and $\boldsymbol{\omega}_0(\mathbf{I}) = (\omega_{0,1}(\mathbf{I}), \ldots, \omega_{0,n}(\mathbf{I}))$ is the n-dimensional frequency vector. Again we aim to construct perturbatively a new set of canonical variables $(\mathbf{J}, \boldsymbol{\varphi})$ which transform the perturbed system to integrable form (i.e., $H(\mathbf{I}, \boldsymbol{\theta}) = K(\mathbf{J})$) by means of a generating function $S = S(\boldsymbol{\theta}, \mathbf{J})$ satisfying the relations

$$\mathbf{I} = \boldsymbol{\nabla}_\theta S(\boldsymbol{\theta}, \mathbf{J}), \qquad \boldsymbol{\varphi} = \boldsymbol{\nabla}_J S(\boldsymbol{\theta}, \mathbf{J}) \tag{3.3.3}$$

S is expanded as an ϵ power series of the form

$$S = \boldsymbol{\theta} \cdot \mathbf{J} + \epsilon S_1 + \epsilon^2 S_2 + \cdots \tag{3.3.4}$$

where the leading term corresponds to the usual identity transformation. The Hamilton–Jacobi equation

$$H_0(\boldsymbol{\nabla}_\theta S) + \epsilon H_1(\boldsymbol{\nabla}_\theta S, \boldsymbol{\theta}) = K_0(\mathbf{J}) + \epsilon K_1(\mathbf{J}) + O(\epsilon^2) \tag{3.3.5}$$

is treated in exactly the same way as before, leading to

$$O(\epsilon^0): \quad H_0(\mathbf{J}) = K_0(\mathbf{J}) \tag{3.3.6a}$$

$$O(\epsilon^1): \quad \boldsymbol{\nabla}_\theta S_1 \cdot \boldsymbol{\nabla}_I H_0(\mathbf{J}) + H_1(\mathbf{J}, \boldsymbol{\theta}) = K_1(\mathbf{J}) \tag{3.3.6b}$$

and so on, where we have again replaced $\mathbf{I}$ by $\mathbf{J}$ in the left-hand side of (3.3.6b). Using (3.3.2b) in (3.3.6b) gives

$$\boldsymbol{\omega}_0(\mathbf{J}) \cdot \boldsymbol{\nabla}_\theta S_1(\boldsymbol{\theta}, \mathbf{J}) = K_1(\mathbf{J}) - H_1(\mathbf{J}, \boldsymbol{\theta}) \tag{3.3.7}$$

The first-order energy correction is again determined by assuming that S_1 is a periodic function of $\boldsymbol{\theta}$ and averaging over all the angle variables, that is,

$$K_1(\mathbf{I}) = \overline{H_1(\mathbf{J}, \boldsymbol{\theta})} \tag{3.3.8}$$

where

$$\overline{H_1}(\mathbf{J}, \boldsymbol{\theta}) = \int_0^{2\pi} d\theta_1 \cdots \int_0^{2\pi} d\theta_n H_1(\mathbf{J}, \boldsymbol{\theta}) \qquad (3.3.9)$$

The real problem arises in trying to solve (3.3.7) for S_1. Expanding both S_1 and the "periodic" part of H_1 in n-dimensional Fourier series, that is,

$$S_1(\boldsymbol{\theta}, \mathbf{J}) = \sum_{\mathbf{m}}' S_{1\mathbf{m}} \, e^{i\mathbf{m}\cdot\boldsymbol{\theta}} \qquad (3.3.10)$$

$$H_1(\boldsymbol{\theta}, \mathbf{J}) = \sum_{\mathbf{m}}' H_{1\mathbf{m}} \, e^{i\mathbf{m}\cdot\boldsymbol{\theta}} \qquad (3.3.11)$$

where $\mathbf{m} = m_1, \ldots, m_n$ and the prime denotes the absence of the term $\mathbf{m} = (0, \ldots, 0)$ in the summation, one easily finds that

$$S_1(\boldsymbol{\theta}, \mathbf{J}) = i \sum_{\mathbf{m}}' \frac{H_{1\mathbf{m}} \, e^{i\mathbf{m}\cdot\boldsymbol{\theta}}}{\mathbf{m} \cdot \boldsymbol{\omega}_0(\mathbf{J})} \qquad (3.3.12)$$

3.3.a Small Divisors

It might appear, in keeping with the results obtained for one-degree-of-freedom systems, that one could proceed in this manner to higher orders in ϵ. However, if the fundamental frequencies $\boldsymbol{\omega}_0(\mathbf{J})$ are commensurable (i.e., $\mathbf{m} \cdot \boldsymbol{\omega}_0(\mathbf{J}) = 0$), the sum in (3.3.12) will obviously diverge. In fact, even if the $\boldsymbol{\omega}_0$ are incommensurable, one can always find some (large) $\mathbf{m}$ such that $\boldsymbol{\omega}_0 \cdot \mathbf{m}$ is arbitrarily small. This is the infamous problem of *small divisors* that so plagued advances in classical mechanics over the past hundred years.

In some cases it is possible that convergence of the series might be maintained by a corresponding smallness of the numerators. However, it was demonstrated a long time ago, by Bruns, that (roughly speaking) the values of the $\boldsymbol{\omega}_0(\mathbf{J})$ for which the series are (absolutely) convergent and those for which the series are not convergent lie arbitrarily close to each other. From this it follows, owing to the $\mathbf{J}$ dependence of the $\boldsymbol{\omega}_0$, that S derived in the way described above is not a continuous function of the $\mathbf{J}$. However, since this continuity is an assumed prerequisite of the whole perturbation procedure (see Eqs. (3.3.2) and (3.3.3)), it follows that the perturbation series, even if they happen to converge for certain $\boldsymbol{\omega}_0(\mathbf{J})$, do not necessarily represent the actual motion. Further research by Poincaré conclusively demonstrated that Fourier-like series (such as (3.3.10) and (3.3.11)) could not provide a convergent representation for perturbed systems. Thus, such fundamental issues as the long time stability of planetary orbits in the solar system could not be satisfactorily resolved. Here, the best that could be obtained were results for a finite time stability based on some truncation of the series.

3.3.b The Fundamental Problem

The best efforts by some of the most eminent mathematicians of the time failed to resolve the small divisor problem. (Poincaré called it the "fundamental problem" of classical mechanics, and it seemed as though there was an insurmountable obstacle to further progress. This state of affairs was nicely summarized by Max Born, who was interested in classical perturbation theory to solve problems in the "old quantum theory," in the statement (Born, 1960):

> It would indeed be remarkable if Nature fortified herself against further advances in knowledge behind the analytical difficulties of the many-body problem.

One opinion held was that the addition of even the smallest nonintegrable perturbation would render the system ergodic; that is, the system would explore the entire energy shell in a statistical manner such that time average equals phase-space average. This was the motivation behind the famous Fermi–Ulam–Pasta calculation—one of the first dynamics calculations every carried out on a computer in the early 1950s. These workers took a chain of harmonic oscillators coupled with a cubic non-linearity and investigated how the energy in one mode would spread among the rest. They expected to see a statistical sharing of energy. Much to their surprise they found it to cycle periodically, implying that the system was much "more" integrable than they had thought. (In fact, it turned out to be integrable beyond their wildest dreams—the continuum limit of their model is a remarkable nonlinear partial differential equation called the *Kortweg–de Vries equation.*†)

The resolution of the "fundamental problem" came in the early 1960s with the announcement of the now famous Kolmogorov–Arnold–Moser (KAM) theorem, which is described in the following sections. However, we conclude this section by remarking that perturbation theory and its many variants, including, for example, the relatively recently developed Lie transform methods, are still a most valuable tool. Many of these techniques are nicely reviewed and illustrated by Lichtenberg and Lieberman (1983).

3.4 THE KOLMOGOROV–ARNOLD–MOSER THEOREM

The big breakthrough came in 1954, when Kolmogorov outlined a theorem that was subsequently proved by Arnold and Moser in the early 1960s (Kolmogorov, 1954; Moser, 1962; Arnold, 1963). Following the notation of Arnold, we assume that an (integrable) Hamiltonian function H_0 is pertur-

†This topic is discussed in Chapter 7.

bed by some function H_1 such that

$$H = H_0(\mathbf{I}) + H_1(\mathbf{I}, \boldsymbol{\theta}) \tag{3.4.1}$$

where H_1 is taken to be periodic in the original angle variables (i.e., $H_1(\mathbf{I}, \boldsymbol{\theta} + 2\pi) = H_1(\mathbf{I}, \boldsymbol{\theta})$) and is required, in some sense, to be "small enough" (i.e., $H_1 \ll 1$). (Think of H_1 being multiplied by some small parameter ϵ, for example.) Hamilton's equations are thus

$$\dot{I}_i = -\frac{\partial H_1}{\partial \theta_i}, \qquad \dot{\theta}_i = \omega_i(\mathbf{I}) + \frac{\partial H_1}{\partial I_i} \tag{3.4.2}$$

where the ω_i are the unperturbed frequencies, that is, $\omega_i = \partial H_0/\partial I_i$.

For most initial data, Kolmogorov sketched the proof that the motion of (3.4.1) remains predominantly quasiperiodic, that is, confined to tori, and that the complement of the quasi-periodic motion (i.e., the chaotic motion) has as small a (Lebesgue) measure as H_1 is small. The KAM theorem is formulated by assuming that the Hamiltonian is analytic in a complex domain (Ω) of phase space and that the unperturbed motion is non-degenerate, that is,

$$\det\left|\frac{\partial \omega_i}{\partial I_j}\right| = \det\left|\frac{\partial^2 H_0}{\partial I_i \partial I_j}\right| \neq 0 \tag{3.4.3}$$

The next step is to identify, in the unperturbed system, a particular torus, call it T_0, through its associated set of frequencies $\boldsymbol{\omega} = \boldsymbol{\omega}(\mathbf{I})$. Specifically, we select an incommensurate frequency vector $\boldsymbol{\omega} = \boldsymbol{\omega}^*$ (i.e., $\boldsymbol{\omega} \cdot \mathbf{k} \neq 0$ for all integers k_i) and state that the equations of the invariant torus $T_0(\boldsymbol{\omega}^*)$ of the unperturbed system are $\mathbf{I} = \mathbf{I}^*$, where $\boldsymbol{\omega}(\mathbf{I}^*) = \boldsymbol{\omega}^*$. That is, the system has frequencies $\boldsymbol{\omega}^*$ on $T_0(\boldsymbol{\omega}^*)$ and $\dot{\boldsymbol{\theta}} = \boldsymbol{\omega}^*$ is the linear flow on the torus T_0.

The stage is now set to state a version of the remarkable KAM theorem:

Theorem (Arnold and Avez (1968), Theorem 21.7). If H_1 is small enough, then for almost all $\boldsymbol{\omega}^*$ there exists an invariant torus $T(\boldsymbol{\omega}^*)$ of the perturbed system such that $T(\boldsymbol{\omega}^*)$ is "close to" $T_0(\boldsymbol{\omega}^*)$.

Moreover, the tori $T(\boldsymbol{\omega}^*)$ form a set of positive measures whose complement has a measure that tends to zero as $|H_1| \to 0$.

The proof of this theorem—which is highly nontrivial, even if it is easily stated—is due to Arnold (1963). The version proved by Moser (1962) deals with an equivalent class of mappings (see Section 3.5 and Chapter 4). We will not give the actual proof here but, instead, will attempt to discuss some of the basic ideas behind it. It is difficult to overstate the profound importance of the KAM theorem since it breaks the age-old deadlock of the small divisor problem in classical perturbation theory and, indeed, provides the starting point for an understanding of the appearance of chaos in Hamiltonian systems.

Notice that the whole philosophy of the KAM theorem is different from that of traditional perturbation theory. Rather than attempting to construct global solutions to the Hamilton–Jacobi equation by building on the unperturbed motion, KAM concentrates on proving the existence of individual tori in the (weakly) perturbed system that satisfy certain conditions. That is, they were able to prove that a given torus $T(\omega^*)$ exists if the frequency ω^* is sufficiently irrational. This approach is analogous to finding a particular root of an (algebraic) equation. The two basic ingredients of the proof are:

(*i*) A "superconvergent" root-finding procedure, which is the function space analogue of the old Newton–Raphson method. This procedure has remarkable convergence properties that can "outstrip" the divergences found in traditional perturbation theory.

(*ii*) A number-theoretic analysis that determines how irrational the frequencies ω^* must be for the "root" (i.e., the torus $T(\omega^*)$) to exist.

3.4.a Superconvergent Perturbation Theory

The superconvergence of the Newton–Raphson technique is easily illustrated by the example of finding the zero of some function $f(x)$ (i.e., $f(x) = 0$), given some initial condition $x = x_0$. Expanding $f(x)$ about $x = x_0$ gives

$$f(x_0 + (x - x_0)) = \sum_{n=0}^{\infty} f_n (x - x_0)^n / n! = 0, \qquad f_n = \left(\frac{\partial^n f}{\partial x^n}\right)_{x = x_0}$$

Writing out the first few terms, we have

$$f_0 + (x - x_0)f_1 + \tfrac{1}{2}(x - x_0)^2 f_2 + \tfrac{1}{6}(x - x_0)^3 f_3 + \cdots = 0$$

or

$$(x - x_0) + \frac{1}{2}(x - x_0)^2 \left(\frac{f_2}{f_1}\right) + \frac{1}{6}(x - x_0)^3 \left(\frac{f_3}{f_1}\right) + \cdots = -\frac{f_0}{f_1}$$

Let $-f_0/f_1 = \epsilon$, and revert the series to express $(x - x_0)$ in powers of ϵ, that is,

$$(x - x_0) = \epsilon + \epsilon^2 \left(\frac{-f_2}{2f_1}\right) + \epsilon^3 \left(2\left(\frac{f_2}{2f_1}\right)^2 - \frac{f_3}{6f_1}\right) + \cdots$$

Thus one can build up a "perturbation series" for finding the "root" x.

Notice this is a power series of the form $\sum c_n \epsilon^n$, where the coefficients c_n are all functions of the "unperturbed" solution, that is, $c_n = c_n(f_i, i = 1, \ldots, n)$, where $f_i = (\partial^i f / \partial x^i)_{x = x_0}$. What we have constructed is, of course, just a regular perturbation expansion of the type introduced in Section 3.1.a.

Now consider Newton's method. Starting at x_0, we obtain the next approximation x_1 from

$$f(x) = f(x + (x - x_0)) \simeq f(x_0) + (x_1 - x_0) f'(x_0) = 0$$

that is,

$$x_1 - x_0 = \frac{-f(x_0)}{f'(x_0)} \equiv \epsilon_1$$

Now, *using x_1 as the initial guess,*

$$f(x) = f(x_1 + (x - x_1)) \simeq f(x_1) + (x_2 - x_1) f'(x_1) = 0$$

we obtain

$$x_2 - x_1 = \frac{-f(x_1)}{f'(x_1)} \equiv \epsilon_2$$

Repeating this procedure at each step leads to

$$x_n - x_{n-1} = \frac{-f(x_{n-1})}{f'(x_{n-1})} \equiv \epsilon_n \tag{3.4.4}$$

We can determine the convergence by estimating ϵ_{n+1} in terms of ϵ_n. Now

$$f(x_n) \simeq f(x_{n-1}) + (x_n - x_{n-1}) f'(x_{n-1}) + \cdots$$

$$= f(x_{n-1}) + \epsilon_n f'(x_{n-1}) + \frac{\epsilon_n^2}{2} f''(x_{n-1}) + \cdots$$

and

$$f'(x_n) = f'(x_{n-1}) + \epsilon_n f''(x_{n-1}) + \frac{\epsilon_n^2}{2} f'''(x_{n-1}) + \cdots$$

Therefore

$$\frac{f(x_n)}{f'(x_n)} \simeq \frac{f(x_{n-1}) + \epsilon_n f'(x_{n-1}) + \frac{1}{2}\epsilon_n^2 f''(x_{n-1}) + \cdots}{f'(x_{n-1}) + \epsilon_n f''(x_{n-1}) + \cdots}$$

$$= \frac{f(x_{n-1})}{f'(x_{n-1})} + \epsilon_n + \frac{\epsilon_n^2}{2} \frac{f''(x_{n-1})}{f'(x_{n-1})} + \cdots = -\epsilon_n + \epsilon_n + \frac{\epsilon_n^2}{2} \frac{f''(x_{n-1})}{f'(x_{n-1})}$$

That is,

$$\epsilon_{n+1} = \frac{-f(x_n)}{f'(x_n)} \simeq -\frac{1}{2}\,\epsilon_n^2\,\frac{f''(x_{n-1})}{f'(x_{n-1})}$$

Hence $\epsilon_{n+1} = O(\epsilon_n^2)$. Thus, the iteration is quadratically convergent, that is,

$$x - x_0 = \sum_n c_n \epsilon_n = \sum_n c_n \epsilon^{2^n} = c_1\epsilon + c_2\epsilon^2 + c_3\epsilon^4 + c_4\epsilon^8 + \cdots \quad (3.4.5)$$

where $c_n = c_n(x_{n-1})$.

The key to this remarkable convergence and its crucial difference to the regular perturbation approach is that at each stage, f is evaluated at the latest solution rather than the original unperturbed solution x_0. It is this that lies at the heart of the convergence properties of the KAM theorem.

3.4.b   Number-Theoretic Properties of the Frequencies

A useful paraphrase of the KAM theorem is that "For sufficiently small perturbation, almost all tori are preserved." The important and delicate issues are the estimates of "sufficiently small" and "almost all." The latter notion is determined by the second ingredient in the theorem, namely, the number-theoretic properties of the ω^*. The theorem excludes tori with rationally related frequencies, that is, $n - 1$ conditions of the form

$$\boldsymbol{\omega} \cdot \mathbf{k} = 0 \qquad\qquad (3.4.6)$$

These tori are, in some sense (to be discussed in detail later), "destroyed." Already there might seem to be a contradiction since we have said, in the context of small divisors, that any irrational number can be approximated arbitrarily closed by a rational number. This would seem to imply that all the tori are, in fact, destroyed. This is where the "almost all" comes in and is related to the number-theoretic problem of just how closely irrational numbers can be approximated by rationals.

For a two-degree-of-freedom system, the condition for closed orbits is just

$$\frac{\omega_1}{\omega_2} = \sigma = \frac{r}{s}, \qquad r, s \text{ integers} \qquad (3.4.7)$$

For quasi-periodic motion the frequency ratio σ is an irrational number†

---

†The discovery of irrational numbers by a student of Pythagoras was suppressed by Pythagoras's followers, and the student was put to death (after he tried to leak the news to the *Daily Papyrus*).

and cannot be written exactly as r/s but can be approximated arbitrarily closely by rational σ.

A simple example of a rational approximation to an irrational number is the approximations of $\pi = 3.141592654$, namely,

$$\sigma \simeq \frac{r}{s} = \frac{3}{1}, \ \frac{31}{10}, \ \frac{314}{100}, \ \frac{3142}{1000}, \ \cdots$$

which are very crude rational approximations to π. For these approximations we obtain

$$\left| \sigma - \frac{r}{s} \right| < \frac{1}{s} \tag{3.4.8}$$

The best rational approximations are obtained by using *continued fractions* namely,

$$\sigma = a_0 + 1 \over a_1 + 1 \over a_2 + 1 \over a_3 + \cdots \tag{3.4.9}$$

The numbers $a_i (i = 0, 1, 2, \ldots)$ can be obtained as follows: a_0 is the integral part of σ, a_1 is the integer part of the reciprocal of $\sigma - a_0$, a_2 is the integer part of the reciprocal of the remainder of a_1, and so on, that is,

$$\pi = 3 + 1 \over 7 + 1 \over 15 + 1 \over 1 + 1 \over 292 + \cdots$$

The successive approximants of a continued fraction are just the sequence

$$\sigma_n = \frac{r_n}{s_n} = a_0 + 1 \over a_1 + 1 \over a_2 + \cdot \atop \cdot \atop \cdot \atop \frac{1}{a_n} \tag{3.4.10}$$

So, the successive approximants to π are

$$\sigma_1 = \tfrac{22}{7} = 3.1429\ldots$$

$$\sigma_2 = \tfrac{333}{106} = 3.14151\ldots$$

$$\sigma_3 = \tfrac{355}{113} = 3.1415929\ldots$$

the last one being known to Tsu-Chung Chi in fifth-century China (later found by A. van Roomen in sixteenth-century Europe). The sequences σ_n always converge, with successive approximants being greater than or less than σ. Moreover, one may show that

$$\left| \sigma - \frac{r_n}{s_n} \right| < \frac{1}{s_n s_{n-1}} \qquad (3.4.11)$$

that is, successive approximants converge at least as fast as $1/s^2$, whereas the decimal truncation only went as $1/s$.

There is an extensive and beautiful theory of continued fractions. Some numbers (e.g., the transcendental numbers like π and e) have very rapidly converging approximants. Quadratically irrational numbers converge most slowly: Lagrange showed that for this class of numbers the continued fraction is periodic, for example,

$$\sqrt{3} = 1 + \cfrac{1}{2 + \cfrac{1}{1 + \cfrac{1}{2 + \cfrac{1}{1 + \cfrac{1}{2 + \cdots}}}}}$$

A very famous irrational number is the golden mean

$$\frac{\sqrt{5} - 1}{2} = 1 + \cfrac{1}{1 + \cfrac{1}{1 + \cfrac{1}{1 + \cdots}}}$$

There is still much that is not known about continued fraction (e.g., the representation of e^3). Continued fractions can also be used to represent functions—Euler used them to find solutions of the Riccati equation. (See, for example, the book by Ince (1978, p. 178).)

KAM showed that the preserved tori satisfy the irrationality condition

$$\left| \frac{\omega_1}{\omega_2} - \frac{r}{s} \right| > \frac{K(\epsilon)}{s^{2.5}} \qquad \text{for all } r, s. \qquad (3.4.12)$$

Not much is known about $K(\epsilon)$, other than $K(\epsilon) \to 0$ as $\epsilon \to 0$. The "des-

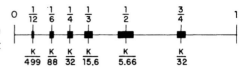

Figure 3.1 Deleted zones of width $k/s^{2.5}$ at rational points in the unit interval.

troyed" tori are the complementary set satisfying

$$\left| \frac{\omega_1}{\omega_2} - \frac{r}{s} \right| < \frac{K(\epsilon)}{s^{2.5}} \tag{3.4.13}$$

This is more restrictive than the strict commensurability condition $n_1\omega_1 + n_2\omega_2 = 0$. However, it is still enough to ensure a finite measure of preserved tori. This may be seen by considering the unit interval and deleting from it zones of width $K/s^{2.5}$ as shown in Figure 3.1. The total length deleted is thus

$$\sum_{s=1}^{\infty} \frac{K}{s^{2.5}} s = K \sum_{s=1}^{\infty} \frac{1}{s^{1.5}} \approx K \tag{3.4.14}$$

which goes to zero as $\epsilon \to 0$. This is only a crude (over) estimate. In fact, any width K/s^{μ}, with $\mu > 2$, would ensure a finite measure of preserved tori.†

3.4.c Other Aspects of the KAM Theorem

The KAM theorem also requires that the perturbation be "sufficiently small." However, there is still no precise estimate, in general, of how small "sufficiently small" should be. Early estimates were of the order of 10^{-48} (any units!). However, what matters is that the theorem does provide us with a proof of existence of tori under (albeit very small) perturbation. Numerical experiments show that tori are preserved under much stronger perturbation than indicated by the theory. We also mention that analyticity of the Hamiltonian (i.e., $H_0 = H_0(\mathbf{I})$) is not strictly necessary. Moser's version of the KAM theorem requires that only the first 333 derivatives exist! Subsequent work has brought that number right down.

The KAM theorem does not say anything about the fate of the "rational tori," that is, the ones that are, in some sense, "destroyed" under perturbation. It is these destroyed tori that provide the "seeds" of chaotic behavior observed in nonintegrable systems. How this comes about will be discussed at length in Chapter 4.

---

†A nice illustration of this idea is given on page 6 of the book by Kac and Uhlenbeck (1958).

3.5 SUMMARY OF KAM THEOREM AND ITS VARIANTS*

The KAM theorem can be generalized to a variety of systems with suitable conditions on the degeneracy and periodicity.†

3.5.a Autonomous Systems

For the Hamiltonian

$$H = H_0(\mathbf{I}) + \epsilon H_1(\mathbf{I}, \boldsymbol{\theta})$$

the nondegeneracy condition

$$\det\left|\frac{\partial^2 H_0}{\partial I_i \partial I_j}\right| \neq 0 \tag{3.5.1}$$

guarantees preservation of most invariant tori under small perturbations, $\epsilon \ll 1$.

In addition, the condition for "*isoenergetic nondegeneracy*," namely,

$$\det\left|\begin{matrix} \dfrac{\partial^2 H_0}{\partial I_i \partial I_j} & \dfrac{\partial H_0}{\partial I_i} \\[2ex] \dfrac{\partial H_0}{\partial I_j} & 0 \end{matrix}\right| \tag{3.5.2}$$

guarantees the existence, on every energy level surface, of a set of invariant tori whose complement has a small measure. If $n = 2$, the isoenergetic nondegeneracy condition also guarantees the stability of the action variables; that is, they remain close to their initial values, for sufficiently small ϵ, for all time.

The idea behind the isoenergetic nondegeneracy condition is to ensure that the ratios of frequencies on a given torus vary smoothly from torus to torus. Consider

$$\frac{\partial}{\partial I_1}\left(\frac{\omega_1}{\omega_2}\right) = \frac{1}{\omega_2^2}\left(\omega_2\frac{\partial \omega_1}{\partial I_1} - \omega_1\frac{\partial \omega_2}{\partial I_1}\right)$$

$$\frac{\partial}{\partial I_2}\left(\frac{\omega_1}{\omega_2}\right) = \frac{1}{\omega_2^2}\left(\omega_2\frac{\partial \omega_1}{\partial I_2} - \omega_1\frac{\partial \omega_2}{\partial I_2}\right)$$

---

†This section is based on Appendix 8 of Arnold (1978).

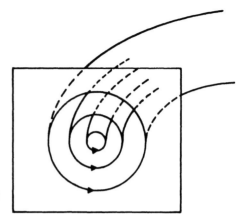

Figure 3.2 Transverse intersection of a family of nested 2-D tori.

These can be combined into an equivalent form:

$$\det \begin{vmatrix} \dfrac{\partial^2 H_0}{\partial I_1^2} & \dfrac{\partial^2 H_0}{\partial I_1 \partial I_2} & \dfrac{\partial H_0}{\partial I_1} \\ \dfrac{\partial^2 H_0}{\partial I_2 \partial I_1} & \dfrac{\partial^2 H_0}{\partial I_2^2} & \dfrac{\partial H_0}{\partial I_2} \\ \dfrac{\partial H_0}{\partial I_1} & \dfrac{\partial H_0}{\partial I_2} & 0 \end{vmatrix} = \begin{vmatrix} \dfrac{\partial \omega_1}{\partial I_1} & \dfrac{\partial \omega_1}{\partial I_2} & \omega_1 \\ \dfrac{\partial \omega_2}{\partial I_1} & \dfrac{\partial \omega_2}{\partial I_2} & \omega_2 \\ \omega_1 & \omega_2 & 0 \end{vmatrix}$$

For a system of two degrees of freedom, we can place a 2-D plane in the 3-D energy level set, transversally intersecting the 2-D family of tori as sketched in Figure 3.2. In this picture, the isoenergetic nondegeneracy condition ensures that the frequency ratio $\sigma = \omega_1/\omega_2$ varies smoothly from circle to circle. Under perturbation, the frequencies of the preserved tori usually depend on the size of ϵ; however, if (3.5.2) is satisfied, the frequency ratios will be preserved under changes in ϵ.

If ω_1/ω_2 is irrational, a given starting point will never exactly return on itself. If, on the other hand, σ is rational—and by our isoenergetic non-degeneracy condition these will be interspersed among the irrationals—a starting point will return to itself. It is these "rational curves" that are "destroyed" under perturbation and their precise fate will be described in the next chapter.

3.5.b Mappings†

From our picture of rotations about circles representing the intersection of a family of tori with a plane (Figure 3.2), it is not difficult to generalize to

†Mappings are discussed in more detail in Chapter 4.

the notion of a "mapping," that is, a transformation that moves one around a $2n$-dimensional annular manifold in discrete steps—that is, a transformation of the form

$$\boldsymbol{\theta}' = \boldsymbol{\theta} + \boldsymbol{\nabla}_{I'} S_0(\mathbf{I}')$$

$$\mathbf{I}' = \mathbf{I} \qquad\qquad (3.5.3)$$

where S_0 is some "generating function". In the case of $n = 1$, this mapping represents the rotation around a circle, of radius I, with "frequency" $\partial S_0 / \partial I$. Under a perturbation $S_1(\mathbf{I}, \boldsymbol{\theta})$, that is,

$$\boldsymbol{\theta}' = \boldsymbol{\theta} + \boldsymbol{\nabla}_{I'} S(\mathbf{I}', \boldsymbol{\theta})$$

$$\mathbf{I}' = \mathbf{I} + \boldsymbol{\nabla}_{\theta} S(\mathbf{I}, \boldsymbol{\theta}) \qquad\qquad (3.5.4)$$

where $S = S_0 + \epsilon S_1$, $\epsilon \ll 1$, the nondegeneracy condition

$$\det\left|\frac{\partial^2 S_0}{\partial I_i \partial I_j}\right| \neq 0 \qquad\qquad (3.5.5)$$

guarantees the preservation of most manifolds under the perturbation.

In the case of $n = 1$ (i.e., mappings of circles into circles), the nondegeneracy condition means that the angle of rotation varies (smoothly) from circle to circle.

Moser's version of the KAM theorem was based on such mappings. Unlike Arnold's version, Moser's version did not require analyticity of the generating function, only that it be a C^{333} function! Later, the theorem was proved requiring the existence of far fewer derivatives.

3.5.c Periodic Systems

For a periodically perturbed system, that is,

$$H = H_0(\mathbf{I}) + \epsilon H_1(\mathbf{I}, \boldsymbol{\theta}, t) \qquad\qquad (3.5.6)$$

where $H_1(\mathbf{I}, \boldsymbol{\theta}, t + \tau) = H_1(\mathbf{I}, \boldsymbol{\theta}, t)$ with τ as the period, one can think of the phase space as being $2n + 1$-dimensional, that is, $(\mathbf{I}, \boldsymbol{\theta}, t) = R^n \times T^{n+1}$ (cf. Figure 4.6). In this situation the invariant tori have dimension $n + 1$, and the nondegeneracy condition

$$\det\left|\frac{\partial^2 H_0}{\partial I_i \partial I_j}\right| \neq 0 \qquad\qquad (3.5.7)$$

guarantees the preservation of most of these tori under small perturbation

$\epsilon \ll 1$. If $n = 1$, the condition also guarantees that the action variable is "stable," that is, for sufficiently small ϵ it remains near its initial value for all t.

3.5.d   Stable Equilibrium Points

In the neighborhood of an elliptic fixed point, the following version of the KAM theorem can be proved. If the linearized frequencies do not have low-order resonances of the form

$$k_1\omega_1 + k_2\omega_2 + \cdots + k_n\omega_n = 0 \tag{3.5.8}$$

where the k_i are integers such that

$$0 < \sum_{i=1}^{n} |k_i| < 4 \tag{3.5.9}$$

then the Hamiltonian function can be reduced to what is known as *Birkhoff normal form* (see Arnold (1978), Appendix 7), namely,

$$H = H_0(\tau) + \text{higher-order terms} \tag{3.5.10}$$

where the $\tau_k = 1, \ldots, n$ correspond to the variables $\tau_k = p_k^2 + q_k^2$ such that $\tau_k = 0$ at the equilibrium point and where

$$H_0(\tau) = \sum_{k=1}^{n} \omega_k \tau_k + \frac{1}{2}\sum_{k,l} \omega_{kl} \tau_k \tau_l \tag{3.5.11}$$

The phrase "higher-order terms" in (3.5.10) denotes terms of degree higher than 4, that is, terms involving the powers τ^N, $N > 4$.

The nondegeneracy condition

$$\det|\omega_{kl}| \neq 0 \tag{3.5.12}$$

guarantees the existence of a set of tori in a sufficiently small neighborhood about the equilibrium position. The isoenergetic nondegeneracy condition

$$\det\begin{vmatrix} \omega_{kl} & \omega_k \\ \omega_l & 0 \end{vmatrix} \neq 0 \tag{3.5.13}$$

further guarantees the existence of such a set, sufficiently close to the equilibrium point, on every energy surface.

SOURCES AND REFERENCES

Arnold, V. I., *Mathematical Methods of Classical Mechanics*, Springer-Verlag, New York, 1978.

Arnold, V. I., and A. Avez, *Ergodic Problems of Classical Mechanics*, Benjamin, New York, 1968.

Bender, C. M., and S. A. Orszag, *Advanced Mathematical Methods for Scientists and Engineers*, McGraw-Hill, New York, 1978.

Berry, M. V., Regular and irregular motion, AIP Conference Proceedings, No. 46, *Topics in Nonlinear Dynamics*, AIP, New York, 1978.

Born, M., *The Classical Mechanics of the Atom*, Ungar, New York, 1960.

Ince, E. L., *Ordinary Differential Equations*, Dover, New York, 1956.

Lichtenberg, A. J., and M. A. Lieberman, *Regular and Stochastic Motion*, Springer-Verlag, New York, 1983.

Markus, L., and K. R. Meyer, Generic Hamiltonian dynamical systems are neither integrable nor ergodic, *Mem. Am. Math. Soc.* **144** (1974).

Percival, I. C., and D. Richards, *Introduction to Dynamics*, Cambridge University Press, Cambridge, 1982.

Poincaré, H. *Les Methods Nouvelles de la Mechanique Celeste*, Gauthier-Villars, Paris, 1892.

Kac, M., and S. Ulam, *Mathematics and Logic*, Praeger, New York, 1968.

Khinchin, A. Ya., *Continued Fractions*, University of Chicago Press, Chicago, 1964.

Siegel, C. L., and J. Moser, *Lectures in Celestial Mechanics*, Springer-Verlag, New York, 1971.

The original KAM papers are:

Arnold, V. I., Small denominators and the problem of stability of motion in classical and celestial mechanics, *Russ. Math. Surv.* **18**, 85–191 (1963).

Kolmogorov, A. N., On the preservation of quasi-periodic motions under a small variation of Hamilton's function, *Dokl. Akad. Nauk. SSSR*, **98**, 525 (1954). An English version can be found in *Proceedings of the 1954 International Congress of Mathematics*, North-Holland, Amsterdam, 1957.

Moser, J., On invariant curves of area-preserving mappings on an annulus, *Nachr. Akad. Wiss. Goettingen Math. Phys.* **K1** 1 (1962).

4

CHAOS IN HAMILTONIAN SYSTEMS AND AREA-PRESERVING MAPPINGS

4.1 THE SURFACE OF SECTION

In the preceding chapters we have talked at length about the evolution of orbits in (multidimensional) phase space, but, apart from the phase-plane pictures of one-degree-of-freedom systems discussed in the first chapter, we have said little about how that motion can be visualized in practice. Clearly, there is the problem of dimension. For a two-degree-of-freedom (Hamiltonian) system, the phase space is four-dimensional and, if conservative, the energy shell is three-dimensional. Even following the motion on this 3-D energy shell is difficult—especially when faced with a 2-D piece of paper on which to plot it! To this end, a most valuable technique, due to Poincaré (1892) and Birkhoff (1932), termed the *surface of section*, has been developed. Although applicable to higher-dimensional systems, it is particularly well suited for conservative Hamiltonian systems of two degrees of freedom. In terms of the development of the subject of nonlinear dynamics, it is worth noting that some of the first—and still highly regarded—surface of section computations for nonintegrable Hamiltonians started to appear in the literature at about the same time as the KAM theorem. Computer studies such as these have played a major role in both complementing and prompting theoretical developments of the subject.

4.1.a Surface of Section for Two-Degree-of-Freedom Hamiltonians

Consider a two-degree-of-freedom conservative Hamiltonian of the form

$$E = H = \frac{1}{2m}(p_x^2 + p_y^2) + V(x, y) \qquad (4.1.1)$$

The study of the motion of the orbits of this system can be reduced to a two-dimensional problem in the following way. On a given energy shell, take a "slice" of the phase space at some given point, say $y = 0$. Now, follow a given orbit (obtained by numerically integrating Hamilton's equations on a computer); every time it passes through the point $y = 0$, note the corresponding values of p_x and x. If the potential $V(x, y)$ supports bounded motion, the orbit will repeatedly pass through this phase-space slice and in this way one can build up a "map" of successive (p_x, x) values as illustrated in Figure 4.1. This is the surface of section, and a point on it defines the state of the system to within a sign. This is easily seen since, given E and $y = 0$, one has

$$p_y = \pm \sqrt{2m\left(E - \frac{1}{2m}p_x^2 - V(x, 0)\right)} \qquad (4.1.2)$$

The surface of section is normally constructed by just keeping one sign of p_y, say $p_y > 0$.

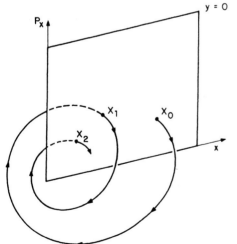

Figure 4.1 Construction of a surface of section.

If we denote the initial conditions of a given orbit (i.e., some p_x, x, E, $y = 0$) on the section as the point X_0, the successive intersections $X_1, X_2, \ldots, X_n$ provide a type of phase-plane "mapping" of the motion. This is an important concept that we shall discuss in detail later on. For now, though, we note that the times between the successive intersections with the surface of section (i.e., the points $X_0, X_1, \ldots$) are not necessarily equal. If we choose an initial condition corresponding to an orbit lying on a torus, the sequence of points $X_0, X_1, X_2, \ldots$ will lie on some smooth curve corresponding to the intersection of that torus with the surface of section (see Figure 4.2). If the chosen torus is one on which the frequency ratio ω_1/ω_2 is irrational, we know that a single orbit covers the torus ergodically. This will be manifested in the surface of section by the (gradual) "filling up" of the smooth curve by the successive iterates X_i. On the other hand, if the frequency ratio is rational, the orbit is closed and there will only be a finite number of intersections X_i $(i = 0, \ldots, n)$, such that $X_0 = X_n$, where n is determined by the rationality of ω_1/ω_2.

As discussed in the previous chapter, the KAM theorem tells us that for a weakly (nonintegrably) perturbed Hamiltonian, most tori are preserved. We have loosely talked about the other tori being, in some sense, "destroyed." Trajectories in these regions are now free to wander over larger regions of the phase space, and this is manifested on the surface of section as a random-looking "splatter" of points through which a smooth curve cannot be drawn. Of course, by eye it is difficult to be objective about what a "random splatter" is, but if a trajectory is run long enough, a pattern clearly different from smooth curves is usually apparent. Eventually, one can even hope to see (small) *areas* of the surface of section being filled up. Furthermore, there are some important computational tests (the computation of power spectra and Lyapunov exponents) to be described in Section 4.5 that enable one to distinguish objectively between the "regular" orbits, which lie on smooth curves in the surface of section, and the "irregular" (or chaotic) ones, which give rise to the random-looking patterns. Nonetheless, the surface of section is an enormously valuable tool and, when computed for a large number of initial conditions on the same

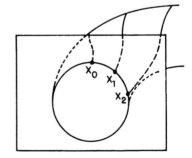

Figure 4.2 Successive intersections of a trajectory on a torus with surface of section.

energy shell, it is able to give an immediate picture of the enormously complicated phase-space structure of nonintegrable systems.

4.1.b The Henon–Heiles Hamiltonian

One of the most famous and enduring surface of section studies, carried out over 20 years ago, is due to Henon and Heiles (1964). Their paper is a model of lucid scientific writing and is a "must" on any reading list. Their Hamiltonian takes the form

$$H = \tfrac{1}{2}(p_x^2 + p_y^2 + x^2 + y^2) + x^2 y - \tfrac{1}{3}y^3 \qquad (4.1.3)$$

and was chosen as a simple model for the motion of a star in a cylindrically symmetric, gravitationally smoothed galactic potential. It can also provide a simple model for a pair of nonlinearly coupled molecular bonds. The potential energy function $V(x, y) = (x^2 + y^2)/2 + x^2 y - y^3/3$, sketched in Figure 4.3, supports bounded motion up to an energy $E = \tfrac{1}{6}$. For small displacements, the motion is nearly linear; however, as the energy increases, the particle "samples" more and more of the nonlinearity in the potential with correspondingly interesting consequences. In Figure 4.4 we show the now-famous surfaces of section computed (for a variety of initial conditions at each energy) at $E = \tfrac{1}{12}$, $\tfrac{1}{8}$, and $\tfrac{1}{6}$. At $E = \tfrac{1}{12}$, the motion is predominantly integrable and (virtually) all the initial conditions studied lead to orbits lying on smooth curves. The self-intersecting curve, where the intersections (points A, B, C in Figure 4.4a) are apparently (to the eye only!) smooth, is a type of separatrix which we shall discuss in more detail later. At $E = \tfrac{1}{8}$, the surface of section has clearly changed. Some of the smooth curves still remain, whereas others have "broken up" in various ways. In the right-hand family of curves, there now appears a "chain" of five "islands." This is generated by a single trajectory which jumps, successively, from island to island, gradually filling up the set of five small curves. On the other hand,

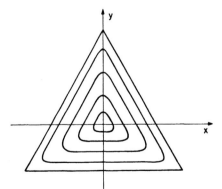

Figure 4.3 Potential energy contours for Henon–Heiles system. Beyond outermost triangle (at energy $E = \tfrac{1}{6}$) motion can become unbounded.

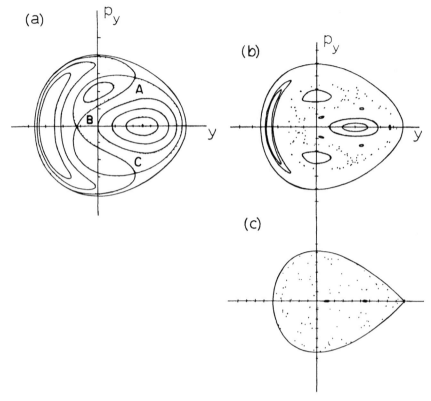

Figure 4.4 Surfaces of section for Henon–Heiles system at (a) $E = \frac{1}{12}$, (b) $E = \frac{1}{8}$, and (c) $E = \frac{1}{6}$. (Reproduced, by permission, from Ford (1975).)

the self-intersecting structure at $E = \frac{1}{12}$ has now disappeared and instead we see random splatter of points (generated by a single orbit) through which a smooth curve cannot be drawn. By $E = \frac{1}{6}$, virtually all smooth curves have disappeared except for a few tiny islands. The set of points, which fills up most of the accessible energy shell, is generated by a single orbit. This remarkable sequence of pictures gives a very clear illustration of the way in which the motion of a nonintegrable Hamiltonian (there is no other first integral apart from the energy) can change from predominantly regular to predominantly chaotic behavior.

4.1.c The Toda Lattice

Before we discuss the nature of this transition in detail, we describe the results of another rather instructive surface of section study. This is the investigation by Ford and co-workers of the three-particle Toda lattice, which consists of three particles on a ring with exponential interactions. The

Hamiltonian takes the form

$$H = \frac{p_1^2}{2m_1} + \frac{p_2^2}{2m_2} + \frac{p_3^2}{2m_3} + e^{-\nu_1(q_1-q_3)} + e^{-\nu_2(q_2-q_1)} + e^{-(q_3-q_2)} - 3 \quad (4.1.4)$$

which, owing to the fact that $p_1 + p_2 + p_3 = 0$ (check this), can be reduced to an equivalent two-dimensional form

$$H = \frac{p_x^2}{2m_x} + \frac{p_y^2}{2m_y} + \frac{1}{24}\{e^{2y+2\sqrt{3}x} + e^{2y-2\sqrt{3}x} + e^{-4y}\} - \frac{1}{8} \quad (4.1.5)$$

For small displacements, the motion is again almost linear, and, in fact, if the exponentials are expanded to third order, the potential-energy term is just the same as that of the Henon–Heiles system.† However, unlike the latter, the motion is bounded for all energies. Working with the *equal* mass case $m_x = m_y$, the surfaces of section were found to be made up entirely of smooth curves at $E = 1$, $E = 256$, and (up to) $E = 56,000$—which was their computer limit! There were absolutely no signs of chaos, and these results strongly suggested that the system (4.1.5) is, in fact, integrable. Motivated by these numerical results, Henon (1974) found the other first integral to be

$$F = 8p_x(p_x^2 - 3p_y^2) + (p_x + \sqrt{3}p_y) \, e^{2y-2\sqrt{3}x}$$
$$+ (p_x - \sqrt{3}p_y) \, e^{2y+2\sqrt{3}x} - 2p_x \, e^{-4y} \quad (4.1.6)$$

which, in the limit of small displacements, tends to

$$F \rightarrow 12(yp_x - xp_y) \quad (4.1.7)$$

which is just the angular momentum of the system. By contrast, a subsequent study by Casati and Ford (1975) for the case of *unequal* masses (i.e., $m_x/m_y \neq 1$) revealed chaotic behavior in the surface of section. Apart from illustrating the nice interplay between numerical experiment and theory, these results again raise the fundamental question of how the integrability of the system (4.1.5) might have been predicted without all that computational effort. A detailed discussion of this problem is postponed to Chapter 8.

4.1.d Surface of Section as a Symplectic Mapping

A fundamental property of the surface of section for Hamiltonian systems is that it corresponds to an *area-preserving* or, to be more precise, *symplectic* mapping. To see what is meant by this, it is first instructive to recall some earlier ideas. Firstly, there is Liouville's theorem (Sections 2.2 and 2.3),

---

†A systematic study of successive truncations of the Toda lattice has been carried out by Contopoulos and Polymilis (1987).

which tells us that phase volume is preserved under the Hamiltonian flow. For a one-degree of freedom system, this is just preservation of area in the (p, q)-phase plane. Thus, for some area A, enclosed by a closed curve $\mathscr{C}$, we can write (using the Stokes' theorem result $\oint_{\mathscr{C}} p\, dq = \int\int_A dp\, dq$)

$$\oint_{\mathscr{C}} p\, dq = \oint_{\mathscr{C}_t} p\, dq \qquad (4.1.8)$$

where $\mathscr{C}_t$ is the shape of the curve after it has evolved under the Hamiltonian flow for a time t. This notion can be extended to many degrees of freedom. Thus, for a tube of trajectories in $2n$-dimensional phase space, encircled by some closed curve $\mathscr{C}$, one again has

$$\oint_{\mathscr{C}} \mathbf{p} \cdot \mathbf{dq} = \oint_{\mathscr{C}_t} \mathbf{p} \cdot \mathbf{dq} \qquad (4.1.9)$$

where $\mathbf{p} = p_1, \ldots, p_n$, $\mathbf{q} = q_1, \ldots, q_n$. Of course, the integrals in (4.1.9) no longer correspond to a simple area as is the case for $n = 1$. Instead, it corresponds to the sum of areas projected onto the set of (p_i, q_i)-planes (Figure 4.5). It is in this sense that we call the preservation *symplectic* (rather than area-preserving), and, in fact, this is just another statement of the property that the Hamiltonian flow itself is a canonical transformation. For conservative *two*-degree-of-freedom systems of the type discussed above one may show (see Appendix 4.1) that an enclosed area on the

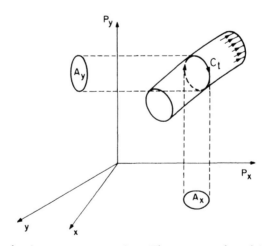

Figure 4.5 Symplectic area preservation. The area enclosed in evolving contour $\mathscr{C}_t$ is the sum of the projections A_x and A_y. This sum is conserved under the Hamiltonian flow.

surface of section (i.e., $\oint_{\mathscr{C}} p_x(x, y = 0, E) \, dx$), will be preserved under the flow.

The surface of section technique can also be used for time-dependent Hamiltonian systems. In the case of periodically driven one-degree-of-freedom systems it is particularly easy to define. For such a Hamiltonian, that is,

$$H(p, q, t + T) = H(p, q, t) \qquad (4.1.10)$$

where T is the period of the time-dependent part, the phase space is just the three-dimensional space of the p, q, and t variables. An area-preserving surface of section is simply obtained by taking stroboscopic "snapshots" of the (p, q)-plane at times nT, $n = 0, 1, 2, \ldots$. The set of points $X_i = (p(t + iT), q(t + iT))$ provides the desired surface of section (Figure 4.6). Obviously in this case, as opposed to the autonomous systems discussed above, the time between successive intersections is precisely equal.

The reader will have noticed that in our discussions of the surface of section technique, the word *mapping* has appeared quite frequently. So far, though, this term has only been used in the rather vague sense of some transformation, generated by the Hamiltonian flow, that takes a phase point X_i to a new phase-space location X_{i+1}, that is, for some mapping T

$$X_{i+1} = TX_i \qquad (4.1.11)$$

Furthermore, this transformation is, in some sense, "area-preserving" or

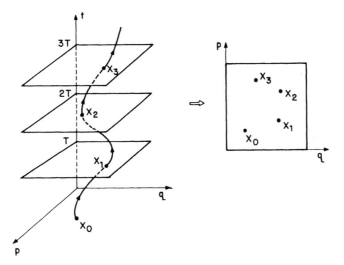

Figure 4.6 Three-dimensional phase space of periodically perturbed system. Surface of section is constructed from stroboscopic slices at $t = nT$.

symplectic. In fact, the notion of an area-preserving mapping is an enormously valuable tool for studying Hamiltonian systems. As will be described, such mappings—even very simple ones—can display all the generic properties of nonintegrable Hamiltonian systems, and eventually we shall discuss these two classes of systems on the same footing. Owing to their relative simplicity, many theorems are more easily proven for mappings than for general Hamiltonians—as well as being much easier to study numerically. So, in order to be able to provide the most detailed discussion of all the phenomena observed in, for example, the Henon–Heiles surfaces of section—such as the break up of tori and the appearance of island chains and chaotic trajectories—we first of all investigate the properties of area-preserving mappings.

4.2 AREA-PRESERVING MAPPINGS

4.2.a Twist Maps

An important class of area-preserving map is the *twist map*. A convenient way to introduce them—and to show their connection with Hamiltonian systems—is to reconsider our previous discussion of surfaces of section. Recall that, for a two-degree-of-freedom system, the surface of section for a trajectory lying on a torus is a sequence of points $X_0, X_1, \ldots$ lying on a smooth curve, which corresponds to the intersection of that torus with the surface of section. Furthermore, if the frequency ratio ω_1/ω_2 is irrational, the sequence X_i fills up the curve ergodically, whereas if ω_1/ω_2 is rational, only a finite sequence of iterates, corresponding to a closed orbit, appears. Now, assuming an integrable system, consider a family of nested tori which, for an isoenergetic nondegenerate system (cf. Eq. (3.5.2)) will have a frequency ratio which varies smoothly, say increases, from torus to torus. Considering just one of these tori, with actions I_1, I_2 (on the energy shell $E = H(I_1, I_2)$), the linear flow on the torus is, of course, just

$$\theta_1(t) = \omega_1 t + \theta_1(0) \tag{4.2.1a}$$

$$\theta_2(t) = \omega_2 t + \theta_2(0) \tag{4.2.1b}$$

where $\omega_1 = \omega_1(I_1, I_2) = \partial H/\partial I_1$ and $\omega_2 = \omega_2(I_1, I_2) = \partial H/\partial I_2$. The time, t_2, it takes for θ_2 to complete a 2π cycle is simply $t_2 = 2\pi/\omega_2$. In this amount of time, the change in θ_1 is thus

$$\theta_1(t + t_2) = \theta_1(t) + \omega_1 t_2$$

$$= \theta_1(t) + 2\pi\omega_1/\omega_2$$

$$= \theta_1(t) + 2\pi\alpha(I_1) \tag{4.2.2}$$

where $\alpha = \omega_1/\omega_2$, which is called the *rotation number*, is written as a function of just I_1 since, on a given energy shell $E = H(I_1, I_2)$, I_2 can always be expressed in terms of I_1, that is, $I_2 = I_2(E, I_1)$. If we now consider the (I_1, θ_1)-plane as a surface of section (see Figure 4.7), the successive intersections of a trajectory (on this torus) with this plane are just the points $X_i = (\theta_1(t + it_2), I_1)$. Changing to the notation $\theta_i = \theta_1(t + it_2)$ and $r = I_1$ the sequence of points $X_i = X_i(r, \theta)$ associated with the flow on a given torus (with a given "radius" I_1) can be represented by the mapping

$$
T: \quad
\begin{aligned}
\theta_{i+1} &= \theta_i + 2\pi\alpha(r_i) && \text{(4.2.3a)} \\
r_{i+1} &= r_i && \text{(4.2.3b)}
\end{aligned}
$$

where α is taken to be a smoothly varying function of r. Such a mapping is called a *twist map*. As it stands, it is rather simple in that all it does is map points, albeit uniformly or discretely, around a given circle—uniformly for irrational α and discretely for rational α. Clearly, one can think of (4.2.3) as an integrable mapping. A given *circle* $\mathscr{C}$ of points will clearly be mapped into itself. We thus say that a twist map maps circles into circles, and we

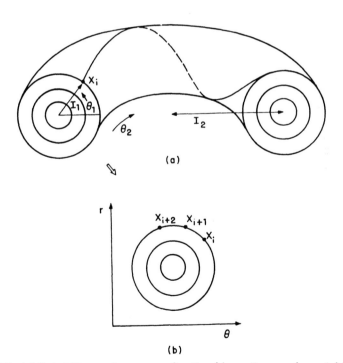

Figure 4.7 (a) Point X_i on a torus coordinatized by action-angle variables I_1, θ_1. (b) Successive points X_i, X_{i+1}, X_{i+2} of corresponding twist map in $(r(= I_1)$, $\theta(= \theta_1))$-plane.

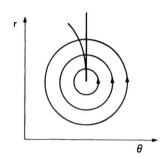

Figure 4.8 Radial line of points twisted under twist-map action.

represent this symbolically as

$$T(\mathscr{C}) = \mathscr{C} \tag{4.2.4}$$

However, since the rotation number $\alpha(r)$ increases with r, a radial line of points will clearly be twisted under T (Figure 4.8). Hence the term *twist* map. The mapping (4.2.3) is obviously area preserving since

$$\frac{\partial(\theta_{i+1}, r_{i+1})}{\partial(\theta_i, r_i)} = 1 \tag{4.2.5}$$

We also comment that, as it stands, it does not really matter whether in Eq. (4.2.3a) we write α as a function of r_i or r_{i+1}. We shall return to this point shortly.

For a nonintegrable system, the KAM theorem tells that tori with rational frequency ratios do not "survive." In terms of the twist map, we can think of the addition of some "nonintegrable" perturbation, that is,

$$T_\epsilon: \quad \begin{aligned} \theta_{i+1} &= \theta_i + 2\pi\alpha(r_i) + \epsilon f(r_i, \theta_i) \tag{4.2.6a} \\ r_{i+1} &= r_i + \epsilon g(r_i, \theta_i) \tag{4.2.6b} \end{aligned}$$

where f and g are chosen to ensure that the area-preserving property (4.2.5) still holds. The natural question to ask about is the preservation of circles under the perturbation. This was Moser's (Moser, 1962) famous contribution to the KAM theorem in which he proved that for sufficiently small perturbation, circles with sufficiently irrational winding numbers are preserved. (See Section 3.5.)

4.2.b Mappings on the Plane

We can also write mappings in cartesian coordinates, that is,

$$T: \quad \begin{aligned} x_{i+1} &= f(x_i, y_i) \tag{4.2.7a} \\ y_{i+1} &= g(x_i, y_i) \tag{4.2.7b} \end{aligned}$$

which will, of course, be area preserving if

$$\frac{\partial(x_{i+1}, y_{i+1})}{\partial(x_i, y_i)} = 1 \qquad (4.2.8)$$

If f and g are polynomials, the mapping is termed an *entire Cremona transformation*. The properties of the mapping depend on the form of f and g. If they are only linear functions, for example,

$$x_{i+1} = x_i \cos \alpha - y_i \sin \alpha \qquad (4.2.9a)$$

$$y_{i+1} = x_i \sin \alpha + y_i \cos \alpha \qquad (4.2.9b)$$

the mapping is just a simple rotation through the angle α. Another simple linear transformation is

$$x_{i+1} = x_i + y_i \qquad (4.2.10a)$$

$$y_{i+1} = y_i \qquad (4.2.10b)$$

which corresponds to a linear shear parallel to the x-axis.

A nonlinearly perturbed version of (4.2.9), that is,

$$T: \begin{array}{ll} x_{i+1} = x_i \cos \alpha - (y_i - x_i^2) \sin \alpha & (4.2.11a) \\ y_{i+1} = x_i \sin \alpha + (y_i - x_i^2) \cos \alpha & (4.2.11b) \end{array}$$

has been the subject of a beautiful study by Henon (1969). (Another piece of obligatory reading!) It is valuable to note that this mapping can be written as the "composition" of two simpler mappings corresponding to a nonlinear shear and a simple rotation (Figure 4.9). That is, one can write

$$T = T_1 T_2 \qquad (4.2.12)$$

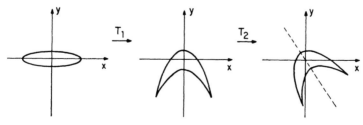

Figure 4.9 Effect on an area element of nonlinear shear T_1 followed by rotation T_2 corresponding to the Henon map.

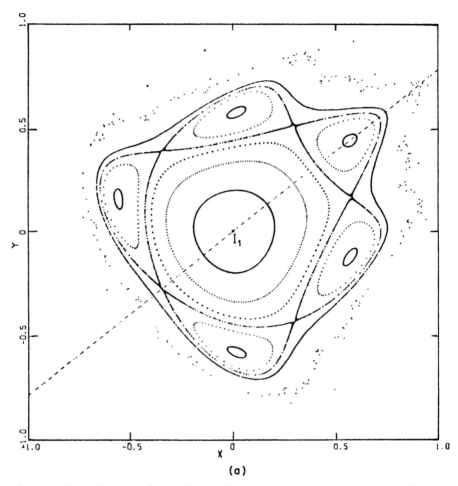

Figure 4.10 (a) Typical phase plane of Henon map with $\alpha = 0.2114$. (b) Blowup of region around right-hand-most hyperbolic point. (Reproduced, by permission, from Henon (1969).)

where

$$T_1: \quad \begin{aligned} x_{i+1/2} &= x_i \\ y_{i+1/2} &= y_i - x_i^2 \end{aligned}$$

$$\qquad (4.2.13a)$$
$$\qquad (4.2.13b)$$

$$T_2: \quad \begin{aligned} x_{i+1} &= x_{i+1/2} \cos \alpha - y_{i+1/2} \sin \alpha \\ y_{i+1} &= x_{i+1/2} \sin \alpha + y_{i+1/2} \cos \alpha \end{aligned}$$

$$\qquad (4.2.14a)$$
$$\qquad (4.2.14b)$$

It is important to note that the mapping (4.2.11) is *invertible*. The inverse transformation T^{-1} takes the form

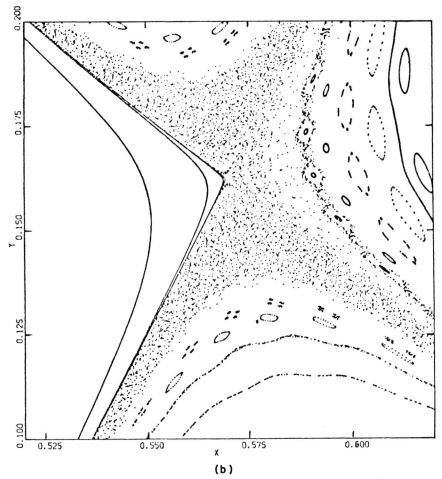

(b)

Figure 4.10 (Continued)

$$T^{-1}: \quad \begin{aligned} x_i &= x_{i+1} \cos \alpha + y_{i+1} \sin \alpha \\ y_i &= - x_{i+1} \sin \alpha + y_{i+1} \cos \alpha + (x_{i+1} \cos \alpha + y_{i+1} \sin \alpha)^2 \end{aligned}$$

Thus some "final" point on an orbit can be uniquely "time-reversed" back to the initial point (x_0, y_0).

A numerical study of Henon's map (4.2.11) is easily performed. For a given rotation angle α, one can study—even on a pocket calculator—the evolving iterates for a variety of different initial conditions (x_0, y_0). Some typical results, as obtained by Henon (1969), are shown in Figure 4.10. They clearly show all the typical features of a Henon–Heiles-like surface of section with families of smooth curves, island chains, and chaotic trajectories. The separatrixlike structure of Figure 4.10a is particularly interesting. On this scale, the intersections look almost smooth; however, when the scale is enlarged, on sees (as shown in the now famous Figure 4.10b) an incredibly rich, fine structure of island chains interspersed in a "sea" of chaos. In the sections that follow, we will (attempt to) explain the origins of this wonderful complexity.

4.2.c Connection between Area-Preserving Maps and Hamiltonians

Although Henon's map displays all the generic features of a nonintegrable Hamiltonian, it is not obviously derivable from one. It is therefore worth asking if area-preserving maps can be derived explicitly from Hamiltonian systems. Consider a simple one-degree-of-freedom Hamiltonian of the form

$$H(p, q) = \tfrac{1}{2}p^2 + V(q) \tag{4.2.15}$$

for which Hamilton's equations are simply

$$\dot{q} = p \tag{4.2.16a}$$

$$\dot{p} = -\frac{\partial V}{\partial q} \tag{4.2.16b}$$

One could try writing the time derivatives on the left-hand sides of (4.2.16) as first-order differences, that is, $\dot{q} = (q_{i+1} - q_i)/\Delta t$, where $q_{i+1} = q(t + \Delta t)$ and $q_i = q(t)$. The discretized version of (4.2.6) would then take the form

$$q_{i+1} = q_i + p_i \Delta t \tag{4.2.17a}$$

$$p_{i+1} = p_i - \Delta t \left(\frac{\partial V}{\partial q_i}\right)_{q=q_i} \tag{4.2.17b}$$

However, this is *not* an area-preserving transformation since

$$\frac{\partial(q_{i+1}, p_{i+1})}{\partial(q_i, p_i)} = \begin{vmatrix} 1 & -\Delta t \dfrac{\partial^2 V}{\partial q_i^2} \\ \Delta t & 1 \end{vmatrix} = 1 + (\Delta t)^2 \frac{\partial^2 V}{\partial q_i^2}$$

and we are assuming that Δt is *finite* (cf. discussion of infinitesimal canonical transformations in Chapter 2). However, if the q dependence of (4.2.17b) is changed from q_i to q_{i+1}, that is,

$$q_{i+1} = q_i + p_i \Delta t \tag{4.2.18a}$$

$$p_{i+1} = p_i - \Delta t \left(\frac{\partial V}{\partial q}\right)_{q=q_{i+1}} \tag{4.2.18b}$$

the mapping is easily seen to become area preserving.

It is now interesting to determine what sort of Hamiltonian would give rise to precisely such equations of motion. Instead of (4.2.15), consider a time-dependent Hamiltonian of the form

$$H(p, q, t) = \begin{cases} \dfrac{p^2}{2\gamma}, & 0 < t < \gamma T \quad (4.2.19a) \\[2ex] \dfrac{V(q)}{1-\gamma}, & \gamma T < t < T \quad (4.2.19b) \end{cases}$$

where $0 < \gamma < 1$. Physically, this corresponds to a situation in which a particle (of unit mass) undergoes free translation for a period γT and then experiences an impulsive force due to the potential $V(q)$, for a period $(1-\gamma)T$—a process that is then repeated periodically. Hamiltonians such as these are used to describe ray propagation in wave guides where the impulse is due to periodically spaced lenses. Integrating Hamilton's equations for (4.2.19) over any one period $t = iT$ to $t = (i+1)T$ yields precisely Eqs. (4.2.18) (with Δt replaced by T). We can also interchange the order of operations in (4.2.19), that is,

$$H(p, q, t) = \begin{cases} \dfrac{V(q)}{\gamma}, & 0 < t < \gamma T \quad (4.2.20a) \\[2ex] \dfrac{p^2}{2(1-\gamma)}, & \gamma T < t < T \quad (4.2.20b) \end{cases}$$

in which case integration of Hamilton's equations yields

$$q_{i+1} = q_i + Tp_{i+1} \qquad (4.2.21a)$$

$$p_{i+1} = p_i - T\left(\frac{\partial V}{\partial q}\right)_{q=q_i} \qquad (4.2.21b)$$

which is again an area-preserving map on the plane.

There is, of course, really no difference between the Hamiltonians (4.2.19) and (4.2.20), although the mapping equations for the latter have a certain nice symmetry about them as follows.

4.2.d Discrete Lagrangians*

For mappings such as (4.2.21), we can introduce the discrete Lagrangian

$$L(q_{i+1}, q_i) = \frac{1}{2}\left(\frac{q_{i+1} - q_i}{T}\right)^2 - V(q_i) \qquad (4.2.22)$$

and the corresponding discrete action function

$$W(q_{i+1}, q_i) = TL(q_{i+1}, q_i) = \frac{1}{2}\frac{(q_{i+1} - q_i)^2}{T} - TV(q_i) \qquad (4.2.23)$$

from which one easily deduces the symmetric pair of generating relations

$$p_i = -\frac{\partial W}{\partial q_i}(q_{i+1}, q_i), \qquad p_{i+1} = \frac{\partial W(q_{i+1}, q_i)}{\partial q_{i+1}} \qquad (4.2.24)$$

Thus one can develop a rather elegant variational formalism for area-preserving maps closely analogous to the standard classical mechanical results. The discrete Lagrangian formalism will be used in our discussion of "quantum maps" in Sections 6.6 and 6.7.

4.2.e The Standard Map

A much-studied mapping is the one obtained by introducing the potential-energy function

$$V(q) = -\frac{k}{(2\pi)^2}\cos(2\pi q) \qquad (4.2.25)$$

into Eqs. (4.2.21), thereby obtaining the equations of motion

$$q_{i+1} = q_i + p_{i+1} \qquad (4.2.26a)$$

$$p_{i+1} = p_i + \frac{k}{2\pi}\sin(2\pi q_i) \qquad (4.2.26b)$$

where we have set $T = 1$. This mapping is usually studied with both p and q taken as periodic variables with period unity; that is, the mapping is confined to the unit torus, and we rewrite (4.2.26) as

$$q_{i+1} = q_i + p_{i+1}, \qquad\qquad \text{mod } q = 1 \qquad (4.2.27a)$$

$$p_{i+1} = p_i + \frac{k}{2\pi}\sin(2\pi q_i), \qquad \text{mod } p = 1 \qquad (4.2.27b)$$

This is the Taylor–Chirikov or "standard" mapping, the latter name stemming from its ubiquity in a variety of theoretical and practical problems. A typical phase plane, for $k = 0.97$, is shown in Figure 4.11. Again we see a wonderfully rich structure of delicately interwoven regular and chaotic motions. The reader may also have noticed that some of the invariant curves (tori) extending across the phase plane are apparently permeated

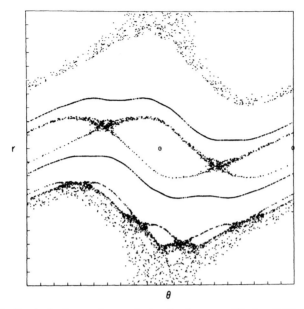

Figure 4.11 Typical phase plane of standard map with $k = 0.97$. (Reproduced, by permission, from Greene (1979).)

with small holes. This is not a numerical artifact but is, instead, a manifestation of the fact that the invariant curves are not really tori but are the so-called *cantori*. This latter term has been invoked to suggest the connection with a Cantor-set-like structure. These exciting notions are outside the scope of these lectures, although we will return to the Cantor-set concept in dynamical systems in the next chapter.

4.3 FIXED POINTS AND THE POINCARÉ–BIRKHOFF FIXED POINT THEOREM

For a mapping T, a fixed point (X^*) of T is a point for which

$$TX^* = X^* \tag{4.3.1}$$

In the case of a periodic orbit which maps out a sequence of n iterates (i.e., $X_0, X_1, \ldots, X_n$) such that $X_{n+i} = X_i$, it is clear that each X_i is a fixed point of T^n, where T^n denotes n successive applications of the mapping T, that is,

$$T^n X_i = X_i \tag{4.3.2}$$

As we found in our discussions in Chapter 1 on phase-plane dynamics, fixed points are convenient centers about which the dynamics can be "organized." Indeed, the fixed-point analysis for mappings is virtually identical to our previous results with the important difference that the area-preserving property of T considerably restricts the allowed types of fixed points.

4.3.a The Tangent Map

Consider some mapping T whose action we denote symbolically as

$$\begin{bmatrix} x_{i+1} \\ y_{i+1} \end{bmatrix} = T \begin{bmatrix} x_i \\ y_i \end{bmatrix} \tag{4.3.3}$$

This could represent a transformation of the sort given in Eqs. (4.2.7). For simplicity, let us assume there is a fixed point of T at the phase-plane origin $(x, y) = (0, 0)$. Linearizing T in the standard way about this point, we have the (linear) mapping, often referred to as the *tangent map*:

$$\begin{bmatrix} \delta x_{i+1} \\ \delta y_{i+1} \end{bmatrix} = \begin{bmatrix} T_{11} & T_{12} \\ T_{21} & T_{22} \end{bmatrix} \begin{bmatrix} \delta x_i \\ \delta y_i \end{bmatrix} \tag{4.3.4}$$

(so, for example, for Eqs. (4.2.7), $T_{11} = (\partial f/\partial x)_{x=y=0}$, etc.). The nature of the fixed point is then determined by the eigenvalues of (4.3.4), that is,

$$\begin{vmatrix} T_{11} - \lambda & T_{12} \\ T_{21} & T_{22} - \lambda \end{vmatrix} = 0 \tag{4.3.5}$$

These eigenvalues are found by solving the quadratic equation

$$\lambda^2 - \lambda(T_{11} + T_{22}) + \lambda(T_{11}T_{22} - T_{12}T_{21}) = 0$$

which we write as

$$\lambda^2 - \lambda(\text{trace}(T)) + \lambda(\det(T)) = 0$$

Since T is an area-preserving transformation (i.e., $\det(T) = 1$), the roots are simply

$$\lambda_{1,2} = \tfrac{1}{2}(\text{trace}(T)) \pm \tfrac{1}{2}\sqrt{(\text{trace}(T))^2 - 4} \tag{4.3.6}$$

There are three possibilities, depending on the value of $\text{trace}(T)$:

1. $|\text{trace}(T)| < 2$; λ_1, λ_2 are a complex conjugate pair lying on the unit circle, that is,

$$\lambda_1 = e^{+i\alpha}, \qquad \lambda_2 = e^{-i\alpha} \tag{4.3.7}$$

2. $|\text{trace}(T)| > 2$; the eigenvalues are real numbers satisfying

$$\lambda_2 = 1/\lambda_1 \tag{4.3.8}$$

3. $|\text{trace}(T)| = 2$; the eigenvalues take on the value

$$\lambda_1 = \lambda_2 = \pm 1 \tag{4.3.9}$$

Using the standard techniques of linear algebra, it is always possible to transform to a representation that diagonalizes (4.3.4), that is,

$$\begin{bmatrix} \xi_{i+1} \\ \eta_{i+1} \end{bmatrix} = \begin{bmatrix} \lambda_1 & 0 \\ 0 & \lambda_2 \end{bmatrix} \begin{bmatrix} \xi_i \\ \eta_i \end{bmatrix}$$

where, given the transformation matrix A,

$$A \begin{bmatrix} \delta x_i \\ \delta y_i \end{bmatrix} = \begin{bmatrix} \xi_i \\ \eta_i \end{bmatrix} \quad \text{and} \quad ATA^{-1} = \begin{bmatrix} \lambda_1 & 0 \\ 0 & \lambda_2 \end{bmatrix}$$

The three different eigenvalue cases are now easily interpreted.

4.3.b Classification of Fixed Points

The first case (i.e., $\lambda_{1,2} = e^{\pm i\alpha}$) is simply the rotation

$$\begin{bmatrix} \xi_{i+1} \\ \eta_{i+1} \end{bmatrix} = \begin{bmatrix} e^{i\alpha} & 0 \\ 0 & e^{-i\alpha} \end{bmatrix} \begin{bmatrix} \xi_i \\ \eta_i \end{bmatrix} \tag{4.3.10}$$

in the neighborhood of the fixed point $(0, 0)$. This obviously corresponds to a stable or elliptic fixed point. Thus in the immediate neighborhood of $(0, 0)$ we expect to find invariant curves (see Figure 1.10c). (Also see the discussion of the KAM theorem for equilibrium points in Section 3.5.)

For the second case (i.e., $\lambda_2 = 1/\lambda_1$), the linearized transformation takes the form

$$\begin{bmatrix} \xi_{i+1} \\ \eta_{i+1} \end{bmatrix} = \begin{bmatrix} \lambda & 0 \\ 0 & 1/\lambda \end{bmatrix} \begin{bmatrix} \xi_i \\ \eta_i \end{bmatrix} \tag{4.3.11}$$

which leads to hyperbolic motion in the neighborhood of $(0, 0)$. The precise behavior of (4.3.11) will be determined by the sign of λ:

Case (a) $\lambda > 0$ gives a regular *hyperbolic* fixed point in which successive iterates of (4.3.11) stay on the same branch of the hyperbola (see Figure 4.12a),

Case (b) $\lambda > 0$ gives a *hyperbolic-with-reflection* fixed point in which successive iterates of (4.3.11) jump backwards and forwards between opposite branches of the hyperbola (see Figure 4.12b). This is easy to see

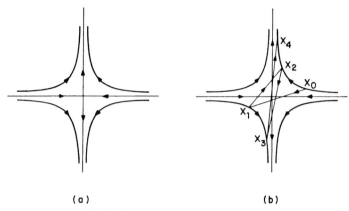

(a)   (b)

Figure 4.12 (a) Hyperbolic fixed point. (b) Hyperbolic-with-reflection fixed point.

since

$$\begin{bmatrix} \delta\xi_1 \\ \delta\eta_1 \end{bmatrix} = \begin{bmatrix} -|\lambda|\delta\xi_0 \\ -\dfrac{1}{|\lambda|}\delta\eta_0 \end{bmatrix}, \quad \begin{bmatrix} \delta\xi_2 \\ \delta\eta_2 \end{bmatrix} = \begin{bmatrix} |\lambda|^2\delta\xi_0 \\ \dfrac{1}{|\lambda|^2}\delta\eta_0 \end{bmatrix}, \quad \begin{bmatrix} \delta x_3 \\ \delta y_3 \end{bmatrix} = \begin{bmatrix} -|\lambda|^3\delta\xi_0 \\ -\dfrac{1}{|\lambda|^3}\delta\eta_0 \end{bmatrix}, \quad \text{etc.}$$

The third case, which is a special case corresponding to $\lambda_1 = \lambda_2 = \pm 1$, is best understood by recognizing that in the original variables $(\delta x_i, \delta y_i)$ the (linearized) transformation (4.3.4) can always be written as (choosing $\lambda_1 = +1$)

$$\begin{bmatrix} \delta x_{i+1} \\ \delta y_{i+1} \end{bmatrix} = \begin{bmatrix} 1 & c \\ 0 & 1 \end{bmatrix} \begin{bmatrix} \delta x_i \\ \delta y_i \end{bmatrix} \tag{4.3.12}$$

where c is any constant, which corresponds to a translation parallel to the x-axis. This is known as a *parabolic* fixed point (see Figure 4.13). For the particular choice $\delta y_0 = 0$, we note that every point on the x-axis is a fixed point of (4.3.12). This is the situation that arises for invariant tori, or

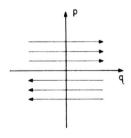

Figure 4.13 Parabolic fixed point.

curves, covered with closed orbits, with every point on the curve (in the phase plane or on the surface of section) being a fixed point of the flow.

4.3.c Poincaré–Birkhoff Fixed-Point Theorem

We are now in the position to tackle, in some detail, the fundamental question of the fate of tori with rational frequency ratios, or curves with rational rotation number, under (small) perturbation. It is convenient to approach this problem from the point of view of twist maps, which we write in the form (see Section 4.3) that makes the connection with two-degree-of-freedom Hamiltonians. Again thinking of the (transversal) intersection of family of tori, we can write the unperturbed twist map as

$$\varphi' = \varphi + \frac{\partial}{\partial I'} S_0(I') \tag{4.3.13a}$$

$$I' = I \tag{4.3.13b}$$

where, in the case of 2-D tori, we note that $\partial S_0/\partial I = 2\pi\omega_1/\omega_2$. (Here the primed variables indicate the $(i+1)$th iterate of the mapping and unprimed variables the ith iterate.) The KAM theorem tells us that for sufficiently small perturbation $\epsilon S_1(I, \varphi)$, that is, the mapping

$$\varphi' = \varphi + \frac{\partial}{\partial I'} S_0(I') + \epsilon \frac{\partial}{\partial I'} S_1(I', \varphi) \tag{4.3.14a}$$

$$I' = I + \epsilon \frac{\partial}{\partial \varphi} S_1(I, \varphi) \tag{4.3.14b}$$

"most" invariant curves will be preserved, provided that the nondegeneracy condition

$$\det \left| \frac{\partial^2 S_0}{\partial I_i \, \partial I_j} \right| \neq 0 \tag{4.3.15}$$

is satisfied, with the term "most" excluding those curves with rational rotation number $\alpha = \omega_1/\omega_2 = r/s$.

We can use these maps to study the precise fate of the rational curves under perturbation. Consider two curves $\mathscr{C}^+$ and $\mathscr{C}^-$ which lie on either side of the curve $\mathscr{C}$ with rational rotation number $\alpha = r/s$, as sketched in Figure 4.14a. We also assume that $\alpha = \alpha(I)$ increases smoothly with increasing I. If we denote the mapping by T, that is,

$$\begin{bmatrix} \varphi' \\ I' \end{bmatrix} = T \begin{bmatrix} \varphi \\ I \end{bmatrix} \tag{4.3.16}$$

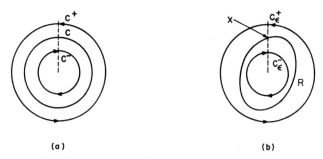

(a)   (b)

Figure 4.14 (a) Invariant curves of unperturbed twist map T with rotation numbers $\alpha < r/s$ for $\mathscr{C}^-$, $\alpha \equiv r/s$ for $\mathscr{C}$, and $\alpha > r/s$ for $\mathscr{C}^+$. (b) Effect of perturbed twist map T_ϵ on these curves. Since relative twists of $\mathscr{C}^+$ and $\mathscr{C}^-$ are preserved, there will be one point X between them whose angular coordinate is preserved. Curve R drawn between $\mathscr{C}^+$ and $\mathscr{C}^-$ is curve of these points.

then *every* point on $\mathscr{C}$ is a fixed point of T^s since

$$T^s\begin{bmatrix} \varphi \\ I \end{bmatrix} = \begin{bmatrix} \varphi + s\left(\dfrac{\partial S_0}{\partial I}\right) \\ I \end{bmatrix} = \begin{bmatrix} \varphi + s2\pi\left(\dfrac{r}{s}\right) \\ I \end{bmatrix} = \begin{bmatrix} \varphi + 2\pi r \\ I \end{bmatrix} = \begin{bmatrix} \varphi \\ I \end{bmatrix} \qquad (4.3.17)$$

Thus, *relative* to $\mathscr{C}$, $\mathscr{C}_+$ rotates anticlockwise and $\mathscr{C}_-$ rotates clockwise under the mapping T^s.

Now consider the (weakly) perturbed mapping T_ϵ. By the KAM theorem, $\mathscr{C}^+$ and $\mathscr{C}^-$ are preserved, albeit in slightly distorted form, say $\mathscr{C}_\epsilon^+$ and $\mathscr{C}_\epsilon^-$. These curves are now invariant curves of T_ϵ, that is,

$$T_\epsilon(\mathscr{C}_\epsilon^+) = \mathscr{C}_\epsilon^+ \quad \text{and} \quad T_\epsilon(\mathscr{C}_\epsilon^-) = \mathscr{C}_\epsilon^- \qquad (4.3.18)$$

Furthermore, we assume that ϵ is sufficiently small such that the relative twists of $\mathscr{C}^+$ and $\mathscr{C}^-$ are preserved under T_ϵ^s. If this is so, then there must be only one point between $\mathscr{C}_\epsilon^+$ and $\mathscr{C}_\epsilon^-$ whose angular coordinate φ is preserved under T_ϵ^s. In fact, along each radius (emanating from the center) there must be one such point, so we can draw a curve R of these points (Figure 4.14*b*). R is not an invariant curve of T_ϵ, but on R must lie the fixed points of T_ϵ^s. (A fixed point has preserved "angle" and "radius"—so far we just have the curve of preserved angles.) These may be found by subjecting R to the mapping T_ϵ^s, that is,

$$R' = T_\epsilon^s R \qquad (4.3.19)$$

The new curve R' will intersect R at an *even* number of points (this follows

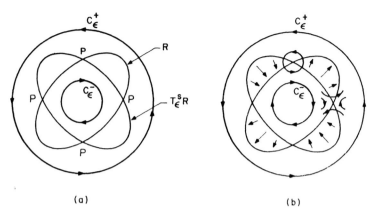

(a)   (b)

Figure 4.15 (a) Mapping of curve of points R into new curve $R' = T_\epsilon^s R$. Points P are the intersection points of these two curves. (b) By following "flow lines," these points are seen to be alternatively hyperbolic and elliptic fixed points.

from simple geometry)—these are the fixed points of T_ϵ^s (see Figure 4.15a). (We exclude from this argument any nongeneric points of tangency between R and R'.) This is the famous *Poincaré–Birkhoff fixed-point theorem*, which states that for the rational curve of an unperturbed system, with rotation number r/s (for which *every* point is a fixed point of T^s), only an even number of fixed points $2ks$ ($k = 1, 2, 3, \ldots$) will remain under perturbation. (We shall soon see that these fixed points are alternately stable and unstable.) That the number of fixed points is an even multiple of s is easy to see. Consider one of the (even number of) fixed points found from the intersection of R and R'. By definition, it is a fixed point of T_ϵ^s. Under T_ϵ the orbit generated is $X, T_\epsilon X, T_\epsilon^2 X, \ldots, T_\epsilon^{s-1} X$. However, each point of this closed orbit is also a fixed point of T_ϵ^s. Hence there are s fixed points associated with each intersection point of R and R'—hence $2ks$ fixed points overall.

Returning to Figure 4.15, all we have to do is follow the "flow lines" to see that the fixed points are *alternately* elliptic and hyperbolic. Thus, under perturbation of the rational curve with $\alpha = r/s$, $2ks$ fixed points of T_ϵ^s remain, of which ks are elliptic and ks are hyperbolic in alternating sequence. Now, around each elliptic fixed-point we will find a family of invariant curves. This family is itself subject to the KAM theorem (see Section 3.5.d), so its rational members will break up according to the Poincaré–Birkhoff fixed point theorem. The same structure must then be repeated about this subsequence of elliptic fixed points, and so on. Thus, around each elliptic fixed point there is a simultaneous application of the Poincaré–Birkhoff fixed-point theorem and KAM theorem which leads to a remarkable self-similar structure on all scales, as sketched in Figure 4.16.

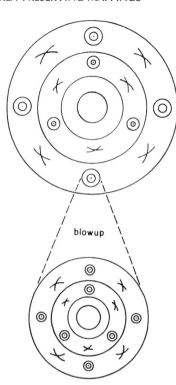

Figure 4.16 Successive applications of KAM and Poincaré–Birkhoff fixed-point theorems leading to self-similar fixed-point structure. (× denotes hyperbolic fixed points.)

4.4 HOMOCLINIC AND HETEROCLINIC POINTS

To complete the picture shown in Figure 4.16, we now have to consider what happens in the neighborhood of the hyperbolic fixed points. As we shall see, the results are very striking. Here we follow the discussion given by Berry (1978).

A hyperbolic fixed point is characterized by four invariant curves or manifolds. These are the two ingoing or *stable manifolds*, H^+, and the two outgoing or *unstable manifolds*, H^-, as shown in Figure 4.17. Points on H^+

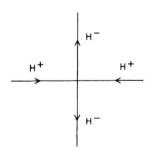

Figure 4.17 Stable manifolds H^+ and unstable manifolds H^- of a hyperbolic fixed point.

approach the fixed point H exponentially slowly, that is,

$$\lim_{s \to \infty} T^s x \to H, \qquad (x \text{ on } H^+) \qquad (4.4.1)$$

whereas points on H^- recede from H exponentially slowly, that is,

$$\lim_{s \to \infty} T^{-s} x \to H, \qquad (x \text{ on } H^-) \qquad (4.4.2)$$

4.4.a The Intersections of H⁺ and H⁻

We will now investigate the way in which H^+ and H^- can "interact" with each other. As we have already seen, for integrable systems the H^+ and H^- manifolds emanating from a hyperbolic fixed point (see, e.g., the pendulum) form a separatrix. In Figure 4.18a we show the case of an outgoing manifold joining up smoothly with an ingoing manifold to form a single smooth loop. Such a curve is sometimes called a *homoclinic orbit*. In Figure 4.18b we show another example in which the H^+ and H^- from a family of three hyperbolic fixed points (i.e., fixed points of T^3) all join up smoothly in the manner shown. This is not dissimilar (only superficially!) to the structure seen in the $E = \frac{1}{12}$ surface of section for the Henon–Heiles system (Figure 4.4a).

This smooth joining of manifolds is the exceptional (i.e., nongeneric) situation that can only arise in integrable systems. The generic situation is far more complex. The manifolds H^+ and H^- are not allowed to intersect themselves but, instead, intersect each other as shown in Figure 4.19. If the intersection point(s) involve the H^+ and H^- manifolds from the same fixed point or from the fixed points of the same family (e.g., the three fixed points of T^3 shown in the inner part of Figure 4.16), they are called *homoclinic*

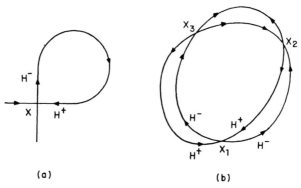

(a)   (b)

Figure 4.18 (a) Smooth joining of H^- to H^+ from same hyperbolic fixed point X leading to homoclinic orbit. (b) Family of three smoothly connected hyperbolic points X_1, X_2, X_3.

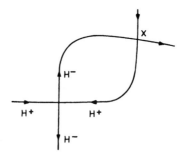

Figure 4.19 Intersection of stable manifold H^+ and unstable manifold H^-, emanating from the same hyperbolic fixed point, to give homoclinic point X. It should be emphasized that the curves drawn here do not correspond to a (single) trajectory but, rather, correspond to the curve(s) drawn through the successive intersections of a trajectory with the plane.

points. If the intersecting H^+ and H^- emanate from the fixed points of different families (e.g., the fixed points of T^3 and T^4 shown in the inner and outer parts of Figure 4.16), they are called *heteroclinic points*. However, these intersections are not at all simple! Consider the homoclinic point X in Figure 4.20a and its adjacent points X' and X''. These two points map, as shown, to TX' and TX'', respectively. The problem is that, since X is "ahead" of both X' and X'', its image TX is required, by the continuity of the mapping, to be "ahead" of both TX' and TX''. This is clearly impossible. However, the contradiction can be resolved by making a loop in the manifold as shown in Figure 4.20b. Now TX is a new intersection (i.e., homoclinic) point. Continuing the argument, TX must map to a new homoclinic point T^2X via a second loop as shown in Figure 4.20c. Furthermore, since T^2X is closer to the hyperbolic point than TX, the distance between T^2X and TX will be less than that between TX and X. By area preservation, the area in the loops between X and TX and T^2X must be the same. This being so, the second loop must be longer and thinner than the first. Further repetition of this argument results in an infinite number of intersections, that is, the whole area becomes dense with homoclinic points with the intervening loops becoming ever longer and thinner. The overall picture (Figure 4.21) is one of incredible complexity. This was appreciated by Poincaré, who wrote about it in his seminal treatise *Les Methodes Nouvelles de la Mechanique Celeste* (Poincaré, 1892) as follows:

> The intersections form a kind of lattice, web or network with infinitely tight loops; neither of the two curves (H^+ and H^-) must ever intersect itself but it must bend in such a complex fashion that it intersects all the loops of the network infinitely many times.

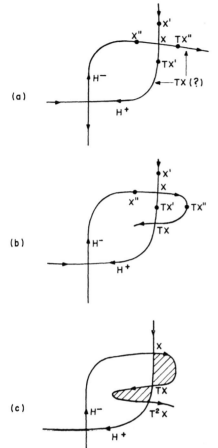

(a)

(b)

(c)

Figure 4.20 (a) Neighboring points X' and X'' being mapped to TX' and TX'' and apparently nonunique image, TX, of homoclinic point X. (b) Creating a unique image TX by inserting a loop in the manifold. (c) Image T^2X of TX creating a longer and thinner loop enclosing the same area.

One is struck by the complexity of this figure which I am not even attempting to draw. Nothing can give us a better idea of the complexity of the three body problem and of all the problems in dynamics where there is no holomorphic integral and Bohlin's series diverge.

We can now fill in, at least approximately, the details of Figure 4.16 and obtain the following picture (Figure 4.22). It is remarkable to think that this structure repeats on all scales and, furthermore, is the generic situation for nonintegrable systems. Thus, as remarkable as the detailed dynamics (shown in Figure 4.10*b*) in the neighborhood of the hyperbolic fixed point of Henon's map might have seemed at first, it should now not be so surprising.

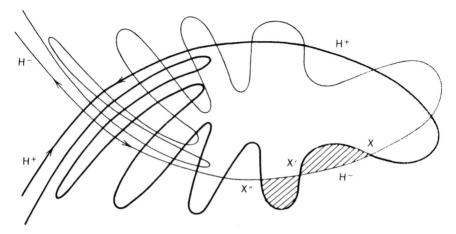

Figure 4.21 Network of intersections of H^+ and H^- leading to dense area of homoclinic points and network of ever longer and thinner loops of the same area (hatched regions).

4.4.b Whorls and Tendrils

In trajectory-by-trajectory studies such as those shown for Henon's map, we see the "seas" of chaos but, of course, we do not see curves corresponding to the wildly intersecting manifolds H^+ and H^-. In order to get a sense of what this structure actually looks like, we have to iterate a whole line element (each point of which corresponds to a different initial condition) under the mapping. Such a computation is shown in Figure 4.23 for a line element in the neighborhood of a hyperbolic fixed point of Henon's map.

Figure 4.22 Typical self-similar network of elliptic and hyperbolic fixed points with associated homoclinic webs. (Reproduced, by permission, from Ford (1975).)

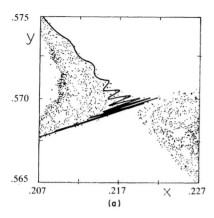

(a)

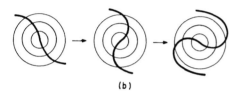

(b)

Figure 4.23 (a) Iteration of line element in neighborhood of hyperbolic fixed point leading to a "tendril." (Reproduced, by permission, from Berry (1978).) (b) Iteration of line element in neighborhood of ellipic fixed point leading to a "whorl."

We call these oscillations of the line element, with their characteristic flailing structure, *homoclinic oscillations*. In fact, in any (strongly) chaotic region (not necessarily the neighborhood of just one hyperbolic fixed point), a line element will evolve in this way—stretching exponentially fast and flailing backwards and forwards. This characteristic feature, found for evolving line elements on the plane, has been termed a *tendril*. This is in contrast to the behavior of a line element evolving in the neighborhood of an elliptic fixed point. On the basis of our discussions of twist maps, it is not difficult to deduce that the line element will form a tightly curling structure that we term a *whorl*. The whorl–tendril features are the "real-world" manifestations of chaos in area-preserving maps. As we will describe in Section 4.8, the lovely patterns that one sees in cream floating on the surface of a cup of coffee, or of gasoline films flowing down the street on a rainy day, can be explained precisely in these terms.

Finally, we make a remark about heteroclinic points. Since these correspond to the intersections of H^+ and H^- from different fixed-point families (e.g., T^3 and T^4 of Figure 4.16), it is likely that some intervening invariant curve (i.e., some torus with (very) irrational winding number) will also have to break down. Clearly, this will require a stronger perturbation. Thus one can expect the appearance of heteroclinic points to herald the onset of fairly widespread chaos. Criteria for the onset of such widespread chaos are important since it is believed that this will, for example, enhance transport processes. A number of techniques to predict this transition have been developed, including the method of overlapping resonances due to

Chirikov and the method of residues due to Greene. These will be discussed in subsequent sections.

4.5 CRITERIA FOR LOCAL CHAOS

4.5.a Lyapunov Exponents

An important characteristic of chaotic motion is the great sensitivity of the motion to small changes in initial conditions. Closely neighboring trajectories are found to diverge *exponentially*, whereas regular trajectories are found to separate only *linearly* in time. (Note, of course, that this divergence cannot go on forever in a bounded phase space.) The rate of divergence can be precisely quantified in terms of *Lyapunov exponents*, which measure the mean rate of exponential separation of neighboring trajectories. In fact, Lyapunov exponents are an extremely useful way of characterizing dynamical systems, and their use is by no means restricted to the Hamiltonian systems considered in this chapter. So, for generality, we will consider some (autonomous) system governed by the differential equations

$$\frac{dx_i}{dt} = F_i(x_1, \ldots, x_n), \qquad i = 1, \ldots, n \qquad (4.5.1)$$

In our examinations of the stability of a given fixed point, we linearized the equations of motion about that point. Now we linearize the equations about any reference *orbit* $\bar{\mathbf{x}} = (\bar{x}_1, \bar{x}_2, \ldots, \bar{x}_n)$ to yield the tangent map

$$\frac{d\delta x_i}{dt} = \sum_{j=1}^{n} \delta x_j \left(\frac{\partial F_i}{\partial x_j}\right)_{\mathbf{x} = \bar{\mathbf{x}}(t)} \qquad (4.5.2)$$

The norm

$$d(t) = \sqrt{\sum_{i=1}^{n} \delta x_i^2(t)} \qquad (4.5.3)$$

provides a measure of the divergence of two neighboring trajectories, that is, the reference trajectory $\bar{\mathbf{x}}$ and its neighbor with initial conditions $\bar{\mathbf{x}}(0) + \delta\mathbf{x}(0)$. The mean rate of exponential divergence is defined as

$$\sigma = \lim_{\substack{t \to \infty \\ d(0) \to 0}} \left(\frac{1}{t}\right) \ln \left(\frac{d(t)}{d(0)}\right) \qquad (4.5.4)$$

where $d(0) = \sqrt{\sum_{i=1}^{n} \delta x_i^2(0)}$. In addition, as will be explained below, it can be shown that there exists a set of n such quantities σ_i, $i = 1, \ldots, n$. These σ_i are called the *Lyapunov characteristic exponents* and they can be ordered by size, that is

$$\sigma_1 \geq \sigma_2 \geq \sigma_3 \geq \cdots \geq \sigma_n \qquad (4.5.5)$$

For regular motion, however, the exponents are zero since $d(t)$ grows only linearly (or possibly algebraically) with time

In order to understand these ideas more fully, it is useful to first of all consider the Lyapunov exponents of mappings. Indeed, the simplest possible case is a *one*-dimensional map of the form

$$x_{i+1} = f(x_i) \qquad (4.5.6)$$

where $f(x)$ is some (simple) nonlinear function of x, for example, $f(x) = 4\lambda x(1-x)$. The remarkable dynamical properties of such a map will be described at greater length in Chapter 5. The tangent map is simply

$$\mathcal{S}x_{i+1} = f'(x_i)\,\delta x_i = \prod_{j=0}^{i} f'(x_j)\,\delta x_0 \qquad (4.5.7)$$

where $f'(x_j)$ is the derivative of $f(x)$ evaluated at each point x_j along the given trajectory. The associated Lyapunov exponent is easily deduced by analogy with (4.5.3) to be

$$\sigma = \lim_{N\to\infty} \frac{1}{N} \ln\left[\prod_{j=1}^{N} f'(x_j)\,\delta x_0 \right]$$

$$= \lim_{N\to\infty} \frac{1}{N} \sum_{j=0}^{N} \ln|f'(x_j)| \qquad (4.5.8)$$

The exponent σ is independent of the initial point x_0 (apart from a set of measure zero initial conditions). In the case of $f(x) = x^2$ the reader should have little difficulty in verifying that $\sigma = \ln 2$.

In the case of multidimensional mappings

$$\mathbf{x}_{i+1} = \mathbf{F}(\mathbf{x}_i)$$

where $\mathbf{x}$ and $\mathbf{F}$ are n-dimensional vectors, there will be a set of n characteristic exponents corresponding to the n eigenvalues of the associated tangent map. Introducing the eigenvalues $\lambda_i(N)$, $i = 1, \ldots, n$, of the matrix

$$(TM)_N = (M(\mathbf{x}_N)M(\mathbf{x}_{N-1})\cdots M(\mathbf{x}_1))^{1/N} \qquad (4.5.9)$$

where $M(\mathbf{x}_i)$ is the linearization of $\mathbf{F}$ at the point $\mathbf{x}_i$, the exponents are defined as

$$\sigma_i = \lim_{N\to\infty} \ln|\lambda_i(N)|, \qquad i = 1, \ldots, n \qquad (4.5.10)$$

It should be clear that for area-preserving maps and Hamiltonian flows, the sum of the exponents must be zero in order to ensure that phase volume is preserved.

We now turn to the case of flows, as governed by Eqs. (4.5.1), and write

(4.5.2) in the vector form

$$\frac{d}{dt}\delta z = M\delta z \qquad (4.5.11)$$

where $\delta z = (\delta x_1, \ldots, \delta x_n)$ and M is the linearized matrix with elements $(M)_{ij} = (\partial F_i/\partial x_j)_{x=x(t)}$. There will exist a set of basis vectors $\hat{e}_i$ $(i = 1, \ldots, n)$ such that $\delta z = \sum_{i=1}^{n} a_i \hat{e}_i$. The stretching (or contracting) rates in each of the directions $\hat{e}_i$ provide us with the set of exponents σ_i $(i = 1, \ldots, n)$, which can be ordered as in (4.5.5). Clearly, as time evolves, a small volume element will be stretched most in the direction $\hat{e}_i$ with the largest exponent. Thus, in practice, (4.5.4) will yield just this exponent (σ_1, according to (4.5.5)). For Hamiltonian systems with n degrees of freedom, the vector δz becomes $2n$-dimensional (i.e., $\delta z = (\delta q_1, \ldots, \delta q_n, \delta p_1, \ldots, \delta p_n)$), and there will be $2n$ exponents. However, now there is a special symmetry between the σ_i, namely,

$$\sigma_i = -\sigma_{2n-i+1} \qquad (4.5.12)$$

Thus any stretching in one "direction" is canceled by contraction in another, thereby ensuring Liouville's theorem. If the exponents are calculated on a given energy shell, the space is $2n - 1$-dimensional. Thus it follows from (4.5.12) that two (or more, depending on the dynamics) of the σ_i must be zero.

The actual computation of exponents for n-dimensional flows (rather than mappings) is nontrivial. Consider, for example, working with the definition (4.5.4). If the norm $d(t)$ increases exponentially, there will be the risk of computer overflow and related errors. Instead, one uses a scheme suggested by Benettin et al. (1976). Here one starts with the initial norm $d(0)$ normalized to unity and computes the divergence over some interval τ, which is then renormalized back to a norm of unity. In this way, one computes (see Figure 4.24) the sequence of quantities

$$d_j = \|\delta x^{(j-1)}(\tau)\| \qquad (4.5.13)$$

where $\| \; \|$ denotes Euclidean norm, and

$$\delta x^{(j)}(0) = \frac{\delta x^{(j-1)}(\tau)}{d_j} \qquad (4.5.14)$$

where $\delta x^{(j)}(\tau)$ is computed from (4.5.2), with the initial values $\delta x^{(j)}(0)$, along the reference trajectory $\bar{x}$ from $\bar{x}(j\tau)$ to $\bar{x}((j+1)\tau)$. By analogy with (4.5.4), one defines

$$\sigma_N = \frac{1}{N\tau} \sum_{j=1}^{N} \ln d_j \qquad (4.5.15)$$

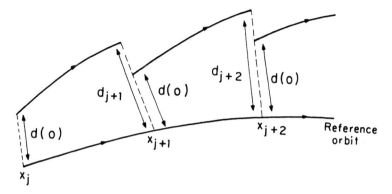

Figure 4.24 Computation of the largest Lyapunov exponent. After each period τ, the distance from the reference orbit is rescaled back to $d(0)$.

Furthermore, for τ not too large, it can be shown that the limit $N \to \infty$ exists and is independent of τ. Indeed, one can show that

$$\lim_{N \to \infty} \sigma_N = \sigma_1 \qquad (4.5.16)$$

where σ_1 is the largest of the set of exponents (4.5.5). The computation of the complete spectrum of Lyapunov exponents $\sigma_1, \ldots, \sigma_n$ requires more sophisticated techniques; a discussion of this important problem is better left to the experts (see Benettin et al. (1980)).

A quantity related to Lyapunov exponents, but more difficult to compute in practice, is the so-called *Kolmogorov entropy* (KS entropy). Formally, it is defined somewhat like entropy in statistical mechanics (i.e., it involves partitions of phase space, etc.) and gives a measure of the amount of information lost or gained by a system as it evolves. A remarkable result of Pesin (1977) shows that it can be computed from the Lyapunov exponents by the relation

$$h_k = \int_P \sum_{\sigma_i > 0} \sigma_i \, d\mu$$

which represents the sum of all positive Lyapunov exponents averaged over some (connected) region of phase space P with measure $d\mu$. A very readable introduction to this idea, which includes computation of h_k for the Henon–Heiles system, has been given by Benettin et al. (1976).

4.5.b Power Spectra

Another valuable characterization of orbits is in terms of their Fourier transform or power spectrum. Indeed, in many experimental systems (e.g., fluid dynamics), the data are often recorded as the Fourier transform rather than as the real time (or space) signal.

Firstly, consider the case of regular motion. Since the trajectories are confined to tori, they can be represented in the standard form

$$\mathbf{q}(t) = \sum_{\mathbf{m}} \mathbf{q}_{\mathbf{m}} \, e^{i\mathbf{m} \cdot (\boldsymbol{\omega}t+\delta)} \tag{4.5.17}$$

where $\mathbf{q}_{\mathbf{m}}$ are the vector of Fourier coefficients associated with the variables $\mathbf{q} = (q_1, \ldots, q_n)$, and $\boldsymbol{\omega} = (\omega_1, \ldots, \omega_n)$ are the frequencies of the associated torus. Clearly, the Fourier transform of (4.5.17)—or rather its squared modulus—will just be a set of δ-functions at the fundamental frequencies $\boldsymbol{\omega}$ and various overtones. The term *power spectrum* is, strictly speaking, defined to be the Fourier transform of the correlation function of a particular variable, say q_i, namely

$$C(t) = \langle q_i(0) q_i(t) \rangle \tag{4.5.18}$$

where $\langle \ \rangle$ denotes some ensemble average. The power spectrum is thus

$$I(\omega) = \frac{1}{2\pi} \int_{-\infty}^{\infty} C(t) \, e^{i\omega t} \, dt \tag{4.5.19}$$

a result which is an example of the Weiner–Khinchine theorem.

For regular motion, a natural choice of ensemble is just the associated torus. Since it is covered by a one-parameter (the initial phase δ) family of trajectories, the ensemble average is just the (phase) average over all δ. In this case it is easy to show that

$$C(t) = \sum_{\mathbf{m}} |q_{\mathbf{m}}^{(i)}|^2 \, e^{i\mathbf{m} \cdot \boldsymbol{\omega}t} \tag{4.5.20}$$

and hence

$$I(\omega) = \sum_{\mathbf{m}} |q_{\mathbf{m}}^{(i)}|^2 \delta(\mathbf{m} \cdot \boldsymbol{\omega} - \omega) \tag{4.5.21}$$

where the $q_{\mathbf{m}}^{(i)}$ are the Fourier coefficients associated with the variable q_i. Furthermore, for tori with incommensurable frequencies, the flow is ergodic on the torus. Thus, phase average equals time average and one can compute $I(\omega)$ just using a single trajectory, that is,

$$I(\omega) = \frac{1}{2\pi} \lim_{T \to \infty} \frac{1}{T} \left| \int_{-T}^{T} q_i(t) \, e^{i\omega t} \, dt \right|^2 \tag{4.5.22}$$

(thus defined, $I(\omega)$ is essentially the spectral line-shape function much used in spectroscopy).

In practice, a time series is only available for a finite time, and the spectrum is really a convolution between the actual motion and the "box-function" $h(t) = 1$, $-T \leq t \leq T$. For example, consider the simple case of the periodic motion

$$q(t) = Ae^{-i\Omega t} \tag{4.5.23}$$

Using (4.5.22), it is easy to show that

$$I(\omega) = \lim_{T \to \infty} \frac{2|A|^2}{\pi} \frac{\sin^2(\omega - \Omega)T}{(\omega - \Omega)^2 T} \tag{4.5.24}$$

which, for finite T, exhibits a maximum at $\omega = \Omega$ and a series of sym-

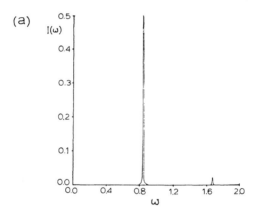

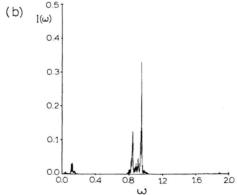

Figure 4.25 Power spectra (of the quantity $(x(t) + y(t))$ for individual trajectories of the Henon–Heiles system at $E = \frac{1}{8}$: (a) regular trajectory; (b) irregular trajectory.

metrical, decaying side-bands. In the limit $T \rightarrow \infty$, (4.5.24) reverts to the expected δ-function at $\omega = \Omega$.

In the chaotic regime, one can still go ahead and compute $I(\omega)$, using a single trajectory, as defined in (4.5.22). (Strictly speaking, this is no longer a "line-shape" function since it is not clear what the ensemble is in (4.5.18).) One finds that the spectrum of an irregular trajectory is much more complicated than for a regular one. Typically, one sees some dominant peaks surrounded by a lot of "grass" (see Figure 4.25). On the basis of numerical evidence alone, it is not clear whether this grassy portion of the spectrum is truly continuous for the irregular trajectories of generic Hamiltonian systems. Nonetheless, the difference in spectrum between regular and irregular motion is usually fairly striking and again provides a valuable characterization of dynamical systems. Indeed, there are a number of important rigorous results available which tell us that a system will only have a discrete spectrum if it is "ergodic," whereas to have a continuous spectrum it must be "mixing." (See Section 4.7 for a brief discussion of these concepts.)

4.6 CRITERIA FOR THE ONSET OF WIDESPREAD CHAOS

So far our discussions of chaos have been of a local nature in that we have concentrated on the chaotic behaviour of individual trajectories as well as the means of identifying them and quantifying their behavior. As useful as these techniques are, it would clearly be valuable if there were also methods for estimating when—as a function of energy or some nonlinear coupling parameter—the bulk of the trajectories become chaotic. This is what we term "widespread" chaos—although we hasten to add that in all of the techniques to be described, the notion of "widespread" is somewhat subjective, and a precise measure of this does not seem to be quantifiable. Nonetheless, estimates of when widespread chaos first appears are most useful. For example, in theories of unimolecular decomposition, it can indicate the validity of statistical, rather than dynamical, theories. In other situations, it can similarly indicate when it becomes valid to think in terms of transport processes. In this section we will briefly describe two such techniques. The first is the method of *overlapping resonances* due to Chirikov (1979), which is capable of giving crude analytical estimates of the onset of widespread chaos. The other is due to Greene (1979) and is able to predict when individual tori will breakdown. More detailed reviews of these and other methods have been given elsewhere (Tabor, 1981; Lichtenberg and Lieberman, 1983).

4.6.a Method of Overlapping Resonances

In order to understand Chirikov's method, we must first understand what is meant by a resonance. Consider some integrable, n-degree-of-freedom

Hamiltonian H_0 perturbed by some H_1, that is,

$$H(\mathbf{I}, \boldsymbol{\theta}) = H_0(\mathbf{I}) + \epsilon H_1(\mathbf{I}, \boldsymbol{\theta}) \qquad (4.6.1)$$

where $\mathbf{I}$, $\boldsymbol{\theta}$ are the n-component action-angle vectors. In the usual way, we write H_1 as a Fourier series in the angle variables, namely,

$$H(\mathbf{I}, \boldsymbol{\theta}) = H_0(\mathbf{I}) + \epsilon \sum_{\mathbf{m}} H_{\mathbf{m}}(\mathbf{I}) e^{i\mathbf{m} \cdot \boldsymbol{\theta}} \qquad (4.6.2)$$

where $H_{\mathbf{m}}(\mathbf{I})$ are the Fourier coefficients and $\mathbf{m} = (m_1, \ldots, m_n)$. The equations for the unperturbed variables are, of course,

$$\begin{aligned} I_i &= I_i(0) \\ \theta_i &= \omega_i(\mathbf{I})t + \theta_i(0) \end{aligned} \qquad (4.6.3)$$

where $\omega_i = \partial H_0 / \partial I_i$ and $I_i(0)$ and $\theta_i(0)$ are the initial values. Now consider the situation in which H_0 is perturbed by only one term in the Fourier sum, that is,

$$H(\mathbf{I}, \boldsymbol{\theta}) = H_0(\mathbf{I}) + \epsilon H_{\mathbf{m}}(\mathbf{I}) \, e^{i\mathbf{m} \cdot \boldsymbol{\theta}} \qquad (4.6.4)$$

In this case the perturbed variables satisfy the equations of motion

$$\dot{I}_i = - i\epsilon m_i H_{\mathbf{m}}(\mathbf{I}) \, e^{i\mathbf{m} \cdot \boldsymbol{\theta}} \qquad (4.6.5a)$$
$$\dot{\theta}_i = \omega_i(\mathbf{I}) + \epsilon H_{\mathbf{m}}'(\mathbf{I}) \, e^{i\mathbf{m} \cdot \theta} \qquad (4.6.5b)$$

where the prime denotes differentiation with respect to I_i. To first order in ϵ, Eqs. (4.6.5) can be integrated by simply substituting, where appropriate, the unperturbed solutions (4.6.3), in which case (4.6.5a) yields

$$I_i \simeq I_i(0) - \frac{\epsilon m_i H_{\mathbf{m}}(\mathbf{I}(0)) e^{i(\mathbf{m} \cdot \boldsymbol{\omega})t + i\delta}}{\mathbf{m} \cdot \boldsymbol{\omega}} \qquad (4.6.6)$$

where δ is a phase factor. Clearly, such an approximation will break down for any $\mathbf{m} \cdot \boldsymbol{\omega}(\mathbf{I}) \leq \epsilon$. This is known as a *resonance*, and the reader will see that this is essentially the same as the small divisor effect discussed in Chapter 3. Indeed, we have already seen an example of a resonance in Chapter 1, in the case of the driven linear oscillator (1.6.4) in which the solution has a denominator proportional to $\omega^2 - \Omega^2$ (1.6.5), where ω is the intrinsic frequency and Ω is the external driving frequency. For a single perturbing term (i.e., Eq. (4.6.4)), such a resonance is not a disaster and can be removed by a suitable canonical transformation as described below. Indeed, the system (4.6.4) is still completely integrable since it is possible to construct new integral of motion of the form

$$F = \hat{\mathbf{m}} \cdot \mathbf{I} \qquad (4.6.7)$$

where $\hat{\mathbf{m}}$ is a vector orthogonal to $\mathbf{m}$ (i.e., $\mathbf{m} \cdot \hat{\mathbf{m}} = 0$). This is easily verified by taking the Poisson bracket of F with H, namely,

$$[F, H] = \sum_{i=1}^{n} \left(\frac{\partial H}{\partial I_i} \frac{\partial F}{\partial \theta_i} - \frac{\partial H}{\partial \theta_i} \frac{\partial F}{\partial I_i} \right)$$

$$= -i\epsilon \sum_{j=1}^{n} \hat{m}_j m_j H_{\mathbf{m}} \, e^{i\mathbf{m} \cdot \boldsymbol{\theta}} = 0 \qquad (4.6.8)$$

Thus, isolated resonances, although they can cause a considerable distortion of tori in their neighborhood, do not introduce any chaos into a system. However, when two or more resonances are simultaneously present, they will render a system nonintegrable. Furthermore, when they are sufficiently "close" to each other, they will result, as is now described, in the appearance of widespread chaos.

In a very readable illustration of the resonance problem, Walker and Ford (1969) took the integrable Hamiltonian

$$H_0(I_1, I_2) = I_1 + I_2 - I_1^2 - 3I_1 I_2 + I_2^2 \qquad (4.6.9)$$

and investigated the effect of adding a $2:2$ resonance and a $3:2$ resonance, that is,

$$H(\mathbf{I}, \boldsymbol{\theta}) = H_0(\mathbf{I}) + \alpha I_1 I_2 \cos(2\theta_1 - 2\theta_2) + \beta I_1^{3/2} I_2 \cos(2\theta_1 - 3\theta_2) \quad (4.6.10)$$

The effect of each resonance in isolation and together is shown in Figure 4.26. At low energies the two resonant zones are well separated. As the energy of the system is increased, the two zones overlap and a "macroscopic zones of instability" appears. By this term, Walker and Ford simply meant a clearly visible splatter of points on the surface of section. The size of this zone increases with increasing energy. The structure is further complicated by the appearance of "secondary" resonant zones as the two principal zones approach each other. By means of a numerical investigation, the authors were able to predict the energy at which the overlap of the resonances first occurred; that is, they were able to predict (successfully) the onset of widespread chaotic motion. The overlapping of resonances would appear to play a key role in such an onset, as well as giving a great deal of physical insight. When principal zones start overlapping, many higher-order resonances are also involved and thus one may be moderately confident that fairly large areas of phase space have had (most of) their tori destroyed and that the ensuing chaos will indeed be "wide-

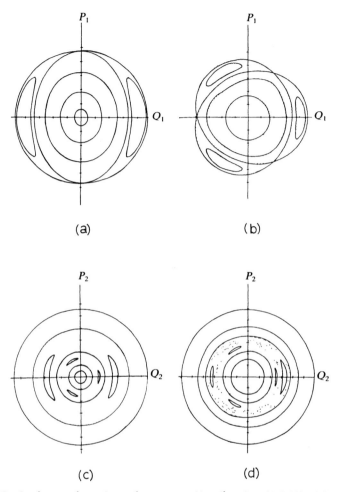

Figure 4.26 Surfaces of section of resonant Hamiltonian (4.6.10): (a) only 2:2 resonance acting ($\beta = 0$); (b) only 2:3 resonance acting ($\alpha = 0$); (c) both resonances present ($\alpha = \beta = 0.02$) but widely separated at $E = 0.18$; and (d) the two resonant zones overlapping at $E = 0.2905$; the erratic splatter of points is generated by a trajectory started in the region of resonance overlap. (Reproduced, by permission, from Walker and Ford (1969).)

spread," since trajectories are now free to wander between regions that were previously separated by nonresonant tori.

A means of predicting approximately when resonant overlap will occur has been proposed by Chirikov (1979). This method works best for forced one-dimensional oscillators, a model very useful in the design of accelerators or in the study of molecular bonds or atoms subjected to radiation fields. We consider a one-dimensional nonlinear oscillator (e.g.,

$H_0 = \frac{1}{2}(p^2 + \frac{1}{2}q^4))$ perturbed by an external periodic force, for example $V = q \cos(\phi)$, where $\phi = \Omega t + \phi_0$ is the external phase. The unperturbed system, being one-dimensional, can always be solved in action-angle variables (I, θ), and we can then express any (reasonable) external field as a Fourier series in these variables, that is,

$$H = H_0(I) + \epsilon \sum_{m,n} V_{mn}(I) e^{i(m\theta + n\phi)} \qquad (4.6.11)$$

In what follows, the results are only significant for nonlinear oscillators. In the linear case, we know that when the external frequency Ω equals the oscillator frequency ω, the motion "blows up." In the nonlinear case, there is also a resonance in the vicinity of $\Omega = \omega$. However, as discussed in Chapter 1, when the amplitude of the oscillator increases, the frequency, which is energy dependent, changes and the system comes out of resonance. In the above case, the oscillator frequency is given by the usual equation, $\omega(I) = \partial H_0/\partial I$, and there is a resonance at those values of $I = I_r$ such that

$$\frac{\omega(I_r)}{\Omega} = \frac{k}{l} \qquad (4.6.12)$$

In this case the corresponding phase (and its harmonics) $l\theta - k\phi$ are slowly varying compared to other terms in the series (4.6.11). By virtue of the nonlinearity of H_0, there will, of course, be other values of I_r giving rise to other resonances, and, generally speaking, the set of resonances is everywhere dense. However, to simplify matters, we start by considering the resonance (4.6.12) in isolation and examine the behavior of the Hamiltonian in its vicinity. In the next few paragraphs we show how to reduce (4.6.11) to a simple form from which a "resonance width" can be estimated.

Drawing on our study of canonical transformations in Chapter 2, we introduce the generating function

$$F = F(J, \theta, \phi) = (l\theta - k\phi)J + \theta I_r \qquad (4.6.13)$$

where J is the new momentum and the term θI_r provides, as we see below, a convenient shift in the origin of the new action variable J. From the generating function, we obtain the relations

$$I = \frac{\partial F}{\partial \theta} = lJ + I_r \qquad (4.6.14)$$

and

$$\psi = \frac{\partial F}{\partial J} = l\theta - k\phi \qquad (4.6.15)$$

where ψ is the new "resonant" phase conjugate to the new momentum J. The time derivative of F is also required, that is,

$$\frac{\partial F}{\partial t} = -k\Omega J \tag{4.6.16}$$

Performing the canonical transformation the Hamiltonian becomes, in terms of the new variables,

$$H = H_0(J) + \epsilon \sum_{m,n} V_{mn}(J) \exp\left[i \frac{1}{l} \{ m\psi + (km + nl)\phi \} \right] - k\Omega J \tag{4.6.17}$$

The transformed Hamiltonian has almost the same form as the original one (4.6.11). However, by performing this transformation, one is effectively putting the observer in a rotating frame in which the rate of change of the new phase measures the slow deviation from resonance. Although it is not necessarily true that $\dot{\psi} \ll \dot{\phi}$, we assume that it is near the resonance and, hence, that during one complete cycle of ψ, ϕ will have passed through many cycles. The average contribution of these rapidly oscillating terms is zero, and this leads us to the next stage, namely, averaging the Hamiltonian over the fast-phase variables, that is,

$$\bar{H}(J, \psi) = \frac{1}{2\pi} \int_0^{2\pi} H(J, \psi, \phi) \, d\phi \tag{4.6.18}$$

The resulting "averaged" Hamiltonian, H, takes the form

$$\bar{H} = H_0(J) + \epsilon \sum_p V_{pl,-pk} \cos(p\psi) - k\Omega J \tag{4.6.19}$$

where we have gone over to real arithmetic (assuming $V_{-l,k} = V_{l,-k}$ and $V_{00} = 0$ as well as absorbing a factor of 2 in the Fourier coefficients $V_{pl,-pk}$). There are still all the harmonics to deal with, but at this stage we make the assumption

$$V_{pl,-pk} \ll V_{l,-k} \qquad \text{for } p = 2, 3, \ldots \tag{4.6.20}$$

and (4.6.19) reduces to

$$\bar{H} = H_0(J) + \epsilon V_{l,-k} \cos \psi - k\Omega J \tag{4.6.21}$$

The last stage is as follows. Having worked on the assumption that $\psi \ll \phi$ in the region of the resonance (4.6.12), we expand (4.6.21) about $I = I_r$ to second order, although we assume that the coefficient $V_{l,-k}(J)$ is only a slowly varying function of I. Thus we obtain

$$\bar{H} = H_0(I_r) + IJ \left(\frac{\partial H_0}{\partial I}\right)_{I=I_r} + I^2 \frac{J^2}{2}\left(\frac{\partial^2 H_0}{\partial I^2}\right)_{I=I_r} + \epsilon V_{l,-k}(I_r)\cos\psi - k\Omega J$$

(4.6.22)

Since $(\partial H_0/\partial I)_{I=I_r} = \omega(I_r)$, the terms that are linear in J conveniently cancel by the resonance condition (4.6.12). Dropping the constant term $H_0(I_r)$, we are left with the "resonant" Hamiltonian

$$H_r = \frac{J^2}{2M} + \epsilon V_{l,-k}\cos\psi$$

(4.6.23)

where the "mass" M is given by

$$M^{-1} = I^2 \left(\frac{\partial^2 H_0}{\partial I^2}\right)_{I=I_r}$$

(4.6.24)

The "resonant" Hamiltonian (4.6.23) has exactly the form of a pendulum Hamiltonian whose separatrix (see Figure 4.27) is determined by the equation

$$J_{sx} = \pm (4M\epsilon V_{l,-k})^{1/2} \cos\left(\frac{\psi}{2}\right)$$

(4.6.25)

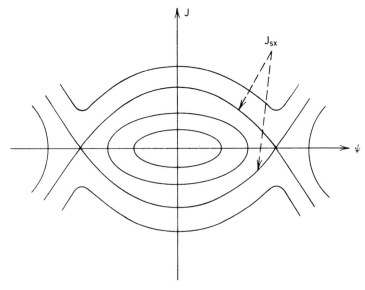

Figure 4.27 Phase plane for pendulum Hamiltonian (4.6.23) showing separatrix.

In the old (I, θ) variables, this is simply

$$I_{sx} = I_r \pm (\Delta I_r) \cos\left(\frac{l\theta - k\phi}{2}\right) \qquad (4.6.26)$$

where

$$(\Delta I_r) = 2l(\epsilon M V_{l,-k})^{1/2} \qquad (4.6.27)$$

The quantity (ΔI_r) is the resonance "half-width," which can also be expressed in terms of frequency, that is,

$$(\Delta \omega_r) = \frac{\partial \omega}{\partial I}(\Delta I_r) = \frac{1}{l^2 M} 2l(\epsilon M V_{l,-k})^{1/2} = \frac{2}{l}\left(\frac{\epsilon V_{l,-k}}{M}\right)^{1/2} \qquad (4.6.28)$$

Notice that since $M^{-1} = l^2(\partial\omega/\partial I)$ the resonance half-width's dependence on the order of the resonance is only in the Fourier coefficients $V_{l,-k}$. From (4.6.27) and (4.6.28), we can see that the effect of a resonant perturbation is of $O(\epsilon^{1/2})$.

It is important to note that several assumptions have gone into the derivation of the resonant Hamiltonian (4.6.23). Nonresonant terms in H are neglected by assuming they are rapidly oscillating, and hence their average value over a cycle of motion is zero. The resonance harmonics have been neglected by assuming $V_{l,-k} \ll V_{pl,-pk}$; furthermore, the higher-order terms in the expansion about I_r have been dropped. Chirikov (1979) has suggested that these assumptions can be summarized in the "moderate nonlinearity condition"

$$\epsilon \ll \alpha \ll \frac{1}{\epsilon} \qquad (4.6.29)$$

where $\alpha = (I/\omega)(\partial\omega/\partial I)$. This condition is discussed further by Chirikov.

So far, though, the "resonant" Hamiltonian (4.6.23) is still integrable, since it consists of only one resonance in isolation. Chirikov's "criterion of overlapping resonances" is obtained by evaluating the width of another (principal) resonance and then finding the coupling strength ϵ at which the two resonances touch; that is, we find the ϵ for which

$$(\Delta \omega_r)_1 + (\Delta \omega_r)_2 = \Delta \Omega \qquad (4.6.30)$$

where $(\Delta \omega_r)_1$ and $(\Delta \omega_r)_2$ are the widths of the two resonances and $\Delta\Omega$ is their separation. The width of each resonance zone is calculated independently of all the others; clearly, this is a major approximation and one simply hopes that the "moderate nonlinearity condition" (4.6.29) will ensure that the error is not too great.

Chirikov (1979) has tested his method out on a number of simple systems. Of particular interest is the one with model Hamiltonian

$$H(I, \theta, t) = \frac{I^2}{2} + K \sum_{n=-\infty}^{\infty} \cos(\theta - nt) \qquad (4.6.31)$$

Physically, it represents a rotor (of unit mass) being acted on by an infinite series of resonances or, alternatively, when written in the equivalent form

$$H(I, \theta, t) = \frac{I^2}{2} + 2\pi K \cos \theta \sum_{m=-\infty}^{\infty} \delta(2\pi m - t) \qquad (4.6.32)$$

as a pendulum subject to a series of "kicks" at times $t = 2\pi m$. The reader may confirm that integration of Hamilton's equations over one period yields precisely the "standard map" (4.2.26) with $p \equiv I$ and $q \equiv \theta$ and $k = K/(2\pi)^2$. Each term in the series of resonances provides us directly with a resonant Hamiltonian (cf. (4.6.23))

$$\bar{H}^{(n)} = \frac{I^2}{2} + K \cos \psi_n \qquad (4.6.33)$$

where ψ_n is the slowly varying phase $(\theta - nt)$. One can immediately write the resonance half-width from (4.6.28), that is,

$$(\Delta \omega_r)_n = 2K^{1/2} \qquad (4.6.34)$$

Resonances occur at every integer value of $\omega = I = I_r = n$ (this is represented schematically in Figure 4.28). The spacing between resonances is unity $(\Delta\Omega = I_r^{(n+1)} - I_r^{(n)} = (n+1) - n = 1)$, and they touch when

$$(\Delta\omega_r) = \frac{\Delta\Omega}{2} = \frac{1}{2} \qquad (4.6.35)$$

This enables one to predict the critical value of the perturbation parameter K at which resonance overlap occurs, that is, since

$$(\Delta\omega_r) = 2K^{1/2} = \tfrac{1}{2} \qquad (4.6.36)$$

this gives

$$K_{\text{crit}} = \tfrac{1}{16} \qquad (4.6.37)$$

Numerical studies of this system suggest that widespread chaos sets in at about $K = \tfrac{1}{40}$; that is, the overlap criterion is out by about a factor $2\tfrac{1}{2}$. Further refinements, such as including higher harmonics of the resonant phases, give an improved estimate of $K \simeq \tfrac{1}{30}$.

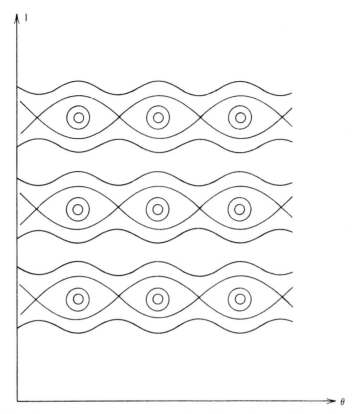

Figure 4.28 Phase plane of resonances for "kicked" pendulum Hamiltonian (4.6.32).

Chirikov's method has been the object of much research activity, and sophisticated versions of it (to deal, for example, with the effect of "secondary" resonances) have been developed which are able to give far more accurate predictions for the onset of widespread chaos. The technique can also be applied to autonomous systems with many degrees of freedom (the theory is "cleanest" for driven one-degree-of-freedom systems). The underlying ideas are the same, but the analysis becomes more complicated; the interested reader is referred elsewhere to follow up this topic.

4.6.b Greene's Method

We now turn (albeit too briefly) to discuss an important method, developed by Greene (1979), for predicting the onset of chaotic motion based on the stability properties of closed orbits. It is based on the hypothesis that the dissolution of an invariant curve (torus) can be associated with the sudden change from stability to instability of nearby closed orbits. To see this more precisely, imagine a weakly perturbed integrable system. According to the KAM theorem, those invariant curves with "sufficiently" irrational winding number are preserved. The neighboring rational (and close-to-rational) curves break up in the manner described previously (the Poincaré–Birkhoff fixed-point theorem), that is, into equal numbers of elliptic (stable) and hyperbolic (unstable) fixed points. Greene's method is based on the observation that when the perturbation is made sufficiently strong (or the energy high enough), the set of stable fixed points also becomes unstable (they become "hyperbolic-with-reflection" fixed points). The contention is that this then signals the dissolution of an invariant curve "close" to that set of fixed points. Recalling the discussion of the KAM theorem in Chapter 3, a rather nice way of estimating the closeness of a closed orbit to a given invariant curve is by expressing that curve's winding number in the form of a continued fraction, that is,

$$\alpha = a_0 + \cfrac{1}{a_1 + \cfrac{1}{a_2 + \cfrac{1}{a_3 + \cdots}}}$$
(4.6.38)

where $a_0, a_1, a_2, \ldots$ are positive integers. Thus the successive trunctions of this representation of an irrational winding number yield the winding numbers of the closed orbits that become ever "closer" to the chosen invariant curve. By following the stability properties of these sequences of closed orbits, as they "close in" on an invariant curve, Greene (1979) was able to predict the breakup of that curve.

The two essential ingredients of this method are (1) finding the closed orbits and (2) determining their stability characteristics. A detailed discussion of the former problem is outside the scope of this chapter. Suffice it to say that there are now a variety of well-developed and efficient methods for finding closed orbits of any desired topology (winding number). Of these we mention the approach described by Greene (1979) and a method developed by Helleman and Bountis (1979). We describe the stability analysis by using the example studied in detail by Greene, namely, the "standard map" (on the unit torus), that is,

$$I_{n+1} = I_n + \frac{k}{2\pi} \sin 2\pi\theta_n, \qquad \text{mod } I = 1 \qquad (4.6.39a)$$

$$\theta_{n+1} = \theta_n + I_{n+1}, \qquad \text{mod } \theta = 1 \qquad (4.6.39b)$$

The parameter k can be regarded as a perturbation parameter; for $k = 0$, the mapping takes the trivial form

$$I_{n+1} = I_n \qquad (4.6.40a)$$

$$\theta_{n+1} = \theta_n + I_{n+1} \qquad (4.6.40b)$$

In this case, the mapping is clearly "integrable," since all the orbits lie on straight lines. These are just the invariant curves of the unperturbed mapping. Returning to Figure 4.11, which shows the standard map computed at $k = 0.97$, we see some strongly irregular orbits, filling up substantial portions of the phase plane, as well as the typical alternating hyperbolic and elliptic fixed-point structure. Notice also that there are still invariant curves remaining that divide the phase space. These curves prevent a trajectory from wandering over the whole phase plane. Clearly, it will not be until these curves are destroyed that the "chaos" will be truly widespread.

The stability of a given closed orbit is determined by evaluating the tangent-space mapping. This corresponds to linearizing the mapping at each iteration. Thus if we denote the "tangent-space" variables as $(\delta I, \delta\theta)$, we have the tangent map

$$\begin{bmatrix} \delta I_{n+1} \\ \delta\theta_{n+1} \end{bmatrix} = M \begin{bmatrix} \delta I_n \\ \delta\theta_n \end{bmatrix} \qquad (4.6.41a)$$

where

$$M = \begin{bmatrix} 1 & -k\cos 2\pi\theta_n \\ 1 & 1 - k\cos 2\pi\theta_n \end{bmatrix} \qquad (4.6.41b)$$

The tangent mapping is, of course, area preserving since

$$\det|M| = 1 \qquad (4.6.42)$$

For an orbit that closes after Q iterations of the mapping, the eigenvalues, $\lambda_{\pm}$, of the 2×2 matrix

$$M^{(Q)} = \prod_{n=1}^{Q} \begin{bmatrix} 1 & -k\cos 2\pi\theta_n \\ 1 & 1 - k\cos 2\pi\theta_n \end{bmatrix} \qquad (4.6.43)$$

give the stability indices, or Floquet multipliers, of the orbit. Denoting the matrix elements of $M^{(Q)}$ by $M_{ij}^{(Q)}$, we have explicitly

$$\lambda_{\pm} = \tfrac{1}{2}(M_{11}^{(Q)} + M_{22}^{(Q)}) \pm \tfrac{1}{2}((M_{11}^{(Q)} + M_{22}^{(Q)})^2 - 4)^{1/2} \qquad (4.6.44)$$

where we have made use of condition (4.6.42), that is,

$$M_{11}^{(Q)}M_{22}^{(Q)} - M_{12}^{(Q)}M_{21}^{(Q)} = 1 \qquad (4.6.45)$$

From our previous discussions we know that if the eigenvalues are complex, the orbits are stable, whereas if the eigenvalues are real, the orbits are unstable. Greene (1979) introduces a quantity called the *residue*, which is defined as

$$R = \tfrac{1}{4}(2 - \text{Tr}(M^{(Q)})) \tag{4.6.46}$$

where Tr denotes trace.

From (4.6.44), it is easy to see that if $0 < R < 1$, the eigenvalues are imaginary and hence the orbit is stable, that is, the fixed points are elliptic. If $R < 0$ or $R > 1$ the eigenvalues are real and hence the orbit is unstable. More precisely, for $R < 0$ the fixed points are hyperbolic and for $R > 1$ they are "hyperbolic-with-reflection." (For parabolic fixed points, $R = 0$.) It can be shown that for an orbit of "length" Q, the residue is proportional to k^Q for both large and small k. (Recall that k is the perturbation parameter for the system studied here.) A quantity, called the *mean residue* f, is then introduced that scales away this exponential dependence on Q, that is,

$$f = \left(\frac{R}{\beta}\right)^{1/Q} \tag{4.6.47}$$

where β is some arbitrary constant introduced for practical convenience. We can now proceed to characterize the stability properties of the sequence of closed orbits converging on a chosen invariant curve. Each successive closed orbit (determined through the successive truncations of the continued fraction representation of the winding number of the chosen curve) has a larger Q, corresponding to increasing topological complexity of that orbit. The remarkable thing is that the corresponding sequence of mean residues is found to converge to some finite value. The rate of convergence seems to be determined by the value of β; for this problem the optimum value was found to be $\beta = \tfrac{1}{4}$. For further discussion of this matter, the reader is referred to Greene's original papers (Greene, 1979). It is then demonstrated (empirically) that when the converged mean residue becomes greater than unity (now assuming $\beta = 1$), the invariant curve associated with that sequence of closed orbits is destroyed.

This criterion then enables one to find the value of the perturbation parameter k at which any chosen invariant curve breaks up. For the above system the method has been found to give very accurate results. Furthermore, Green (1979) has made an ingenious extension of his method to predict the onset of widespread chaos. It is based on the conjecture that the more closely an irrational curve can be approximated by a sequence of rationals, the smaller the perturbation (k) required to destroy it. Thus one might reasonably assume that the last invariant curve to be destroyed will be the one whose winding number is least closely approximated by a sequence of rationals. This is the invariant curve whose winding number has the continued fraction representation

$$\alpha = 1 + \cfrac{1}{1 + \cfrac{1}{1 + \cfrac{1}{1 + \cfrac{1}{1 + \cdots}}}}$$

$$= \frac{(\sqrt{5} - 1)}{2} \tag{4.6.48}$$

which is the famous "golden mean." Thus by the time k is sufficiently large so that this invariant curve breaks up, one may fairly confidently assume that all the other curves have also been destroyed. There will then be no impediment to an irregular trajectory wandering over the whole of the phase plane, and widespread chaos will ensue. The critical value of k corresponding to the breakup of the golden mean curve was found to be about unity, in close agreement with the observed onset of widespread chaos.

4.7 STATISTICAL CONCEPTS IN STRONGLY CHAOTIC SYSTEMS

We have already mentioned the concept of ergodicity several times, and in this section we will enlarge the discussion to include other, related concepts that are useful in understanding the properties of strongly chaotic systems. Excellent, introductory discussions of these ideas have been given by Lebowitz and Penrose (1973) and Zaslavskii and Chirikov (1972).

4.7.a Ergodicity

A simple illustration of ergodicity is provided by recalling the case, given in Section 2.5, of flow on a torus with irrationally related frequencies. Con-

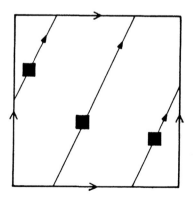

Figure 4.29 Ergodicity on the torus: Small-area element exploring the torus by uniform translation.

sider a two-dimensional torus, for which we write the flow as

$$\phi_1 = 2\pi\omega_1 t + \phi_1(0), \qquad \text{mod } \phi_1 = 1 \qquad (4.7.1a)$$

$$\phi_2 = 2\pi\omega_2 t + \phi_2(0), \qquad \text{mod } \phi_2 = 1 \qquad (4.7.1b)$$

where we have introduced the variables $\phi_i = \theta_i/2\pi$ which have period 1. This 2-D torus is topologically equivalent to the unit square with identified edges, and, as already described, it is easy to see that the flow is ergodic—and hence time average equals phase average—if ω_1/ω_2 is irrational. Notice (of course) that a small-area element only explores the torus by uniform translation without itself ever undergoing any distortion (see Figure 4.29). It should be clear from this example that ergodicity does not imply chaotic behavior.

4.7.b Mixing

Chaotic behavior is associated with the exponential divergence of nearby trajectories and hence positive Lyapunov exponents. In this case a small-area element will clearly undergo considerable distortion as it evolves—this leads to the concept of *mixing*. A simple system which exhibits mixing is the famous "cat map" of Arnold (linear automorphism of the unit torus), which is nothing more than the linear, area-preserving transformation T

$$T: \quad \begin{bmatrix} x_{n+1} \\ y_{n+1} \end{bmatrix} = \begin{bmatrix} 1 & 1 \\ 1 & 2 \end{bmatrix} \begin{bmatrix} x_n \\ y_n \end{bmatrix} \quad \begin{array}{l} \text{mod } x = 1 \\ \text{mod } y = 1 \end{array} \qquad (4.7.2)$$

As illustrated in Figure 4.30 after only two iterations of the map, an area element undergoes considerable distortion. The difference between (4.7.1) and (4.7.2) is that the latter has a shearing component and it is this plus the periodicity of the torus that lead to the observed effect of both stretching and *mixing*. By contrast with (4.7.1), a small-area element not only translates around the torus but rapidly becomes a long thin filament. Clearly mixing implies ergodicity, but ergodicity does not imply mixing.

The eigenvalues of T are easily computed to be

$$\lambda_{\pm} = \frac{3 \pm \sqrt{5}}{2} \qquad (4.7.3)$$

with $\lambda_+\lambda_- = 1$ since the transformation is area preserving. The two real eigenvalues λ_+ and λ_- lead to exponential stretching and contracting, respectively—the former in the direction of the eigenvector

$$\xi_+ = \begin{bmatrix} 1 \\ \dfrac{1+\sqrt{5}}{2} \end{bmatrix} \qquad (4.7.4a)$$

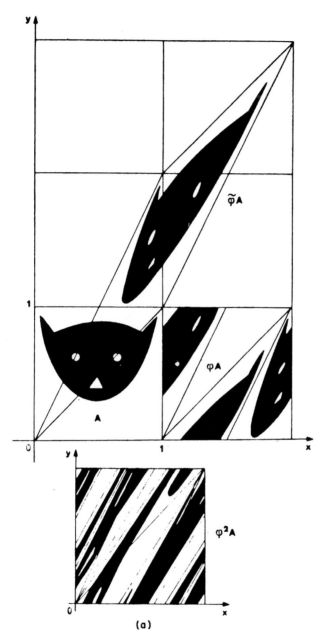

Figure 4.30 (a) Two iterations of the Arnold cat map. (Reproduced, by permission, from Arnold and Avez (1968).) (b) Mixing on the torus: Small-area element exploring the torus by translation and stretching.

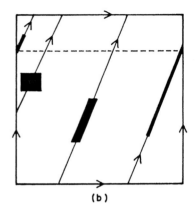

Figure 4.30 (*Continued*)   **(b)**

and the latter in the direction

$$\boldsymbol{\xi}_- = \begin{bmatrix} 1 \\ \dfrac{1 - \sqrt{5}}{2} \end{bmatrix}$$ (4.7.4b)

For this simple linear mapping, it is easy to see that the (positive) Lyapunov exponent is just $\sigma = \ln[(3 + \sqrt{5})/2]$.

The mapping T^n has many fixed points. They are easily determined by

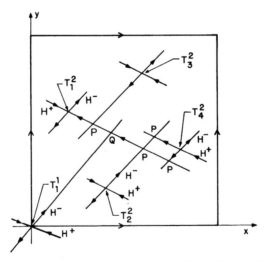

Figure 4.31 Schematic representation of homoclinic (P) and heteroclinic (Q) points of the cat map. T_i^2 ($i = 1, \ldots, 4$) are the four fixed points of T^2, and T_1^1 is the fixed point of T. The homoclinic points P are due to intersections of the H^+ and H^- manifolds of the T_i^2, and the heteroclinic point Q is due to intersection of H^- manifold of T_1^1 and H^+ manifold of T_1^2.

solving the equation

$$\begin{bmatrix} x \\ y \end{bmatrix} = T^n \begin{bmatrix} x \\ y \end{bmatrix} - \begin{bmatrix} k \\ l \end{bmatrix} \tag{4.7.5}$$

where k and l are the integers required to mod the iterates of T^n back onto the unit square. Obviously, the only fixed point of T itelf is just $(x, y) = (0, 0)$. For T^2 the fixed points are $(1/5, 3/5)$, $(2/5, 1/5)$, $(3/5, 4/5)$, and $(4/5, 2/5)$, which are all hyperbolic with eigenvalues $(\lambda_\pm)^2 = (7 \pm 3\sqrt{5})$. It is also rather easy to identify the homoclinic and heteroclinic points of the cat map. For example, the fixed point $(0, 0)$ of T has stable (H^+) and unstable (H^-) manifolds which wrap around the torus in irrational directions, given by ξ_- and ξ_+, respectively, and therefore intersect each other (but never themselves) infinitely often. The same argument also applies to the stable and unstable manifolds of the fixed points of T^2, but these will also intersect the H^+ and H^- of the fixed point of T, thereby leading to an infinity of heteroclinic points as well (see Figure 4.31).

4.7.c The Baker's Transformation and Bernoulli Systems

Another simple transformation with striking properties is the so-called *Baker's transformation*, which can be written as the following mapping on the unit square:

$$\begin{bmatrix} x_{n+1} \\ y_{n+1} \end{bmatrix} = \begin{bmatrix} 2x_n \\ y_n/2 \end{bmatrix}, \qquad 0 \le x_n < \tfrac{1}{2}$$

$$= \begin{bmatrix} 2x_n - 1 \\ y_n/2 + \tfrac{1}{2} \end{bmatrix}, \qquad \tfrac{1}{2} \le x_n < 1 \tag{4.7.6}$$

which corresponds to repeated doublings in the x direction and halvings in the y direction. The mapping is completely reversible; and if it is run backwards, the doubling occurs in the y direction and the halving occurs in the x direction. As Figure 4.32 shows, this area-preserving transformation is reminiscent of a baker rolling out dough. Again it is clear that just a few iterations of the mapping will lead to rapid mixing.

The strongly random nature of this simple transformation can be seen by representing the iterates (x_n, y_n) in binary notation, that is, as strings of zeros and ones. Simple examples of binary numbers are $\tfrac{1}{16} = 0.0001000\ldots$, $\tfrac{1}{8} = 0.001000\ldots$, $\tfrac{1}{4} = 0.01000\ldots$, and so on. Less trivial numbers, such as irrationals, consist of infinite, nonrepeating strings of zeros and ones. Note, however, the basic property that the doubling of a number corresponds to moving the decimal point one place to the right and halving a number corresponds to moving it one place to the left. This is ideal for the baker's transformation. An initial condition $X_0 = (x_0, y_0)$ is represented by the strings

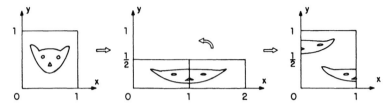

Figure 4.32 Baker's transformation.

$$x_0 = . \, a_1 a_2 a_3 \ldots a_i \ldots \tag{4.7.7a}$$

$$y_0 = . \, b_1 b_2 b_3 \ldots b_i \ldots \tag{4.7.7b}$$

where the a_i and b_i are all either 0 or 1. The position of this point in the unit square can be conveniently represented by putting these two strings back to back, that is,

$$X_0 = \ldots b_i \ldots b_3 b_2 b_1 . \, a_1 a_2 a_3 \ldots a_i \ldots \tag{4.7.8}$$

Since (forward) iteration of the map corresponds to doubling of x and halving of y, X_1 is obtained by simply moving the decimal point (4.7.8) one place to the right, that is,

$$X_1 = \ldots b_i \ldots b_3 b_2 b_1 a_1 . \, a_2 a_3 \ldots a_i \ldots \tag{4.7.9}$$

and so on, for the successive X_i. This process is known as a *Bernoulli shift*. Now consider a more "coarse-grained" description of the motion in which an orbit (or some function of the orbit) is identified by 0 if $0 \le x_n < \frac{1}{2}$ and by 1 if $\frac{1}{2} \le x_n < 1$. This identification simply corresponds to taking just the first digit in the (full) binary representation of the x_n. In this way, the coarse-grained history of the motion (i.e., the sequence of iterates $\bar{X}_0, \bar{X}_1, \bar{X}_2, \ldots, \bar{X}_i, \ldots$) will just be the sequence $a_1, a_2, a_3, \ldots, a_i, \ldots$, and so on. If the map is run from $n = -\infty$ to $n = +\infty$, the history is then the doubly infinite sequence

$$\ldots b_i \ldots b_3 b_2 b_1 a_1 a_2 a_3 \ldots a_i \ldots \tag{4.7.10}$$

(Recall that in the reverse direction (i.e., $-\infty \le n \le 0$), x and y deformations are interchanged.) The crucial point is that for typical, irrational, initial coordinates (x_0, y_0), the associated binary representations (4.7.7) are infinite nonrepeating sequences of zeros and ones and hence the doubly infinite sequence (4.7.10) is as random as that which would be obtained by a fair coin toss (1 for heads, 0 for tails). Thus a completely deterministic dynamical system (4.7.6) generates motion that appears to be completely random! Such a system is known as a *Bernoulli system* and represents the ultimate in randomness. A most significant result in dynamical systems theory is the demonstration that near any homoclinic point of a mapping, the motion can be locally represented by a mapping with the Bernoulli property. We have already seen that homoclinic points are dense in the neighborhood of hyperbolic fixed points in nonintegrable Hamiltonian systems; so the above result reinforces the idea that the chaotic, yet deterministic, trajectories observed in these systems have a truly random nature. Some nice illustrations of this concept have been given by Berry (1978).

4.7.d Hierarchies of Randomness

From the above examples it is clear that different types of dynamical system can have greater or lesser degrees of randomness. An approximate classification of this hierarchy is as follows

 (i) *Ergodic System.* This is the "weakest" type of behavior in which phase average equals time average, that is,

$$\lim_{T \to \infty} \int_{-T}^{T} f(x, t) \, dt = \langle f(x, t) \rangle \tag{4.7.11}$$

where $\langle \ \rangle$ denotes ensemble average over the manifold being considered. Simple examples are irrational flow on a torus or, for one-dimensional systems, flow on the energy shell (see Section 2.5). It is important to recall that generic n-dimensional Hamiltonians $(n > 1)$ are not ergodic over the entire energy shell, since this is typically divided by (surviving) tori.

 (ii) *Mixing Systems.* As illustrated, this is a much stronger property than ergodicity. By contrast with (4.7.11), mixing implies that

$$\lim_{t \to \infty} f(x, t) = \langle f(x, t) \rangle \tag{4.7.12}$$

that is, no time averaging is required to achieve "equilibrium." It may be shown that the spectrum of mixing systems is continuous, whereas that of ergodic systems is discrete. (There is, in fact, an intermediate state termed *weakly mixing* which is sufficient for a continuous spectrum.)

 (iii) *K-Systems.* These are systems with positive Kolmogorov entropy. This means that a connected neighborhood of trajectories must exhibit a positive average rate of exponential divergence.

 (iv) *C-Systems* (also called *Anosov Systems*). These are systems which are *globally* unstable, that is, every trajectory has a positive, Lyapunov exponent. The cat map is an example of a C-system.

(v) *Bernoulli Systems.* These are systems whose motion is as random as a fair coin toss (e.g., the baker's transformation).

Any member of the hierarchy also exhibits the properties of "lower" members. So, for example, the cat map, which is a C-system, also has the properties of a K-system (in this case the KS entropy is $\ln[(3 + \sqrt{5}/2)]$ and exhibits mixing, which, in turn, implies ergodicity).

4.8   HAMILTONIAN CHAOS IN FLUIDS

So far, all our discussions of chaos have been illustrated by computer studies of simple model systems. Furthermore, we have done little more than pay lip-service to the physical contexts in which these models might arise (e.g., nonlinear oscillations, accelerators, wave guides, molecules in radiation fields, etc.). Clearly, the reader would be much happier if there was some "real-life" physical situation in which he could actually *see* chaos—possibly even a surface of section—with the naked eye. As it turns out, this wish can be fulfilled for certain classes of fluid dynamical problems, which we shall now describe.

4.8.a   Fluid Mechanical Background

The first step is to understand the difference between the "Eulerian" and "Lagrangian" descriptions of fluid dynamics. The former specifies the velocity field, $\mathbf{u} = (u, v, w)$, of a fluid with respect to a fixed coordinate frame, that is,

$$u = u(x, y, z, t) \tag{4.8.1a}$$

$$v = v(x, y, z, t) \tag{4.8.1b}$$

$$w = w(x, y, z, t) \tag{4.8.1c}$$

where (u, v, w) are obtained (in principle) by solving the fluid dynamical equations (see Section 5.1) subject to the specified boundary conditions. If the velocity field is explicitly dependent on time, it is termed *unsteady*, in contrast to the time-independent *steady* fields. On the other hand, the Lagrangian description involves the following individual fluid "particle" trajectories. Thus for a given velocity field, one follows a particle by solving the set of ordinary differential equations

$$\dot{x} = u(x, y, z, t) \tag{4.8.2a}$$

$$\dot{y} = v(x, y, z, t) \tag{4.8.2b}$$

$$\dot{z} = w(x, y, z, t) \qquad (4.8.2c)$$

subject to the initial conditions $(x(0), y(0), z(0))$.

For an incompressible two-dimensional fluid, one has

$$u_x + v_y = 0 \qquad (4.8.3)$$

which tells us that there must be an exact differential $d\psi$ such that

$$u = \frac{\partial \psi}{\partial y} \qquad (4.8.4a)$$

$$v = -\frac{\partial \psi}{\partial x} \qquad (4.8.4b)$$

The function $\psi = \psi(x, y, t)$ is termed the *stream function*. If one is using the Lagrangian description, one can then write the equations of motion (4.8.2) in the form

$$\dot{x} = \frac{\partial \psi}{\partial y}(x, y, t) \qquad (4.8.5a)$$

$$\dot{y} = -\frac{\partial \psi}{\partial x}(x, y, t) \qquad (4.8.5b)$$

which has a Hamiltonian structure, with ψ playing the role of the Hamiltonian and x and y being the canonical variables. It is worth emphasizing that this Hamiltonian structure stems from the incompressibility condition (4.8.3) and is valid whether the fluid is viscous or not. Thus, in two dimensions, one is able to determine the fluid particle paths by following the phase-space dynamics of Eqs. (4.8.5).

The precise dynamics will, of course, depend on the nature of ψ. For steady flows, ψ is time independent and Eqs. (4.8.5) reduce to the autonomous system

$$\dot{x} = \psi_y(x, y) \qquad (4.8.6a)$$

$$\dot{y} = -\psi_x(x, y) \qquad (4.8.6b)$$

which we know, from chapter 1, to be completely integrable with the trajectories lying on smooth curves—termed *stream lines* in fluid dynamics—in the (x, y) phase plane. However, for unsteady flows (i.e., time-dependent ψ), we know that there is the possibility of chaotic behavior. In order to actually follow the fluid-particle trajectories, they must be tagged in some "passive" way, that is, in such a way that they still follow the dynamics of (4.8.5) without affecting the velocity field. This can usually be achieved with dye, but of course in practice this leads to a distribution ("passive scalar distribution") of tagged fluid particles (e.g., a streak or blob being tagged rather than an individual particle). Thus one is then observing the phase-space evolution of a whole family of trajectories. The deformations of such a family (line element or curve) were discussed in Section 4.4—namely, that line elements on the plane can develop two principal structures, "whorls" and "tendrils," corresponding to the presence of elliptic and hyperbolic fixed points, respectively. The appearance of tendrils is thus a manifestation of fluid-particle chaos in the fluid—a phenomenon that is also referred to as *chaotic advection* or *Lagrangian turbulence*, the latter term being invoked to imply the idea of chaos within the Lagrangian picture of fluid dynamics. (Notions of *Eulerian turbulence* will be briefly discussed in Chapter 5.) Finally we mention that whereas in two dimensions an unsteady flow is needed for the appearance of chaos, in three dimensions chaos can occur for steady flows (i.e., Eqs. (4.8.2) with time independent right-hand sides).

4.8.b The Model System

In practice, we would like to find a fluid dynamical system that is two-dimensional, unsteady, *and* one for which the associated stream function is known explicitly—thereby enabling us to perform computer studies to compare with laboratory experiments. Although this might sound somewhat like a tall order, such a system can be constructed and was the subject of a recent, detailed study by Chaiken et al. (1986). The system in question is closely related to what is known as a *journal bearing*. This corresponds to two cylinders, one placed within the other with the gap between them filled with a viscous fluid (traditionally a heavy lubricant). Both cylinders can rotate independently about their axes, which are typically set not to coincide with each other (i.e., the cylinders are eccentric). If the fluid depth is sufficiently large compared to the (outer) cylinder radius, the system can be considered as approximately two-dimensional. If the fluid viscosity is sufficiently high and the cylinder rotation rates are sufficiently low (leading to the so-called *Reynolds number* being very small), the fluid equations can be solved in the Stokes' approximation, that is,

$$\nu \nabla^2 \mathbf{u} = \nabla p \qquad (4.8.7)$$

where $\mathbf{u} = (u, v)$ is the two-dimensional velocity field, ν is the kinematic viscosity, and p is the pressure. The boundary conditions are that the component of fluid velocity normal to inner and outer cylinder boundaries must equal the corresponding component of the associated cylinder velocities. Provided that the cylinders (one or both) are rotating at a *steady*

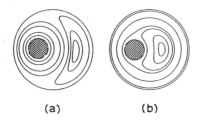

(a)   (b)

Figure 4.33 Typical streamlines for (a) inner-cylinder-only rotation and (b) outer-cylinder-only rotation. Curves plotted are exact solutions to Eq. (4.8.8) with appropriate boundary conditions. Hatched regions correspond to inner cylinder.

(angular) velocity, (4.8.7) is a steady-state problem. Furthermore, by noting that the curl operation kills gradients (i.e., $\nabla \times (\nabla p) = 0$) and that $\nabla \times \mathbf{u} = -\nabla^2 \psi$, (4.8.7) can be reduced to the biharmonic equation

$$\nabla^4 \psi = 0 \tag{4.8.8}$$

For the particular geometry of the system discussed here, it has been known for some time that this equation admits an exact, closed form solution in bipolar coordinates. Some typical results are shown in Figure 4.33 for the cases of inner-cylinder-only and outer-cylinder-only rotation. What is plotted here are curves of constant ψ in the (x, y)-plane—streamlines in fluid dynamical terms, invariant tori in dynamical terms. For these steady-state solutions, the fluid particles behave in an entirely regular way following their respective streamlines.

We now need to introduce some (controlled) unsteadiness into the system. This can be achieved by modulating the cylinder rotations. The simplest case of this is to rotate the cylinders alternately in time. The solution to (4.8.8) is then a piecewise linear combination of the solutions corresponding to the separate cylinder motions. Thus as one cylinder

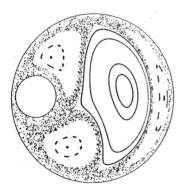

Figure 4.34 Typical surface of section computed from Eqs. (4.8.5). (Reproduced, with permission, from Chaiken et al. (1986).)

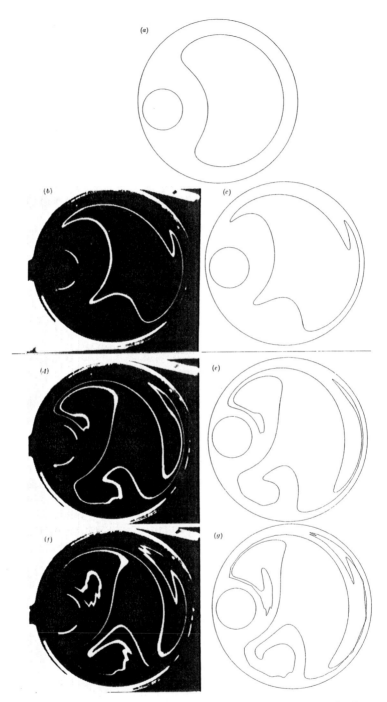

Figure 4.35 (a) Initial contour generated by rotation of outer cylinder. Comparison of experiment and computation after: 7 periods (b) and (c); 15 periods (d) and (e); 21 periods (f) and (g). (Reproduced, with permission, from Chaiken et al. (1986).)

rotates, a fluid particle follows its streamline and then jumps (virtually instantaneously in a Stokes' flow) to a different streamline as the cylinder motions are switched. It is this mechanism that can lead to chaos as the particles evolve in the three-dimensional phase space (x, y, t). The motion is conveniently followed by constructing a surface of section by taking *strobosopic* "snapshots" of the (x, y) phase plane—the time interval being the sum of the two cylinder rotation times (see Figure 4.6). A typical such surface of secton, generated by solving Eqs. (4.8.5) numerically, is shown in Figure 4.34 exhibiting the generic mixture of regular and irregular motions.

4.8.c Experimental Results

The formation of whorls and tendrils can be used to identify regions of regular and irregular motion, respectively. For example, in the above case an evolving line element (i.e., a streak of dyed fluid) will form whorls in the neighborhood of the main three-island chain and tendrils in the intervening chaotic regions. In Figure 4.35 we show the evolution of an initial curve embracing these regions. As the sequence clearly shows, three large whorls are formed on which are superimposed small homoclinic oscillations (tendrils). The computer simulations are compared directly with the laboratory experiment in which a matching curve of dyed fluid (glycerine in this experiment) placed on the fluid surface evolves under exactly the same cylinder motions (alternate piecewise rotations) as used in the computer studies. (The apparatus is shown in Figure 4.36.)

As the cylinder parameters (i.e., rotation rates and eccentricity) are varied, phase planes exhibiting different degrees of chaos are found. In Figure 4.37 we show the case of a strongly chaotic phase plane with only a few tiny island regions left. In this situation we would expect to see the formation of giant tendrils. This is demonstrated in the sequence in Figure 4.38. Computer simulations become difficult in this case since the exponential separation of neighboring points leads to a rapid breakdown of numerical resolution. (To overcome this the initial curve would have to be packed with an enormous number of initial points.) By contrast, the laboratory experiment has no such problems and in Figure 4.39 we show the striking result of running the experiment for several periods longer than can be followed numerically. As it turns out, the laboratory experiment can also be used to construct surfaces of section. (The reader is referred to the original paper (Chaiken et al., 1986) for the technical details.) In Figure 4.40 we show a laboratory surface of section obtained at the parameters corresponding to the numerical simulation shown in Figure 4.34. Quite good agreement is obtained, with the laboratory result clearly showing the correct elliptic and hyperbolic regions. This is one of the few examples of a surface of section of a chaotic Hamiltonian flow being directly visualized in the laboratory.

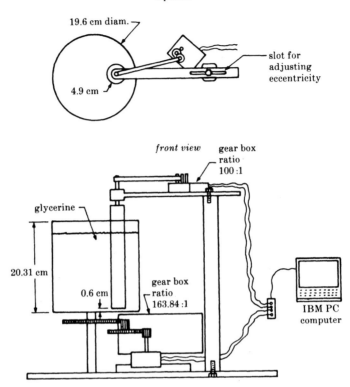

Figure 4.36 Schematic diagram of experimental apparatus. Contours of dyed fluid are placed on surface of glycerine, and their evolution is followed under controlled cylinder motions. (Reproduced, with permission, from Chaiken et al. (1986).)

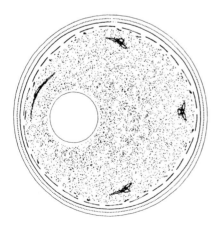

Figure 4.37 Strongly chaotic surface of section computed from Eqs. (4.8.5). (Reproduced, with permission, from Chaiken et al. (1986).)

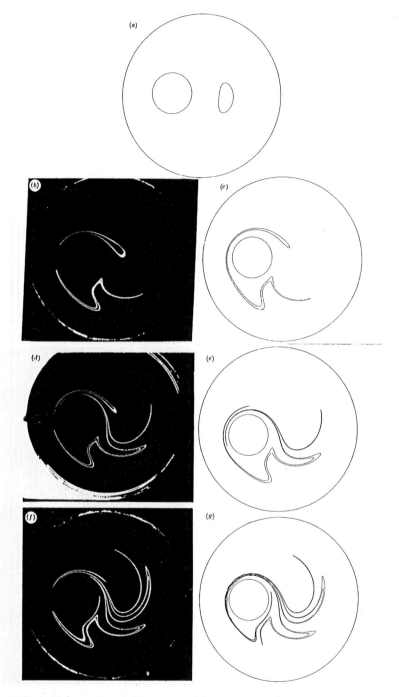

Figure 4.38 Initial contour (a) generated by rotation of outer cylinder. Comparison of experiment and computation after: 4 periods (b) and (c); 5 periods (d) and (e); 6 periods (f) and (g). (Reproduced, with permission, from Chaiken et al. (1986).)

Figure 4.39 Same experiment as in Figure 4.38 after 12 periods. (Reproduced, with permission, from Chaiken et al. (1986).)

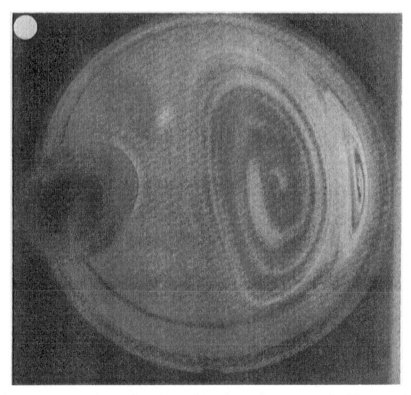

Figure 4.40 Experimental surface of section to be compared with numerically obtained result shown in Figure 4.34. (Reproduced, with permission, from Chaiken et al. (1986).)

APPENDIX 4.1 SURFACE OF SECTION AS A SYMPLECTIC MAPPING

The special "area-preserving" property of the surface of section for two-degree-of-freedom, conservative systems can be seen most elegantly in the geometric language introduced in Appendix 2.2.

From the invariance properties of the Poincaré–Cartan 1-form $\sum_{i=1}^{n} p_i \, dq_i - H \, dt$ we have (cf. Eq. (2.A.29))

$$\oint_{\mathscr{C}} \sum_{i=1}^{n} p_i \, dq_i - H \, dt = \oint_{\mathscr{C}'} \sum_{i=1}^{n} p_i \, dq_i - H \, dt \qquad (4.A.1)$$

where $\mathscr{C}$ and $\mathscr{C}'$ are any two closed curves enclosing the same tube of phase-space trajectories. If $\mathscr{C}$ and $\mathscr{C}'$ are curves taken at constant time slices, $t = 0$ for curve $\mathscr{C}$ and $t = T$ for curve $\mathscr{C}'$ (now denoted as $\mathscr{C}_T$), there is no contribution from the $H \, dt$ terms and we obtain the result (4.1.9) (or Eq. (2.A.31)), namely,

$$\oint_{\mathscr{C}} \sum_{i=1}^{n} p_i \, dq_i = \oint_{\mathscr{C}_T} \sum_{i=1}^{n} p_i \, dq_i \qquad (4.A.2)$$

Now consider the two-degree-of-freedom case with the 1-form $p_x \, dx + p_y \, dy$ and define an "initial" curve $\mathscr{C}$ as a set of initial conditions on the (p, x) plane at $y = 0$ at a fixed energy E. That is, $\mathscr{C}$ is a curve of points on the surface of section. Under the Hamiltonian flow these points will evolve, forming a "tube" of trajectories in the "extended" phase space $(p_x, p_y, x, y; t)$. For bounded motion this tube will eventually pass back through the surface of section at $y = 0$. However, there is no reason to assume that each point on the tube will pass through $y = 0$ at the same time. Thus, although the reintersection of the tube with the surface of section will be some closed curve, say $\mathscr{C}'$, it will not be of the "fixed time slice" type $\mathscr{C}_T$ used in (4.A.2). Thus we must consider the Poincaré–Cartan invariant in (4.A.1), namely,

$$\oint_{\mathscr{C}} p_x \, dx + p_y \, dy - H \, dt = \oint_{\mathscr{C}'} p_x \, dx + p_y \, dy - H \, dt \qquad (4.A.3)$$

However, for a family of trajectories on a given energy shell $E = H =$ constant we note that

$$\oint_{\mathscr{C}} H \, dt = \oint_{\mathscr{C}'} H \, dt = 0 \qquad (4.A.4)$$

Furthermore, since the curves $\mathscr{C}$ and $\mathscr{C}'$ are defined on planes with fixed y

$(y = 0)$,

$$\oint_{\mathscr{C}} p_y \, dy = \oint_{\mathscr{C}'} p_y \, dy = 0 \tag{4.A.5}$$

Thus we are left with the result

$$\oint_{\mathscr{C}} p_x \, dx = \oint_{\mathscr{C}'} p_x \, dx \tag{4.A.6}$$

namely, that the area on a surface of section is conserved under the Hamiltonian flow. That is why we can speak of the surface of section, for a two-degree-of-freedom system, as an *area*-preserving map.

For more than 2 degrees of freedom the same idea holds but now the "surface of section" is a $(2n - 2)$-dimensional surface embedded in the $(2n - 1)$-dimensional energy shell (of a conservative n-degree-of-freedom Hamiltonian). If the "surface" is defined at $q_1 = $ constant, then (4.A.3) generalizes to

$$\oint_{\mathscr{C}} \sum_{i=2}^{n} p_i \, dq_i + p_1 \, dq_1 - H \, dt = \oint_{\mathscr{C}'} \sum_{i=2}^{n} p_i \, dq_i + p_1 \, dq_1 - H \, dt \tag{4.A.7}$$

By analogy with (4.A.4) and (4.A.5) we have vanishing contributions from the terms $\oint p_1 \, dq_1$ and $\oint H \, dt$ about both $\mathscr{C}$ and $\mathscr{C}'$ and are hence left with the result

$$\oint_{\mathscr{C}} \sum_{i=2}^{n} p_i \, dq_i = \oint_{\mathscr{C}'} \sum_{i=2}^{n} p_i \, dq_i \tag{4.A.8}$$

These integrals no longer correspond to simple areas but rather the projected areas on the various (p_i, q_i) planes (see (2.A.50)), namely,

$$\oint_{\mathscr{C}} \sum_{i=2}^{n} p_i \, dq_i = \sum_{i=2}^{n} \iint_{A_i} dp_i \, dq_i$$

$$= \oint_{\mathscr{C}'} \sum_{i=2}^{n} p_i \, dq_i = \sum_{i=2}^{n} \iint_{A_i'} dp_i \, dq_i \tag{4.A.9}$$

where the A_i are the various projections of $\mathscr{C}$ and A_i' the projections of $\mathscr{C}'$. Thus we see that the surface of section is a symplectic mapping.

SOURCES AND REFERENCES

Texts and General Review Articles

Arnold, V. I., and A. Avez, *Ergodic Problems of Classical Mechanics*, Benjamin, New York, 1968.

Berry, M. V., Regular and irregular motion, AIP Conference Proceedings, No. 46, *Topics in nonlinear dynamics*, AIP, New York, 1978.

Birkhoff, G. D., *Dynamical Systems*, American Mathematical Society, Providence, RI, 1927.

Ford, J., The statistical mechanics of analytical dynamics, in E. D. G. Cohen, Ed., *Fundamental Problems in Statistical Mechanics*, Vol. 3, North-Holland, Amsterdam, 1975.

Helleman, R. H. G., Self generated chaotic behaviour in nonlinear mechanics, in E. D. G. Cohen, Ed., *Fundamental Problems in Statistical Mechanics*, Vol. 5, North-Holland, Amsterdam, 1980.

Lichtenberg, A. J., and M. A. Lieberman, *Regular and Stochastic Motion*, Springer-Verlag, New York, 1983.

MacKay, R. S., and J. D. Meiss, *Hamiltonian Dynamical Systems*, Adam Hilger, Bristol, 1987. This volume contains a valuable compilation and extensive bibliography of many fundamental research papers in Hamiltonian dynamics.

Moser, J., *Stable and Random Motions in Dynamical Systems*, Annals of Mathematical Studies, Vol. 77, Princeton University Press, Princeton, NJ, 1973.

Poincaré, H., *Les Methods Nouvelles de la Mechanique Celeste*, Gauthier-Villars, Paris, 1892.

Section 4.1

Casati, G., and J. Ford, Stochastic transition in the unequal-mass Toda lattice, *Phys. Rev.*, **A12**, 1702 (1975).

Contopoulos, G., and C. Polymilis, Approximations of the 3-particle Toda lattice, *Physica*, **24D**, 328 (1987).

Ford, J., S. D. Stoddard, and J. S. Turner, On the integrability of the Toda lattice, *Prog. Theor. Phys.*, **50**, 1574 (1973).

Henon, M., and C. Heiles, The applicability of the third integral of motion: some numerical experiments, *Astron. J.*, **69**, 73 (1964).

Henon, M., Integrals of the Toda lattice, *Phys. Rev.*, **B9**, 1921 (1974). See also accompanying paper by H. Flaschka.

Section 4.2

Henon, M., Numerical study of quadratic area preserving mappings, *Quart. Appl. Math.*, **27**, 291 (1969).

MacKay, M. S., J. D. Meiss, and I. C. Percival, Transport in Hamiltonian systems, *Physica*, **13D**, 55 (1984). Cantori and their applications are discussed.

MacKay, R. S., Introduction to the dynamics of area preserving maps, *Proceedings of the Spring College on Plasma Physics, Trieste, 1985,* World Scientific, Singapore, 1985. A good geometrical introduction to area preserving maps is given.

Percival, I. C., Variational principles for invariant tori and cantori, AIP Conference Proceedings, No. 57, *Nonlinear Dynamics and the Beam–Beam Interaction,* AIP, New York, 1980. Discrete Lagrangians are introduced.

Section 4.5

Benettin, G., L. Galgani, and J. M. Strelcyn, Kolmogorov entropy and numerical experiments, *Phys. Rev.,* **A14,** 2338 (1976).

Benettin, G., L. Galgani, A. Giorgilli, and J. M. Strelcyn, Lyapunov characteristic exponents for smooth dynamical systems and for Hamiltonian systems; a method for computing all of them, *Meccanica,* **15,** 9 (1980).

Pesin, Ya. B., Characteristic Lyapunov exponents and smooth ergodic theory, *Russ. Math. Surveys,* **32**(4), 55 (1977).

Section 4.6

Chirikov, B., A universal instability of many dimensional oscillator systems, *Phys. Reports,* **52,** 263 (1979).

Escande, D. F., Stochasticity in classical Hamiltonian systems: universal aspects, *Phys. Reports,* **121,** 165 (1985). Further developments of the Chirikov method are described.

Greene, J. M., A method for determining a stochastic transition, *J. Math. Phys.,* **20,** 1183 (1979).

Helleman, R. H. G., and T. Bountis, Periodic solutions of arbitrary period, variational methods, Lecture Notes in Physics, Vol. 93, *Stochastic Behavior in Classical and Quantal Hamiltonian Systems,* Springer-Verlag, New York, 1979.

Tabor, M., The onset of chaos in dynamical systems, *Adv. Chem. Phys.,* **46,** 73 (1981).

Walker, G. H., and J. Ford, Amplitude instability and ergodic behavior for conservative nonlinear oscillator systems, *Phys. Rev.,* **188,** 416 (1969).

Section 4.7

Lebowitz, J., and O. Penrose, Modern ergodic theory, *Phys. Today,* **26,** 23 (1973).

Zaslavskii, G. M., and B. Chirikov, Stochastic instability of nonlinear oscillations, *Sov. Phys. Usp.,* **14,** 549 (1972).

Section 4.8

Chaiken, J., R. Chevray, M. Tabor, and Q. M. Tan, Experimental study of Lagrangian turbulence in a Stokes' flow, *Proc. R. Soc. London A,* **408,** 165 (1986).

5

THE DYNAMICS OF DISSIPATIVE SYSTEMS

5.1 DISSIPATIVE SYSTEMS AND TURBULENCE

In this chapter we turn our attention from Hamiltonian systems and area-preserving maps to the dynamics of dissipative systems. These have already been discussed in Chapter 1 in the context of various damped oscillator systems which can display limit point and limit cycle behavior. However, as was already hinted at in Section 1.6, dissipation does not always damp out interesting dynamics and can, under the right circumstances, result in the appearance of chaotic behavior.

Much of the current research on dissipative dynamical systems has been motivated by the desire to explain various phenomena observed in real fluid dynamical experiments, and some examples of these will be discussed in the next section. There is, furthermore, the possibility that the properties of these systems may provide some insights into the enormously complicated issue of fluid dynamical *turbulence*, although, to date, success in this direction has been rather limited.† To set the stage for our discussion of the successes (and failures) of this approach, we first give a very brief discussion of the relevant fluid dynamical concepts.‡

---

†At a meeting of the British Association in London in 1932 the noted fluid dynamicist Sir Horace Lamb is reputed to have said: "I am an old man now, and when I die and go to Heaven there are two matters on which I hope for enlightenment. One is quantum electrodynamics and the other is the turbulent motion of fluids. And about the former I am really rather optimistic." (The author would like to thank Prof. P. C. Martin for his help with finding this quotation.)
‡Clearly, the reader should consult the standard texts, such as the excellent *Fluid Mechanics* by Landau and Lifschitz (1959), for more details.

5.1.a The Navier–Stokes Equation

The fundamental equations of motion for an incompressible fluid are the Navier–Stokes equations, namely,

$$\frac{\partial \mathbf{u}}{\partial t} + (\mathbf{u} \cdot \nabla)\mathbf{u} - \nu \nabla^2 \mathbf{u} = -\frac{1}{\rho}\nabla p + \mathbf{f} \qquad (5.1.1a)$$

$$\text{div } \mathbf{u} = 0 \qquad (5.1.1b)$$

$$\mathbf{u} = 0 \text{ on } \mathscr{D} \qquad (5.1.1c)$$

where $\mathscr{D}$ is the region bounding the fluid, $\mathbf{u}$ is the fluid velocity field ($\mathbf{u} = \mathbf{u}(x, y, z, t)$), p is the pressure, ρ is the fluid density, $\mathbf{f}$ is the external forces (if any), and ν is the kinematic viscosity. It is the presence of the viscous term $\nu \nabla^2 \mathbf{u}$ that provides the mechanism by which energy can be dissipated. Equation (5.1.1a) is a three-dimensional partial differential equation for the velocity field $\mathbf{u}$ with respect to a fixed coordinate frame. (From our discussion in Section 4.8, we recall that this corresponds to the Eulerian description of the fluid.) Equation (5.1.1b) is the incompressibility condition, and Eq. (5.1.1c) is the boundary condition. The sad truth is that even today, little is known about the formal properties (let alone finding exact solutions) of the Navier–Stokes equation in three dimensions; for example, there is still no proof of existence of solutions for all times. (Such results are, however, available for the two-dimensional equations.)

However, as far as experimental observations of fluid flows are concerned (i.e., the physical reality that the Navier–Stokes equations represent), much is known. These studies were pioneered by Reynolds in the 1880s. An important part of his work was the introduction of the dimensionless parameter, now termed the *Reynolds number*,

$$R = \frac{UL}{\nu}$$

where U and L are, respectively, typical velocity and length scales of the system under study, and ν is the kinematic viscosity. When the Navier–Stokes equations are written in nondimensional form, that is, scaling the velocities by U and the coordinates by L (the time scale is thus L/U), one obtains (dropping the force term in (5.1.1a))

$$\frac{\partial \mathbf{u}}{\partial t} + (\mathbf{u} \cdot \nabla)\mathbf{u} - \frac{1}{R}\nabla^2 u = -\nabla p \qquad (5.1.2)$$

where the pressure has been rescaled as $p \to p/\rho U^2$. Reynolds' fundamental observation was that as R is increased, the flow can make a transition from smooth, regular behavior—*laminar flow*—to an erratic, chaotic behavior—*turbulent flow*.

5.1.b The Concept of Turbulence

In fluid dynamics the term *turbulence* is generally accepted to mean a state of *spatiotemporal chaos*; that is, the fluid exhibits chaos on all scales in both space *and* time. A satisfying mathematical description of this state has proved to be one of the great challenges of applied mathematics. In recent years there has been some hope that the chaos exhibited by simple dynamical systems (o.d.e.'s and mappings) may help in this understanding. However, it is most important to emphasize that these simple systems *only exhibit temporal chaos*. Thus although there is little chance, at this stage, that they can explain fully developed turbulence, they may provide some insight into the *onset* of turbulence, that is, the situation in which the velocity fields start to fluctuate randomly in time but still stay fairly well organized spatially. It is important to compare what we are about to discuss in this chapter with the ideas described in Section 4.8. There we talked about Lagrangian turbulence, that is, the phase space chaos exhibited by individual (or distributiuons thereof) fluid-particle trajectories. We were able to show that very simple velocity fields—even at vanishingly small Reynolds numbers—were able to exhibit chaotic behavior; we also showed how this chaos was able to explain certain spatially complicated patterns seen in a fluid. By contrast, turbulence in the Eulerian description (Eulerian turbulence) corresponds to the appearance of spatiotemporal chaos of the associated velocity fields—a situation that arises at high Reynolds numbers. At this stage there is little understanding of the connection (if any) between Lagrangian and Eulerian turbulence.

In this chapter we will be considering the onset of turbulence (in the Eulerian description) using models involving simple, dissipative dynamical systems. The study of such models, which may, in reality, have only a tenuous link to real hydrodynamic situations, has, nonetheless, led to some profound new ideas, and, despite their sometimes abstract nature, their choice has been motivated by the desire to interpret real experiments. So to start with, we shall describe some experimental studies that exhibit the onset of turbulence.

5.1.c A Hamiltonian Digression*

In the absence of viscosity the Navier–Stokes equation is replaced by Euler's equation:

$$\frac{\partial \mathbf{u}}{\partial t} + (\mathbf{u} \cdot \nabla)\mathbf{u} = -\frac{1}{\rho} \nabla p \tag{5.1.3}$$

The required boundary condition corresponds to the fact that the fluid cannot penetrate the boundary $\mathcal{D}$, that is,

$$\partial_n \mathbf{u} = 0 \text{ on } \mathcal{D}$$

where ∂_n corresponds to the derivative normal to the boundary (here assumed stationary). Note this is in contrast to the boundary condition (5.1.1c) for the Navier–Stokes equation in which the velocity (as opposed to its normal derivative) is zero on $\mathcal{D}$.

In two dimensions the Euler equations are, in component form ($\mathbf{u} = (u, v)$),

$$\frac{\partial u}{\partial t} + uu_x + vu_y = -\frac{1}{\rho}\frac{\partial p}{\partial x} \tag{5.1.4a}$$

$$\frac{\partial v}{\partial t} + uv_x + vv_y = -\frac{1}{\rho}\frac{\partial p}{\partial y} \tag{5.1.4b}$$

Furthermore, the equation of continuity (incompressibility condition) is

$$\text{div } \mathbf{u} = u_x + v_y = 0 \tag{5.1.5}$$

From this condition follows, as already described in Section 4.8, the existence of an exact differential $d\psi$ such that

$$u\,dy + v\,dx = d\psi \tag{5.1.6}$$

which thus enables one to write

$$u = \frac{\partial \psi}{\partial y}, \qquad v = -\frac{\partial \psi}{\partial x} \tag{5.1.7}$$

The function ψ is called the *stream function*.†

An important fluid dynamical quantity is the *vorticity*:

$$\boldsymbol{\xi} = \nabla \times \mathbf{u} \tag{5.1.8}$$

Taking the curl of the Euler equations kills the pressure term and yields

$$\frac{\partial \boldsymbol{\xi}}{\partial t} + \nabla \times (\mathbf{u} \times \boldsymbol{\xi}) = 0 \tag{5.1.9}$$

---

†We emphasize that the existence of a stream function in two dimensions is a consequence of the incompressibility condition (5.1.5)—it is not affected by the presence (or absence) of viscosity.

However, in two dimensions the vorticity becomes a simple scalar field, that is,

$$\xi = v_x - u_y = -\nabla^2 \psi \qquad (5.1.10)$$

in which case Euler's equation becomes

$$\frac{\partial}{\partial t}(\nabla^2 \psi) + J(\psi, \nabla^2 \psi) = 0 \qquad (5.1.11)$$

where

$$J(\psi, \nabla^2 \psi) = \frac{\partial(\psi, \nabla^2 \psi)}{(x, y)} = \psi_x(\nabla^2 \psi)_y - \psi_y(\nabla^2 \psi)_x$$

A common approach in fluid dynamical studies is to expand the vorticity ξ as a sum of "point vortices," that is,

$$\xi = \sum_{j=1}^{n} \Gamma_j \delta(\mathbf{x} - \mathbf{x}_j), \qquad \mathbf{x} = (x, y) \qquad (5.1.12)$$

where the Γ_j are the "strengths" or circulations of each point vortex located at $\mathbf{x}_j$. One may then easily show that

$$\psi = \frac{1}{2\pi} \sum_{j=1}^{n} \Gamma_j \log |\mathbf{x} - \mathbf{x}_j| \qquad (5.1.13)$$

The equations of motion for the ith point vortex moving in the field of the remaining $n - 1$ point vortices are

$$\frac{dx_i}{dt} = u_i = \frac{\partial \psi}{\partial y_i} = -\frac{1}{2\pi} \sum_{j \neq i}^{n} \Gamma_j \frac{(y_i - y_j)}{r_{ij}^2} \qquad (5.1.14a)$$

$$\frac{dy_i}{dt} = v_i = -\frac{\partial \psi}{\partial x_i} = \frac{1}{2\pi} \sum_{j \neq i}^{n} \Gamma_j \frac{(x_i - x_j)}{r_{ij}^2} \qquad (5.1.14b)$$

where $r_{ij}^2 = |\mathbf{x}_i - \mathbf{x}_j|^2$. If one now defines the function

$$H = -\frac{1}{4\pi} \sum_{ij} \Gamma_i \Gamma_j \log |\mathbf{x}_i - \mathbf{x}_j| \qquad (5.1.15)$$

then the equations of motion can be written in Hamiltonian form:

$$\Gamma_i \frac{dx_i}{dt} = \frac{\partial H}{\partial y_i}, \qquad \Gamma_i \frac{dy_i}{dt} = -\frac{\partial H}{\partial x_i} \qquad (5.1.16)$$

Is such a system integrable? To be so would require $N-1$ integrals of motion. For three point vortices, one can find integrals associated with the rotational and translational invariances (Noether's theorem), thereby reducing the system to one that can always be integrated. For four point vortices, one can reduce the Hamiltonian to one of two degrees of freedom. Numerical studies (e.g., surfaces of section and Lyapunov exponents) show that this is a nonintegrable system displaying the generic mixture of regular and irregular motion (see, for example, the review by Aref (1983)). Interestingly enough, this does not necessarily imply that the system remains nonintegrable as $N \to \infty$, and little is known about the formal properties of Eqs. (5.1.16) in this limit.

5.2 EXPERIMENTAL OBSERVATIONS OF THE ONSET OF TURBULENCE

5.2.a Couette Flow

The first experiment we shall consider is a Couette flow. This situation corresponds to the flow of a fluid between concentric cylinders with either the inner or outer (or both) cylinders rotating with some known angular velocity Ω (see Figure 5.1). The Reynolds number can be defined as

$$R = \frac{r_i \Omega (r_0 - r_i)}{\nu} \qquad (5.2.1)$$

As the Reynolds number is increased past a critical value R_c, the initial, purely azimuthal flow makes a transition to flow consisting of horizontal toroidal vortices superimposed on the azimuthal flow as sketched in Figure 5.2. This transition was first examined in detail and successfully predicted by G. I. Taylor in 1923.

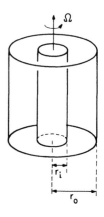

Figure 5.1 System of concentric cylinders used in Taylor–Couette flow experiment.

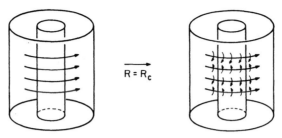

Figure 5.2 Transition at critical Reynolds number R_c from azimuthal flow to toroidal vortices (Taylor vortices).

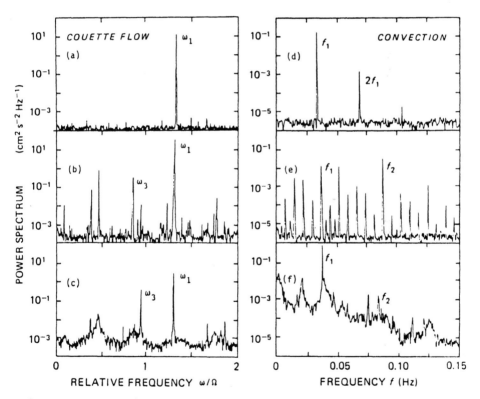

Figure 5.3 (a–c) Velocity power spectra for Taylor–Couette flow at (a) $R/R_c = 9.6$, (b) $R/R_c = 11.0$, and (c) $R/R_c = 18.9$. (d–f) Velocity power spectra for Rayleigh–Bénard convection at relative Rayleigh number (d) 35.6, (e) 41.9, and (f) 43.0. The flow is periodic in (a) and (d), quasi-periodic in (b) and (e), and broad band (chaotic) in (c) and (f). (Reproduced, with permission, from Swinney (1978)).

As the Reynolds number is increased beyond R_c, further transitions occur corresponding to the superposition of transverse waves on the horizontal vortices. Detailed experimental investigations of these processes have been carried out in the last few years. In particular, we shall refer to the work of Swinney, Gollub, and co-workers (Swinney and Gollub, 1978), using what is called *laser Doppler velocimetry*, a technique that determines the fluid velocity at a given point in space without perturbing the flow. They examined the radial velocity component at a series of Reynolds numbers. The results are conveniently expressed in the form of the power spectrum of the fluid velocity.

The results show a series of distinct regimes (the described power spectra are shown in Figure 5.3a–c):

(i) Periodic Flow (Figure 5.3a). The power spectrum consists of a single intense line at frequency ω_1.

(ii) Quasiperiodic Flow (Figure 5.3b). A second frequency component ω_3 is present (appearing above $R/R_c \simeq 10$). Other lines in the spectrum are various combinations and overtones of the two fundamental frequencies.

(iii) Chaotic Flow (Figure 5.3c). The spectrum contains significant broad band portions (well above experimental noise levels), implying a chaotic component in the motion. At the R/R_c value shown here, the sharp frequencies are still present but disappear at higher R/R_c, leaving an essentially continuous spectrum.

Thus, overall, we see that the spectrum of the flow seems to go through just a few well-defined transitions, corresponding to periodic flows of increasing complexity, to an essentially continuous spectrum.

5.2.b Rayleigh–Benard Convection

In this system the fluid is contained between horizontal, thermally conducting plates heated from below. The distance from equilibrium is measured in terms of the so-called Rayleigh number, R_a, which is proportional to the temperature difference between the plates. Rayleigh, in 1916, showed that at a critical value of R_a the pure conduction state would become unstable to perturbation, forming "convection rolls" as sketched in Figure 5.4. This instability is usually called the *Rayleigh–Benard instability* (similar convection cells had been studied by Benard in 1900). We shall discuss the underlying theory in more detail later.

Experimentally, this problem has also been studied by the same techniques used for Couette flow. We show the results (in Figure 5.3d–f) for water, at 70°C, contained in a small rectangular box. The results are again in the form of power spectra. Overall, there seems to be a similar sequence of events to the Couette flow. Namely, beyond the critical Rayleigh number there is, at first, only one frequency and its harmonics in

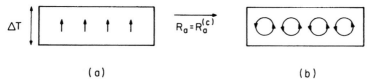

(a)   (b)

Figure 5.4 Transition at critical Rayleigh number $R_a^{(c)}$ from conductive mode to convective rolls.

the spectrum. At some higher R_a a second line appears. All other lines in the spectrum are identifiable as overtone and combination lines. Finally, as R_a is increased further, the sharp components essentially vanish, leaving an approximately continuous spectrum.

It turns out that the geometry of the container can be important. For example, studies by Ahlers and Behringer (1978) on the convection of liquid helium in cylindrical cells show similar results for small aspect ratio (height-to-radius ratio), but for large aspect ratio the spectrum becomes noisy almost immediately after the initial onset of convection; that is, there appear to be no periodic regimes at all.

For both the Couette flow and Rayleigh–Benard convection, the primary instabilities were successfully predicted (by Taylor and Rayleigh, respectively) on the basis of linear stability theory. However, there are other flows for which linear stability analyses are unsuccessful and in which the onset of turbulence undergoes different sequences of events, for example, flow past a cylinder and flow down a pipe (Poiseuille flow). In the latter case, for example, linear stability analysis yields a critical Reynolds number of infinity, whereas in practice it lies between 2000 and 100,000. Here the turbulence can result from irregularities at the entrance region, in the pipe walls, or from other causes.

These examples should illustrate just how complicated the onset of turbulence can be—every case can be a little bit different. However, in the Couette and Rayleigh–Benard problems, there can be common features, namely, initially periodic regimes of increasing complexity making a rather sharp transition to an essentially continuous spectrum corresponding to chaotic behavior. With these observations in mind, we now turn to a discussion of some of the theories that have been put forward to explain the onset of turbulence.

5.3 THEORIES AND CONCEPTS IN THE ONSET OF TURBULENCE

In this section we discuss a number of theories—and the concepts they invoke—that have been devised to explain the onset of turbulence as observed, for example, in the experiments just described. This discussion will provide the background for our investigations of the model systems on which they can be·tested.

5.3.a   Landau–Hopf Theory

This theory was developed in the early 1940s. The idea here is that the solution (to the Navier–Stokes equation) is quasi-periodic, but as the Reynolds number is increased the solution "picks up" more and more frequency components, that is,

$$u(t) = f(\omega_1 t) \rightarrow f(\omega_1 t, \omega_2 t) \rightarrow f(\omega_1 t, \omega_2 t, \omega_3 t) \rightarrow \cdots \qquad (5.3.1)$$

or more generally

$$\mathbf{u}(\mathbf{x}, t) = \sum_{\mathbf{m}} A_{\mathbf{m}}(\mathbf{x}) \, e^{\, i m (\omega t + \delta)} \qquad (5.3.2)$$

where $\omega = \omega_1, \omega_2, \ldots, \omega_k$ and $\mathbf{m} = m_1, m_2, \ldots, m_k$ with $k \rightarrow \infty$ as $R \rightarrow \infty$.

The rate at which new frequencies are picked up becomes faster as $R \rightarrow \infty$. The frequencies $(\omega_1, \ldots, \omega_k)$ are all assumed to be irrationally related so that as the number of components increases the spectrum becomes (rapidly) so complicated that it is essentially indistinguishable from the continuous spectrum of truly chaotic motion.

We can already see that this picture of perpetually quasi-periodic "turbulence" is at variance with the experimental observations, for example, Couette flow and Rayleigh–Benard convection. However, before we dismiss this model, we should consider its implications a little further.

First of all, what is the meaning of a solution "picking up" new frequencies? Our solution $\mathbf{u}(\mathbf{x}, t)$ can be thought of as following a trajectory in some (infinite-dimensional) phase space. However, unlike the phase space of Hamiltonian systems, the presence of dissipation causes the contraction of phase-space volume. This means that the trajectory will eventually end up on some limiting manifold (of measure zero in the whole of the phase space). For example, in a linear oscillator subject to simple frictional damping, the phase-space trajectory will spiral into a *limit point*. The same oscillator subject to a more sophisticated damping mechanism—for example, the Van de Pohl oscillator—will spiral onto a *limit cycle*. In a similar way, our phase-space trajectory corresponding to the state of the fluid flow will end up on some limiting manifold.

Clearly, when there is just one frequency component present (i.e., $u(t) = f(\omega_1 t)$), this means that the solution has settled down onto a simple *periodic orbit* (limit cycle). The appearance of a second, independent frequency (i.e., $u(t) = f(\omega_1 t, \omega_2 t)$) then implies that the solution has settled onto a two-dimensional *torus*. (An n-dimensional torus, T^n, can be defined as the product of n independent periodic cycles.) Thus with the

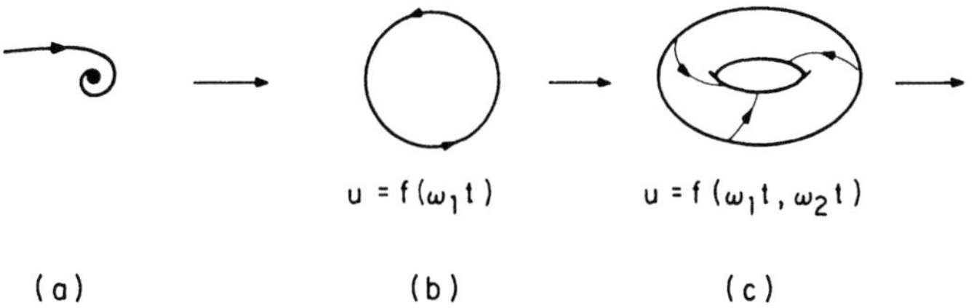

$$u = f(\omega_1 t) \qquad u = f(\omega_1 t, \omega_2 t)$$

(a) (b) (c)

Figure 5.5 Transition from (a) limit point to (b) limit cycle (one frequency) to (c) limit 2-torus (two frequencies).

appearance of each new frequency component, the solution trajectory makes a transition to a new torus with the same dimensionality as the number of independent frequencies (see Figure 5.5). The next question to ask is how the trajectory makes the transition from one limiting manifold to the next; that is, why does the original manifold become unstable and a new stable one of higher dimensionality appear?

5.3.b Hopf Bifurcation Theory

The basic ideas governing these transitions are provided by the celebrated Hopf bifurcation theory, which we shall now discuss briefly. A full account can be found in the text by Marsden and McCracken (1976). We assume that our equations of motion are governed by a "flow" of the form

$$\frac{d\mathbf{x}}{dt} = F_\mu(\mathbf{x}), \qquad \mathbf{x} = x_1, \ldots, x_k \tag{5.3.3}$$

where μ is some system parameter (e.g., the Reynolds number). The Hopf theorem is concerned with the stability of the solutions of the above equation as a function of the parameter μ. The critical points of Eq. (5.3.3) are those points $\mathbf{x} = \mathbf{x}^*$ for which

$$\frac{d\mathbf{x}^*}{dt} = 0, \quad \text{i.e.,} \quad F_\mu(\mathbf{x}^*) = 0 \tag{5.3.4}$$

Their stability is determined by examining the eigenvalues, $\lambda = \lambda(\mu)$, of the associated tangent map of $F_\mu(\mathbf{x})$, evaluated at the critical points. Provided that the λ all lie in the left half-plane, that is, all have strictly negative real parts, the critical point is a simple limit point (Figure 5.6a). The Hopf bifurcation occurs if one can show that a complex conjugate pair of eigenvalues cross, at nonzero speed, from the left to the right half-plane, that is, acquire positive real parts. Provided that certain complicated conditions involving the first, second, and third derivatives of $F_\mu(\mathbf{x}^*)$ are satisfied at the critical value of $\mu = \mu_c$ (the value of μ at which the λ cross

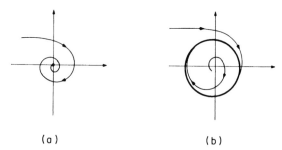

(a)   (b)

Figure 5.6 (a) Orbits spiraling in to limit point. (b) Orbits spiraling on to a stable limit cycle.

to the right half-plane), it can be shown that the limit point has bifurcated into a *stable* periodic orbit (limit cycle), that is, all nearby orbits are attracted to it (Figure 5.6b). This transition is known as a *normal Hopf bifurcation* (sometimes called a *supercritical bifurcation*). A simple way to visualize the transition from stable limit point to stable limit cycle is, as sketched in Figure 5.7, to imagine the motion of a particle in a single minimum potential which changes to a double minimum potential as a function of μ. Another situation can also arise, namely, that no attracting (stable) solution is formed at μ_c. In this case, one finds that for $\mu < \mu_c$ there exists an *unstable* periodic orbit which gradually shrinks down to an unstable limit point as μ passes μ_c, as sketched in Figure 5.8. This situation is known as an *inverted* (or *subcritical*) *Hopf bifurcation*. It should be clear that the Landau–Hopf picture for the onset of turbulence requires a sequence of normal Hopf bifurcations. One might ask, assuming normal bifurcations, how the higher transitions actually occur (e.g., the transition from periodic orbit to 2-torus). Pictorially, one could imagine the situation sketched in Figure 5.9. To demonstrate this explicitly is not so easy. Briefly,

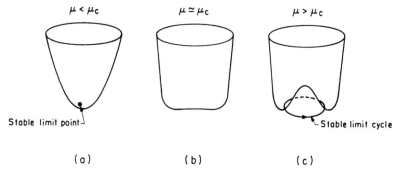

$\mu < \mu_c$ $\mu \simeq \mu_c$ $\mu > \mu_c$

Stable limit point   Stable limit cycle

(a)   (b)   (c)

Figure 5.7 Schematic representation of normal Hopf bifurcation. Stable limit point at $\mu < \mu_c$ (a) making transition through $\mu = \mu_c$ (b) to unstable limit point for $\mu > \mu_c$ (c).

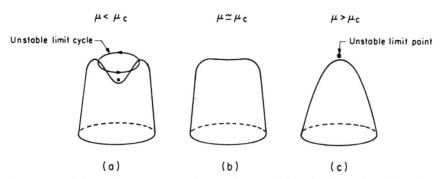

$\mu < \mu_c$ $\mu \simeq \mu_c$ $\mu > \mu_c$

Unstable limit cycle ⌐   ⌐ Unstable limit point

(a)   (b)   (c)

Figure 5.8 Schematic representation of inverted Hopf bifurcation. Unstable limit cycle at $\mu < \mu_c$ (a) making transition through $\mu \simeq \mu_c$ (b) to unstable limit point for $\mu > \mu_c$ (c).

this is done by looking (Figure 5.10) at the motion on a *Poincaré map* (which is just like the Poincaré surface of section used in the study of Hamiltonian systems). If one can prove the existence of stable invariant circles on the map, one then has a torus. Note that by examining the transformation on the Poincaré map, the original flow (differential equation) is now seen in the form of a diffeomorphism (smooth invertible mapping). Formally, one thus needs a Hopf bifurcation theorem for diffeomorphisms. Such theorems are available.

However, as we saw at the beginning of our discussion of the Lan-dau–Hopf theory, it is not in agreement with observation. Clearly, a more sophisticated theory is required—this is the Ruelle–Takens theory.

5.3.c Ruelle–Takens Theory

This theory, which was announced in 1971, is initially like the Landau–Hopf theory in that there are normal bifurcations to invariant tori of successively higher dimensionality. However, Ruelle and Takens (1971) suggested that by the time T_4 is reached, the motion can become trapped on manifolds that are no longer smooth tori but are of a complex topology which they called *strange attractors*. Later work, with Newhouse (Newhouse et al., 1978), has shown that strange attractors can exist by the time T_3 is reached.

Figure 5.9 Schematic representation of limit cycle making transition to 2-torus.

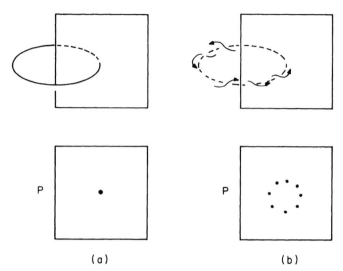

Figure 5.10 (a) Poincaré map *P* of limit cycle is a single point. (*b*) As cycle makes transition to 2-torus, Poincaré map displays circle of points.

These strange attractors are manifolds that do not have a simple integer dimension; that is, they lie somewhere between, say, a surface and a volume. This notion of noninteger dimensionality has been studied extensively by Mandelbrot (1983) in the context of *fractals*, which will be discussed in Section 5.3.e.

The crucial point of the Ruelle–Takens–Newhouse theory is that in the case of the attractor satisfying certain conditions such that it is an "axiom A attractor" (it turns out in practice that this class is rather restrictive), the motion on it is *chaotic*. That is, the motion is very sensitive to initial conditions, time correlations show exponential decay (proved only in certain cases), and the average behavior is described by a measure $d\mu$ with nonzero entropy.

These properties of strange attractors provide the crucial distinction between the Landau–Hopf theory and the Ruelle-Takens theory. In the former, the motion is always assumed to be intrinsically *quasi-periodic*. In the latter, by invoking the existence of a strange attractor, there is now a mechanism by which the motion of entirely deterministic dissipative dynamical systems can display intrinsically *chaotic* behavior. Thus, chaos is possible in a dissipative system without the imposition of external noise.

At first sight, the idea of chaos in a dissipative system would seem to be counterintuitive since, as we learned in the context of Hamiltonian systems, sensitivity to initial condition implies divergence of nearby trajectories (and hence positive Lyapunov exponents), whereas dissipation implies attraction of trajectories. Certainly, for flows (differential equations) on the plane (i.e.,

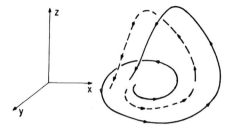

Figure 5.11 Two nearby initial conditions (dashed line and solid line) in x–y plane spiraling out exponentially (stretching) and being attracted back (folding) in the z-direction.

in a two-dimensional phase space), it would seem (and is) topologically impossible to reconcile these two opposing concepts. However, if there is a third (or more) dimension to the phase space, the paradox can be resolved as follows. Imagine trajectories diverging in a plane by spiraling out from some unstable spiral point, leaving the plane and then "folding over" and returning (i.e., being attracted) back to the center of the spiral. This is illustrated in Figure 5.11. There are clearly two basic processes at work (i) *stretching*, which gives the sensitivity to initial conditions, and (ii) *folding*, which gives the attraction. Since trajectories in phase space cannot cross the repeated stretching and folding operations must result in an object of great topological complexity.

To obtain some idea of this complexity, we appeal to a simple model transformation, known as the *Smale horseshoe mapping* (Smale, 1963), which consists of a sequence of stretching and folding operations as follows. The sequence begins (see Figure 5.12) by taking some initial rectangle and stretching it out in (say) the x-direction and contracting it by a somewhat

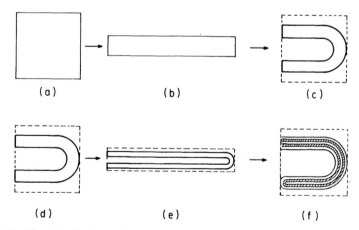

(a) (b) (c)

(d) (e) (f)

Figure 5.12 The Smale horseshoe mapping. Initial square (a) is stretched to double its length to rectangle of reduced area (b) and folded over into horseshoe (c). In (d), (e), and (f), the process is repeated, giving doubly folded horseshoe in (f).

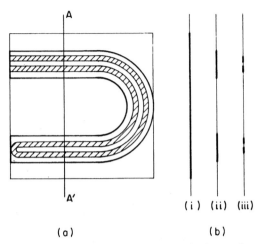

Figure 5.13 (a) Typical cross section A–A' of Smale horseshoe. (b) Successive cross sections (i), (ii), and (iii) show appearance of Cantor-set-like structure.

greater amount in the *y*-direction. In this way the original rectangle is stretched out into a long thin one of slighter reduced area (i.e., the area has been contracted as behoves a dissipative system. The next step is to fold the resulting rectangle over into the shape of a horseshoe.† The sequence of operations is now repeated. The horseshoe is stretched out into a longer and thinner one and then folded over, giving a type of double hairpin structure. As this transformation is iterated further, a highly complex structure is clearly obtained. If one looks at a cross section of successive images of the transformation (see Figure 5.13), one sees a sequence of segments doubling in number with each iteration (i.e., two segments after the first iteration, four segments after the second, and so on). As is described in Section 5.3.e, this structure corresponds to what is known as a *Cantor set*, which is a simple example of a fractal object.

5.3.d   Other Scenarios

Although the Ruelle–Takens–Newhouse theory is a great improvement over the previous theories, it is still not quite the whole story. It has now been realized that there are, in fact, many different "routes to turbulence"—the Ruelle–Takens–Newhouse "scenario" just being one of them. A particularly interesting route is one that occurs via a sequence of *period doubling bifurcations*; this route is discussed in detail in Section 5.5.

---

†This sequence is not dissimilar to the Baker's transformation described in Section 4.7—however, with the important difference that that transformation is area preserving.

Another important route is that of *intermittency*. These scenarios have been nicely reviewed by Eckman (1981) and are discussed at length in the excellent book by Bergé et al. (1984).

5.3.e Fractals

A well-known example of a simple fractal is the *Koch snowflake* shown in Figure 5.14. This leads to a curve which has infinite length but which encloses a finite area. Unlike a simple curve that has (topological) dimension one, this curve has a dimension, known as the *Hausdorf–Besicovitch* dimension (D_H), lying between one (a curve) and two (a surface). In fact, one may show for this case that

$$D_H = \frac{\log 4}{\log 3} = 1.2618\ldots \tag{5.3.5}$$

A fundamental property of such fractals—as is readily apparent in this simple example—is that they are *self-similar*; that is, at whatever scale they are perceived, they have the same appearance. A fractal object may have some largest scale, but, in principle, there should be no smallest scale at which the basic structure is not reproduced. (In practice, there may well be a lower limit too.) One often says that these structures have a Cantor-set-like structure. The original Cantor set was constructed by dividing the unit interval into thirds, removing the central third and repeating the process on each remaining segment, and so on (see Figure 5.15). In this case the Hausdorf–Besicovitch dimension is

$$D_H = \frac{\log 2}{\log 3} = 0.6309\ldots \tag{5.3.6}$$

For such geometrically simple objects as the snowflake and the Cantor set, D_H is easily computed from the following considerations. Here we consider the latter case. Define the total measure of points lying on a line segment of length l to be some function $\mu = \mu(l)$. If, in the Cantor set construction of dividing a line into two separate thirds, we assume the total measure of points to be preserved, it follows that

$$\mu(3l) = 2\mu(l) \tag{5.3.7}$$

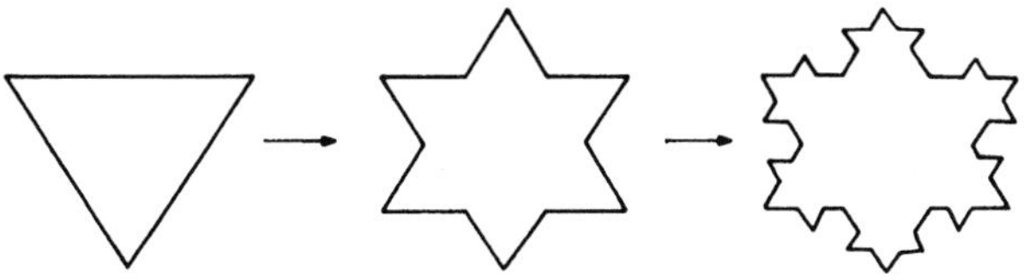

Figure 5.14 First stages in construction of Koch snowflake.

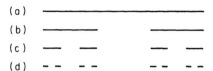

Figure 5.15 Construction of Cantor set: (a) unit interval; (b) removal of central $\frac{1}{3}$rd portion; (c) removal of central $\frac{1}{9}$th portions; (d) removal of central $\frac{1}{27}$th portions, and so on.

If one then assumes that $\mu(l)$ scales as a power of l, that is,

$$\mu(l) \sim l^{\delta} \tag{5.3.8}$$

the exponent δ is easily determined from (5.3.7) to be just $\log 2/\log 3$. This exponent is then identified as the "fractal dimension" D_H. In many cases the fractal dimension of a self-similar object has to be computed numerically. At the time of writing, this has become a most active area of research. To date, one of the most practical algorithms for computing dimension is that due to Grassberger and Procaccia (1983). In any such algorithm the computed quantity is usually not D_H itself but is, instead, some related dimension. In the case of the Grassberger–Procaccia method, based on the properties of a certain correlation function, the computed dimension is a lower bound to D_H.

Finally, while on the subject of fractals, in theories of fully developed turbulence it is believed that vortex sheets fold themselves up into infinitely complicated fractal structures.

5.4 MATHEMATICAL MODELS OF STRANGE ATTRACTORS

We now turn to an examination of some simple mathematical models that exhibit some form of strange attractor.

5.4.a The Lorenz System

This remarkable system was investigated by Lorenz in 1963—all the more remarkable since it was constructed long before strange attractors were known about. Its purpose was to provide a simplified model of atmospheric convection to determine whether long-range weather forecasting was possible. In more recent years, this system has been investigated in detail by many workers. Many of these results are reviewed in the book by Sparrow (1982).

One considers a fluid of uniform depth H with imposed temperature different ΔT. In the case where all the motions are parallel to the $(x–z)$ plane and uniform in the y-direction, the governing equations of motion can be expressed in the two-dimensional form

$$\frac{\partial}{\partial t}(\nabla^2 \psi) = \frac{\partial \psi}{\partial z}\frac{\partial}{\partial x}(\nabla^2 \psi) - \frac{\partial \psi}{\partial x}\frac{\partial}{\partial z}(\nabla^2 \psi) + \nu\nabla^2(\nabla^2 \psi) + g\alpha\frac{\partial \theta}{\partial x}$$

$$\frac{\partial \theta}{\partial t} = \frac{\partial \theta}{\partial z}\frac{\partial \psi}{\partial x} - \frac{\partial \theta}{\partial x}\frac{\partial \psi}{\partial z} + \kappa\nabla^2\theta + \frac{\Delta T}{H}\frac{\partial \psi}{\partial x}$$

(5.4.1)

where ψ is the stream function for the two-dimensional motion; that is, the velocity components ($\mathbf{u} = (u, w)$) are given by

$$u = \frac{\partial \psi}{\partial z}, \qquad w = -\frac{\partial \psi}{\partial x}$$

(5.4.2)

and θ is the temperature field measuring the deviation from the equilibrium state. The various coefficients are

$g =$ gravitational constant,
$\alpha =$ coefficient of thermal expansion,
$\nu =$ kinematic viscosity, and
$\kappa =$ thermal conductivity.

Rayleigh found that solutions of the form

$$\psi = \psi_0 \sin\left(\frac{\pi a x}{H}\right)\sin\left(\frac{\pi z}{H}\right)$$

$$\theta = \theta_0 \cos\left(\frac{\pi a x}{H}\right)\sin\left(\frac{\pi z}{H}\right)$$

(5.4.3)

would grow if the Rayleigh number, that is, the quantity

$$R_a = \frac{g\alpha H^3 \Delta T}{\nu \kappa}$$

(5.4.4)

is greater than the critical value

$$R_a^{(c)} = \frac{\pi^4(1 + a^2)^3}{a^2}$$

(5.4.5)

The minimum value of $R_a^{(c)}$ occurs for $a^2 = \frac{1}{2}$, giving

$$R_a^{(c)} = \frac{27\pi}{4} = 657.511$$

(5.4.6)

Solution of the problem numerically requires the integration of the pair of two-dimensional partial differential equations (5.4.1). This is always an unpleasant task. An alternative approach to direct numerical integration is to expand the functions ψ and θ in a basis set. In this case by taking periodic boundary conditions in both directions, one has

$$\psi(x, z, t) = \sum_{m=1}^{\infty} \sum_{n=1}^{\infty} \psi_{mn}(t) \sin\left(\frac{m\pi x a}{H}\right) \sin\left(\frac{m\pi z}{H}\right)$$

$$\theta(x, z, t) = \sum_{m=0}^{\infty} \sum_{n=1}^{\infty} \theta_{mn}(t) \cos\left(\frac{m\pi x a}{H}\right) \sin\left(\frac{n\pi z}{H}\right) \quad (5.4.7)$$

Substitution of these expansions into the partial differential equations yields an *infinite* set of *ordinary* differential equations. Actual integration requires a *finite* truncation of the infinite set.

Lorenz (1963) took the most extreme truncation possible by including only the coefficients ψ_{11} (denoted by X), $\theta_{11}(Y)$, and $\theta_{02}(Z)$. With various scalings, this yields the system of three coupled ordinary differential equations:

$$\dot{X} = \sigma(Y - X)$$
$$\dot{Y} = -XZ + rX - Y \quad (5.4.8)$$
$$\dot{Z} = XY - bZ$$

where $\sigma = \dfrac{\nu}{\kappa} =$ Prandtl number

$r = \dfrac{R_a}{R_a^{(c)}} =$ (normalized) Rayleigh number

$b = 4/(1 + a^2) =$ geometric factor

and the time variable is scaled as $\tau = \pi^2(1 + a^2)\kappa t/H^2$. In the set of equations (5.4.8), usually referred to as the *Lorenz equations*, the variables X, Y, and Z can be assigned simple physical interpretations, that is,

$X \propto$ intensity of convective motion,

$Y \propto$ temperature difference between ascending and descending currents, and

$Z \propto$ distortion of vertical temperature profile from linearity.

We now examine some of the properties of the equations (5.4.8):

(a) *Divergence*

$$D = \frac{\partial \dot{X}}{\partial X} + \frac{\partial \dot{Y}}{\partial Y} + \frac{\partial \dot{Z}}{\partial Z} = -(b + \sigma + 1) \quad (5.4.9)$$

which is negative since b and σ are positive. Denoting a typical element of phase-space volume by $\Gamma(t)$, we thus have a contraction of the form

$$\Gamma(t) = \Gamma(0)e^{-(b+\sigma+1)t} \tag{5.4.10}$$

Hence all trajectories will ultimately become confined to some form of limiting manifold.

(b) *Critical Points.* The points satisfying the condition

$$\dot{X} = \dot{Y} = \dot{Z} = 0$$

are

(i) $X = Y = Z = 0$, corresponding to the state of *no* convection, that is, pure conduction.

(ii) $X = Y = +\sqrt{b(r-1)}$, $Z = (r-1)$ and $X = Y = -\sqrt{b(r-1)}$, $Z = (r-1)$, corresponding to the states of *steady* convection. Note that the steady convective states only exist for $r > 1$.

(c) *Stability Properties.* The linearized transformation is of the form

$$\frac{d}{dt}\begin{bmatrix} \delta X \\ \delta Y \\ \delta Z \end{bmatrix} = \begin{bmatrix} -\sigma & \sigma & 0 \\ (\sigma - Z) & -1 & -X \\ Y & X & -b \end{bmatrix} \begin{bmatrix} \delta X \\ \delta Y \\ \delta Z \end{bmatrix} \tag{5.4.11}$$

from which one can deduce the stability properties of the critical points.

(i) $(X, Y, Z) = (0, 0, 0)$: For $r < 1$ this is stable, that is, all eigenvalues have negative real parts; for $r > 1$, one eigenvalue acquires a positive real part. The critical point is unstable and hence convection will start on infinitesimal perturbation. Note that the stability of the critical point depends only on the value of the Rayleigh number.

(ii) $(X, Y, Z) = (\pm\sqrt{b(r-1)}, \pm\sqrt{b(r-1)}, r-1)$: For $r > 1$, the eigenvalues consist of one real negative root and a pair of complex conjugate roots. This pair of critical points can be shown to become unstable if

$$r = \frac{\sigma(\sigma + b + 3)}{(\sigma - b - 1)} \tag{5.4.12}$$

a condition that can only be satisfied for positive r if

$$\sigma > (b + 1)$$

Note that the stability of these critical points no longer depends on just the value of the Rayleigh number.

In his study, Lorenz (1963) chose the parameter values

$$b = 8/3 \quad \text{and} \quad \sigma = 10$$

With this choice, the steady (convective) states become unstable at

$$r = \frac{470}{19} \simeq 24.74 \ldots$$

and the contraction rate is $D = -13.67$, which is, in fact, extremely fast.

We now summarize what happens to the solutions of the Lorenz equations as r is gradually increased.

(i) $0 < r < 1$: The origin is a globally attracting stationary solution, and all trajectories (i.e., all different initial conditions) eventually spiral into it.

(ii) $1 < r < 24.74$: The origin becomes unstable and bifurcates into a pair of locally attracting stationary solutions $C = (\sqrt{b(r-1)}, \sqrt{b(r-1)}, r-1)$ and $C' = (-\sqrt{b(r-1)}, -\sqrt{b(r-1)}, r-1)$. Virtually all trajectories converge to either C or C'. The exceptions are the (measure zero) set of trajectories that stay in the vicinity of the origin. At around $r \simeq 13.926$, the origin develops into a homoclinic point. Beyond this r value, the "basins of attraction" around C and C' are no longer distinct and trajectories can cross backward and forward between the two before settling down.

(iii) $r \simeq 24.74$: As mentioned, this is the critical value at which the steady states C and C' become unstable. However, the Hopf analysis shows that as this critical value of r is passed, there is an inverted bifurcation; that is, the limit points C and C' do not become stable limit cycles.

(iv) $r > 24.74$: Trajectories integrated in this regime show remarkable behavior. In Lorenz's original paper (Lorenz, 1963), he studied the trajectory with initial conditions $(X, Y, Z) = (0, 1, 0)$ (i.e., a slight deviation from equilibrium) at a value of $r = 28$. For this value of r, the unstable steady states are $C = (6\sqrt{2}, 6\sqrt{2}, 27)$ and $C' = (-6\sqrt{2}, -6\sqrt{2}, 27)$. His computations showed that once certain transient oscillations had died away, the motion became highly erratic. This is a result of the solution spiraling around one of the fixed points, C or C', for some arbitrary period and then jumping to the vicinity of the other fixed point, spiraling around that for a while and then jumping back to the other, and so on. This combination of spiraling out (along the unstable manifold) and returning (along the stable manifold) gives rise to the stretching and folding mechanism discussed earlier and results in a highly complex manifold, namely, a form of

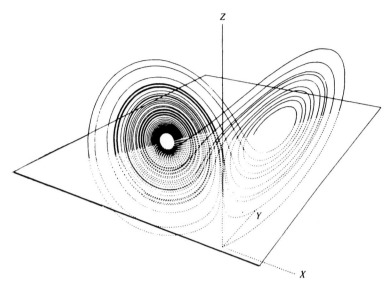

Figure 5.16 Solution of the Lorenz equation computed at $r = 28$. The horizontal plane is at $z = 27$. (Reproduced, with permission, from O. Lanford, *Turbulence Seminar*, Lecture Notes in Mathematics, Vol. 615, Springer-Verlag, New York, 1977.)

strange attractor. A typical orbit on this attractor is shown in Figure 5.16. (The apparent regularity of this structure in the figure is deceptive—the attractor is highly complex.) The power spectrum of the trajectory is essentially continuous, indicating highly chaotic motion.

The Lorenz system has a three-dimensional phase space, and one might ask if there is some compact way of representing the motion—perhaps analogous to the surface-of-section technique used for Hamiltonian systems. Once transients have died out and the system has "settled" on the strange attractor, the individual variables X, Y, and Z all show chaotic behavior when plotted as a function of time. With great perspicacity, Lorenz (1963) followed the values of the successive maxima of $Z(t)$—that is, the maximum value attained by Z (call it Z_n)—as it spiraled about one of the fixed points C or C' before jumping over to the other fixed point and attaining its next maximum value (call it Z_{n+1}), and so on. The plot of Z_{n+1} versus Z_n takes the form shown in Figure 5.17. Remarkably enough, this one-dimensional "mapping" contains the essence of the Lorenz attractor dynamics. The properties of such one-dimensional maps will be discussed in detail in Section 5.5.

The nature of the Lorenz attractor has been investigated in great detail. It is certainly strange, although apparently not an axiom A attractor. Notice

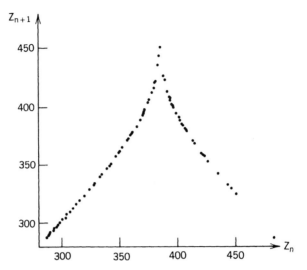

Figure 5.17 Maxima Z_{n+1} versus previous maxima Z_n for Lorenz equations at $r = 28$. (Reproduced, with permission of the American Meteorological Society, from E. N. Lorenz, *J. Atmos. Sci.*, **20**, 130–141 (1963).)

that although the motion clearly becomes chaotic beyond $r \simeq 24.74$, the sequence of events leading up to this chaos does not include any periodic regimes; that is, this is not the complete Reulle–Takens scenario for the onset of turbulence.

It is also of interest to ask what happens at very large values of r. Investigations by various workers reveal that there are alternating regimes of turbulent and periodic behavior. That is, as r is increased above 28, the strange attractor changes into a periodic limit cycle (around $r = 145$ to $r = 148$). This lives on for a while as r is increased and then changes back into a strange attractor. At higher r this then changes back into another limit cycle (around $r = 210$ to $r = 234$). As the limit cycle regime changes into a chaotic regime, one can detect an effect called *intermittency*, that is, turbulent "bursts" interrupting the otherwise periodic motion. Various types of intermittency can occur; these are all discussed in detail in the book by Bergé et al (1984). Finally we remark that as interesting as all this high-r behavior is (i.e., alternating chaotic and periodic regimes), it makes the Lorenz model rather unrealistic as far as "real" turbulence is concerned.

5.4.b Variations on the Lorenz Model

One of the major deficiencies of the Lorenz model is that it is such an extreme truncation. Clearly, it is important to examine what happens as more modes are added on. A study of a 14-mode version of the Lorenz model has been made by Curry (1978). We give a very brief summary of his results. There seem to be a number of distinct regimes. (Note that Curry's *r* parameter is slightly different from Lorenz's, so absolute values should not be compared.)

(i) $1 < r < 43.48$: The motion converges to the stable fixed points (there are two) of the system.

(ii) $r \approx 43.70$. A stable limit cycle appears; that is, there has been a normal Hopf bifurcation. At around $r \approx 44.07$, the period of this limit cycle doubles.

(iii) $44.6 < r < 45.10$. There is substantial evidence for the existence of a stable 2-torus.

(iv) $r > 45.10$. Here a strange attractor appears which looks remarkably similar (in a suitable projection) to the Lorenz attractor.

Although it is encouraging that a strange attractor is still found for this system, a word of caution is now in order. A more systematic study of Lorenz-like models with ever-increasing numbers of modes certainly reveals very rich dynamics. However, the most sensitive parameter in this family of systems seems to be the number of modes itself! In other words,

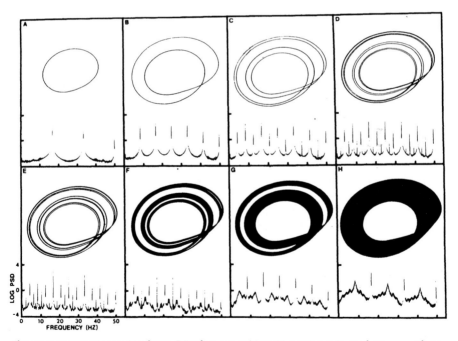

Figure 5.18 Trajectories from Rössler equations (5.4.13) projected on x–y plane with associated power spectra. Computed at (*A*) $c = 2.6$, (*B*) $c = 3.5$, (*C*) $c = 4.1$, (*D*) $c = 4.18$, (*E*) $c = 4.21$, (*F*) $c = 4.23$, (*G*) $c = 4.30$, (*H*) $c = 4.60$. (Reproduced, with permission, from Crutchfield et al. (1980).)

there does not seem to be a unique behavior to which these systems converge. This strongly suggests that small-mode truncations of partial differential equations may not give a reliable picture of the "true" behavior (whatever that is!) when many spatial scales are physically significant. Indeed, an active area of current research is devoted to proving the existence of finite-dimensional attractors in various partial differential equations; this is known as "inertial manifold" theory (see, e.g., Doering et al. (1988)).

Nonetheless, simple models such as the Lorenz equations are interesting in their own right for their rich dynamical behavior. Another interesting system, for example, is the *Rossler model*:

$$\dot{X} = -(Y + Z)$$
$$\dot{Y} = X + 0.2\,Y \qquad\qquad (5.4.13)$$
$$\dot{Z} = 0.2 + XZ - cZ$$

As the parameter c is increased, the motion undergoes a distinctive sequence of period-doubling bifurcations until a type of strange attractor-like behavior is attained (shown in Figure 5.18). Another model that shows a similar behavior is the Duffing oscillator (1.6.11) introduced in Chapter 1.

5.4.c The Henon Map

Naturally, one would also like to find simple algebraic mappings that can display strange attractors. It will come as no surprise by now that such a mapping has been constructed by Henon. It takes the very simple form

$$T: \quad \begin{aligned} x_{i+1} &= y_i - ax_i^2 + 1 \qquad\qquad &(5.4.14a)\\[1ex] y_{i+1} &= bx_i &(5.4.14b) \end{aligned}$$

It is a contraction mapping since

$$\frac{\partial(x_{i+1},\, y_{i+1})}{\partial(x_i,\, y_i)} = -b \qquad\qquad (5.4.15)$$

T can be thought of as the product of three simpler mappings, that is,

$$T = T'T''T''' \qquad\qquad (5.4.16)$$

where

$$\begin{aligned} T': \quad & x' = x, \qquad y' = y + 1 - ax^2\\[1ex] T'': \quad & x'' = bx', \qquad y'' = y'\\[1ex] T''': \quad & x''' = y'', \qquad y''' = x'' \end{aligned} \qquad\qquad (5.4.17)$$

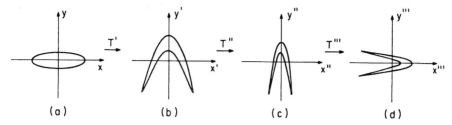

Figure 5.19 Sequence of transformations in Henon map. An area element (a) is folded (b), contracted (c), and rotated (d).

which corresponds to bending, contraction, and rotation, respectively (see Figure 5.19). It is this composition of transformations that provides the necessary ingredients of stretching and folding required for the generation of chaotic behavior. It should also be noted that the mapping, like its area-preserving cousin (4.2.11), is invertible, the inverse taking the form

$$T^{-1}: \quad x_i = y_{i+1}/b \tag{5.4.18a}$$

$$y_i = x_{i+1} - 1 + ay_{i+1}^2/b^2 \tag{5.4.18b}$$

Thus some final point can, in principle, be uniquely "time-reversed" back to its initial point (x_0, y_0).

Overall, the numerical results show that within some fairly large region of the x–y phase plane, all trajectories land up on a complicated looking manifold—the "Henon attractor." Outside this region the trajectories escape to infinity. In a beautiful sequence of computer-generated pictures (see Figure 5.20), Henon (1976) clearly shows that this attractor has a Cantor-set-like structure. Power spectra of the Henon map have been computed and found to be suitably noisy. (Although the Henon attractor is clearly strange, it apparently does not fall within the rather restrictive axiom A category.)

Finally, we mention an amusing variant of the Henon mapping, known as the *Lozi mapping*:

$$T: \quad x_{i+1} = y_i + 1 - a|x_i| \tag{5.4.19}$$

$$y_{i+1} = bx_i$$

which also exhibits a strange attractor made up of straight line segments. For this system it is possible to prove that the attractor is of the axiom A type.

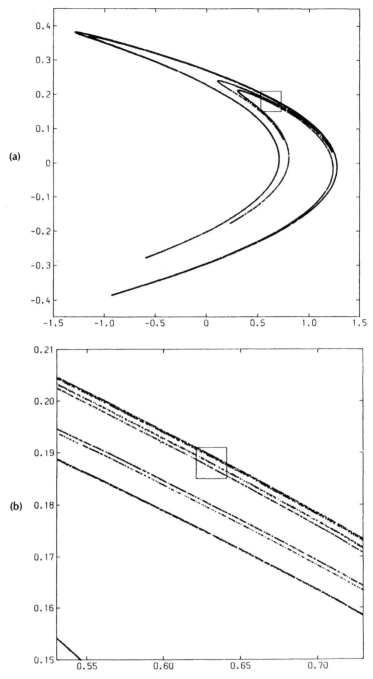

Figure 5.20 (a) Henon attractor computed from (5.4.19) with $a = 1.4$, $b = 0.3$. (b) Blowup of boxed region in (a) showing Cantor-set-like structure of attractor. Further blowups of boxed region in (b) show same structure. (Reproduced, with permission of Springer-Verlag, from M. Henon, *Comm. Math. Phys.*, **50**, 69–77 (1976).)

5.5 PERIOD-DOUBLING BIFURCATIONS

As mentioned earlier, an important "route to turbulence" is that of period-doubling bifurcations. Remarkably, this complicated behavior is exhibited by one of the simplest nonlinear mappings possible, namely, one-dimensional mappings of the form

$$x_{n+1} = f(x_n) \qquad (5.5.1)$$

if $f(x)$ satisfies certain conditions. Of these mappings, one of the most famous is the so called *logistic* map

$$x_{n+1} = 4\lambda x_n(1 - x_n), \qquad 0 < x < 1 \qquad (5.5.2)$$

where λ is an adjustable parameter. One of its earlier applications was in biology as a simple model of population dynamics (hence its name), and the reader is recommended to read the beautiful and seminal article by May (1976) which describes some of the basic phenomenology of this system as the parameter λ is varied. For small λ, one finds that all iterates (providing $x_0 \neq 0$) converge onto a single limit point. This behavior persists until λ passes 0.75. The single fixed point, for larger λ, bifurcates into a pair of fixed points—that is, a period-2 limit cycle. As λ is further increased, the period-2 limit cycle bifurcates into a period-4 cycle, which subsequently bifurcates into a period-8 cycle, and so on. The λ values at which the bifurcations occur $(\lambda_1, \lambda_2, \lambda_3, \ldots)$ become ever closer, converging geometrically to a critical value λ_∞ (about 0.892). At this point, the orbit becomes aperiodic. Beyond this point, both chaotic orbits and odd-period limit cycles start to appear. At $\lambda = 1$, the motion is formally ergodic on the unit interval $(0, 1)$; beyond $\lambda = 1$, all orbits escape to ∞.

A most significant observation is due to Feigenbaum (1978) who noticed a geometric convergence of the period-doubling sequence, that is,

$$\lambda_\infty - \lambda_n \propto \delta^{-n} \qquad (5.5.3)$$

On defining

$$\delta_n = \frac{\lambda_{n+1} - \lambda_n}{\lambda_{n+2} - \lambda_{n+1}} \qquad (5.5.4)$$

he found numerically that δ_n (as $n \to \infty$) converged to the value $\delta = 4.6692016$. His remarkable discovery was that other nonlinear maps, of the form (5.5.1) with different forms of $f(x)$ (within a certain class), all had this same convergence rate. Thus δ became dubbed a *universal* number. Although this particular "universality" is limited to a relatively small class of maps and flows (these include the Henon map, the Lorenz model, and some small-mode truncations of the Navier–Stokes equations), it has generated some beautiful theory and has been observed in a variety of experiments.

5.5.a The Period-Doubling Mechanism

For a mapping of the form

$$x_{n+1} = f(x_n) \tag{5.5.5}$$

the sequence of iterates, starting at x_0, is

$$x_1 = f(x_0)$$
$$x_2 = f(x_1) = f(f(x_0))$$
$$\vdots$$
$$x_{n+1} = f(x_n) = f(f(\ldots f(x_0))) = f^n(x_0)$$

where $f^n(x_0)$ denotes the nth *functional iteration* of $f(x)$ (not the nth power of $f(x)$). If $f(x)$ is a linear function of x, these iterates are trivial, but if $f(x)$ is nonlinear, for example,

$$f(x) = 4\lambda x(1 - x) \tag{5.5.6}$$

the successive iterates become polynomials of ever-higher order (for (5.5.6), $f^n(x)$ is a polynomial of order 2^n). Despite this complexity, the iterates x_i can be followed by the simple graphical procedure of plotting $y = f(x)$ and the line $y = x$ on the same graph (Figure 5.21) and moving successively vertically and horizontally between these two curves. As shown, this yields the successive iterates $x_0, x_1, x_2, \ldots$. The point at which the curve and the line intersect (i.e., $y = f(x) = x$) must correspond to a fixed point of the iteration sequence. This fixed point is easily determined by solving the equation

$$x^* = 4\lambda x^*(1 - x^*) \tag{5.5.7}$$

which has the two roots

$$x^* = 0 \quad \text{and} \quad x^* = 1 - \tfrac{1}{4}\lambda$$

For iterates of $f(x)$ to remain in the interval $(0, 1)$, we require that $0 < \lambda < 1$ and $0 < x < 1$. Thus for $\lambda < \tfrac{1}{4}$ only the fixed point $x^* = 0$ lies in the bounded interval; but for $\tfrac{1}{4} < \lambda < 1$, both fixed points can be reached. These points

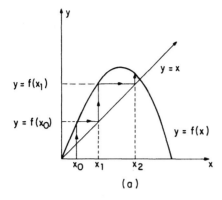

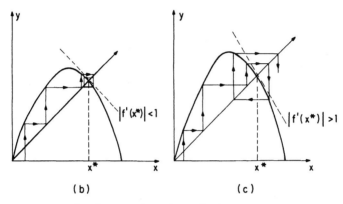

Figure 5.21 (a) Graphical procedure for calculating successive iterates x_i of logistic map. (b) Iterates spiraling onto stable fixed point x^*. (c) Iterates spiraling away from unstable fixed point x^*.

will be attracting or repelling (stable or unstable) depending on the slope of $f(x)$. Consider the fixed point x^* (i.e., $x^* = f(x^*)$) and set $x_n = x^* + \epsilon_n$; then by expanding to first order, one finds that

$$x_{n+1} = f(x^* + \epsilon_n) \simeq f(x^*) + \epsilon_n f'(x^*)$$
$$\equiv x^* + \epsilon_n f'(x^*)$$

Now $x_{n+1} = x^* + \epsilon_{n+1}$, so we obtain

$$\frac{\epsilon_{n+1}}{\epsilon_n} = f'(x^*) \tag{5.5.8}$$

Clearly, the subsequent iterates of ϵ_n will only converge if $|f'(x^*)| < 1$. In summary, we say that

if $|f'(x^*)| < 1$,   then x^* is stable,
if $|f'(x^*)| = 1$,   then x^* is marginally stable, and
if $|f'(x^*)| > 1$,   then x^* is unstable.

One can also observe these basic stability properties from the graphical procedure (sketched in Figure 5.21). For the choice of mapping (5.5.2), one may easily deduce the stability of the two fixed points $x^* = 0$ and $x^* = 1 - 1/4\lambda$. One finds that:

$x^* = 0$ is stable for $0 < \lambda < \frac{1}{4}$ and unstable for $\lambda > \frac{1}{4}$
$x^* = 1 - 1/4\lambda$ is unstable for $0 < \lambda < \frac{1}{4}$ and stable for $\frac{1}{4} < \lambda < \frac{3}{4}$.

Thus, at least for $0 < \lambda < \frac{3}{4}$, the entire behavior of (5.5.2) is known. For $\lambda < \frac{1}{4}$, all iterates (in the interval $(0, 1)$) converge to $x = 0$; and for $\frac{1}{4} < \lambda < \frac{3}{4}$, all iterates converge to $x = 1 - 1/4\lambda$.

For $\lambda > \frac{3}{4}$ it might appear, at first glance, that there are no attracting fixed points. What in fact happens is that the stable fixed point at $x = 1 - 1/4\lambda$, which becomes unstable at $\lambda = \frac{3}{4}$ (at this point $f'(x^*) = -1$), bifurcates into a stable *2-cycle*. The crucial point is that these two fixed points (which make up the 2-cycle) are now the stable fixed points of the composed function

$$f^2 = f(f(x)) \tag{5.5.9}$$

The appearance of these points can be understood by examining graphs of f and f^2 as shown in Figure 5.22. The function f is a single humped function (symmetric about $x = \frac{1}{2}$), whereas the function f^2 is a symetric double-humped function (it is a polynomial of degree 4). If x^* is a fixed point of f, then it is also a fixed point of f^2. For example, for $\lambda < \frac{3}{4}$, x^* is the only (stable) fixed point of f and f^2. Its stability can be deduced as follows: Define $x_2 = f^2(x_0) = f(x_1)$, where $x_1 = f(x_0)$, and use the chain rule to compute

$$\frac{d}{dx} f^2(x) \bigg|_{x=x_0} = \frac{d}{dx} f(x_1) \bigg|_{x=x_0} = \frac{d}{dx_1} f(x_1) \frac{dx_1}{dx} \bigg|_{x=x_0} = \frac{d}{dx_1} f(x_1) \frac{d}{dx} f(x) \bigg|_{x=x_0} \tag{5.5.10}$$

a result that is easily generalized to

$$\frac{d}{dx} f^n(x_0) \bigg|_{x=x_0} = f'(x_0)f'(x_1)f'(x_2) \ldots f'(x_{n-1}) \tag{5.5.11}$$

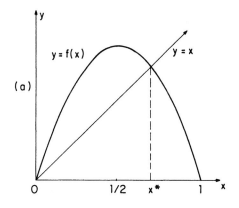

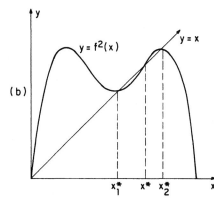

Figure 5.22 (a) For $\lambda > \frac{3}{4}$, x^* is an unstable fixed point of $f(x)$. (b) The composed function $f^2(x) = f(f(x))$. In (b), x_1^* and x_2^* are the stable fixed points of $f^2(x)$. The slope of $f^2(x)$ is the same at both of these points.

For the fixed point $x^* = x_0$, clearly $x_2 = x_1 = x^*$ and hence

$$f^{2\prime}(x^*) = f'(x^*)f'(x^*) = (f'(x^*))^2 \qquad (5.5.12)$$

Thus if $|f'(x)| < 1$, then $|f^{2\prime}(x)| < 1$; and if $|f'(x)| > 1$, then $|f^{2\prime}(x)| > 1$. In other words, if x^* is a stable limit point of f, it is also a stable limit point of f^2; and if x^* is an unstable limit point of f, it is also an unstable limit point of f^2.

For $\lambda > \frac{3}{4}$, x_1^* and x_2^* are (stable) fixed points of f^2. However, since they are not fixed points of f, they are mapped into each other under f, that is,

$$x_1^* = f(x_2^*) \quad \text{and} \quad x_2^* = f(x_1^*)$$

Thus the pair of points x_1^* and x_2^* make up the *2-cycle*, and an initial point will eventually settle into the sequence $x_1^*, x_2^*, x_1^*, x_2^*, \ldots$. An important point to note is that the slopes of f^2 at x_1^* and x_2^* are the same, that is,

$$f^{2\prime}(x_2^*) = f'(x_1^*)f'(x_2^*) \quad \text{and} \quad f^{2\prime}(x_1^*) = f'(x_2^*)f'(x_1^*)$$

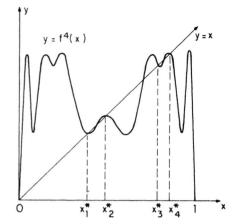

Figure 5.23 The four stable fixed points x_i^* $(i = 1, \ldots, 4)$ of the composed function $f^4(x) = f^2(f^2(x))$. The slope of $f^4(x)$ is the same at each of the x_i^*.

since x_1^* and x_2^* map into each other. Thus the two fixed points x_1^* and x_2^* are simultaneously stable for $\lambda > \frac{3}{4}$ and then become simultaneously unstable (when $f^{2\prime}(x^*) = -1$) at some larger value of λ. At this value of λ (call it λ_2, where $\lambda_1 = \frac{3}{4}$), these two points each bifurcate into two new fixed points to give a stable *4-cycle* corresponding to the fixed points of

$$f^4 = f^2(f^2(x)) \tag{5.5.13}$$

as shown in Figure 5.23. As described at the beginning of this section, this period-doubling sequence continues with the associated λ values converging geometrically.

This period-doubling sequence (or cascade) is universal for mappings of the form (5.5.5), provided that $f(x)$ has a single, locally quadratic, maximum.† The reasons for this universality are deep and have led to the development of *functional renormalization group* theory with close analogies to the renormalization group techniques used in statistical mechanics. Excellent accounts of this have been given by Feigenbaum (1978) and in the book by Collet and Eckman (1980).

5.5.b The Bifurcation Diagram

The detailed dynamics of the mapping (5.5.2) are easily followed on a computer. The basic numerical experiment (ideal for a PC) is to study the iterates of $f(x)$ for successive values of λ. The diagram shown in Figure

†The precise condition is that the so-called Schwarzian derivative of $f(x)$, that is, the quantity

$$\frac{f'''(x)}{f'(x)} - \frac{3}{2}\left(\frac{f''(x)}{f'(x)}\right)^2$$

is negative in the bounded interval.

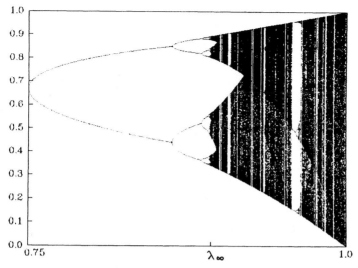

Figure 5.24 Bifurcation diagram computed for logistic map from first period-doubling bifurcation at $\lambda = \frac{3}{4}$ to ergodic limit at $\lambda = 1$. λ_∞ marks the period-doubling accumulation point.

5.24 is easily obtained: For each λ value, one iterates the map until all "transients" have died out and the orbit settles onto its "asymptotic state" (e.g., 2-cycle, 4-cycle, 2^n-cycle, ..., 6- or 3-cycle or aperiodic attractor, etc.) Notice the following basic features:

(i) The successive bifurcations of $2^n (n = 1, 2, ...)$ cycles become ever more compressed in λ space.
(ii) Chaotic bands appear beyond λ_∞.
(iii) Odd cycles (e.g., 3-cycles appear in the chaotic regime.†
(iv) Completely "uniform" chaotic behavior occurs at $\lambda = 1$. (We shall later see that here the motion is truly ergodic on the (0, 1) interval.)

The structure beyond the period-doubling accumulation point λ_∞ is incredibly rich. Some of the current terminology (introduced by Yorke and co-workers (Grebogi et al., 1983)) for the observed phenomena are as follows: The sudden collapse of a chaotic band is called a *subduction*, and its subsequent broadening is called an *internal crisis*; and the complete broadening at $\lambda = 1$ is called a *crisis*.

Returning to the period-doubling bifurcations, their successive

---

†The appearance of odd-period cycles seems to be intimately related to the appearance of chaos in mappings and has led to the Li–Yorke conjecture "period three implies chaos" (Li and Yorke, 1975).

parameter values are tabulated below:

n	Cycle Type	λ_n (λ Value at Which Cycle Appears)
1	2-cycle	0.75
2	4-cycle	0.86237
3	8-cycle	0.88602
4	16-cycle	0.89218
5	32-cycle	0.8924728
.		
.		
.		
∞	Aperiodic attractor	0.892486418

Of course, one would like to know if there are any "real" experiments (as opposed to non-physical numerical models) in which such a sequence can be observed. Clearly, resolving the high-order bifurcations is experimentally very difficult. However, some beautiful experiments by Libchaber et al. (1983) on Rayleigh–Benard convection for mercury in a magnetic field show at least four period doublings with a universal scaling that agrees to within 5% of Feigenbaum's number.

5.5.c Behavior beyond λ_∞

Although the attractor at λ_∞ is aperiodic (has infinite period) and looks chaotic, it does not, in fact, display sensitive dependence to initial conditions; that is, the aperiodic attractor should not be thought of as some sort of strange attractor. This can be demonstrated by computing the associated Lyapunov exponent. For one-dimensional maps (cf. Section 4.5), this exponent is just

$$\lim_{n \to \infty} \frac{1}{n} \log |Df^n(x_0)| \qquad (5.5.14)$$

where x_0 is the initial condition and Df^n denotes the quantity df^n/dx defined in Eq. (5.5.11). The orbit at λ_∞ is found to have zero Lyapunov exponent. As λ is increased beyond λ_∞, the orbits are found to have (overall) increasing exponent values except in the neighborhood of those points where the odd-period cycles appear (at which point the exponent drops to zero).

The maximum Lyapunov exponent is found at $\lambda = 1$, at which point the map can be shown to have an invariant ergodic measure. At this λ value, the map (5.5.2) is easily transformed to the form

$$y_{n+1} = 1 - 2y_n^2 \qquad (5.5.15)$$

under the change of variable $x = \frac{1}{2}(y + 1)$. The transformed map is now bounded on the interval $-1 \leq y \leq 1$. Under the further change of variable $y = \sin \theta$, we obtain

$$\theta_{n+1} = \sin^{-1}(\cos 2\theta_n) \qquad (5.5.16)$$

This is just a linear transformation in θ (i.e., $\theta_{n+1} = 2\theta_n + \pi/2$), and, except for a set of measure zero, all initial conditions will iterate uniformly over the interval $0 \leq \theta \leq \pi$; that is, the probability measure $P(\theta)$ is constant. In the corresponding y interval, $-1 \leq y \leq 1$, the measure $P(y)$ must therefore be related to $P(\theta)$ by

$$P(y) \, dy = \frac{1}{\pi} \, d\theta$$

where the factor $1/\pi$ is just the normalization $\int_0^\pi P(\theta) \, d\theta = \pi$. Thus

$$P(y) = \frac{1}{\pi} \frac{d\theta}{dy} = \frac{1}{\pi \sqrt{1 - y^2}} \qquad (5.5.17)$$

It is an easy exercise to verify this distribution numerically by iterating (5.5.15) for very large n and placing the iterates in small bins of width Δy. It is also not difficult to show that the Lyapunov exponent of (5.5.15) is $\ln 2$.

The fact that orbits beyond λ_∞ can have positive Lyapunov exponents implies that they show sensitive dependence to initial conditions. How can this be so for one-dimensional mappings? Up to now it would appear that the relevant stretching and folding mechanisms required to produce chaotic behavior require a three-dimensional phase space for flows (differential equations) and two dimensions for diffeomorphisms, that is, smooth, invertible mapping (which can be thought of as the Poincaré maps of some higher-dimensional flow). The crucial point is that, unlike the two-dimensional mappings considered previously (e.g., Henon's quadratic map (4.2.11) or attractor map (5.4.14)), the one-dimensional map (5.5.2) is *noninvertible*. Thus, as is easily seen from Figure 5.25, an iterate x_{n+1} has two possible pre-images. It is this multivaluedness that gives rise to the desired stretching and folding mechanism. Consider the line interval $0 \leq x \leq 1$. As shown in Figure 5.25, the portion $0 \leq x \leq \frac{1}{2}$ is stretched out to double its length (double for $\lambda = 1$ and less than double for $\lambda < 1$) to cover the interval $0 \leq x \leq 1$. The remaining portion $\frac{1}{2} \leq x \leq 1$ is also doubled in length but is folded back onto the interval $1 \geq x \geq 0$. This suggests why the mapping of Z_{n+1} versus Z_n constructed by Lorenz (Figure 5.17) for orbits on his attractor is able to capture the essence of the dynamics. The mapping is clearly noninvertible, implying that the motion can be chaotic.

Another important feature of the motion beyond λ_∞ is the appearance of period-3 (and other odd) cycles. The appearance of these period-3 cycles

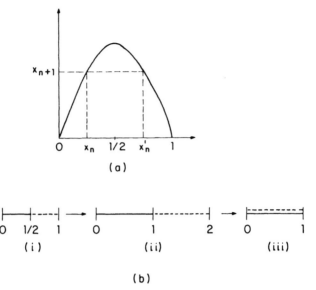

(a)

(b)

Figure 5.25 (a) The logistic map is noninvertible since each iterate x_{n+1} has two possible pre-images x_n and x_n'. (b) The map stretches the unit interval (i) to double its length (ii) and folds it back onto the unit interval (iii).

can be understood by examining the properties of $f^3(x)$. For small λ, the only fixed points of $f^3(x)$ are the fixed points of $f(x)$ (unstable for f^3, of course). As λ increases, a special value, $\lambda_*(>\lambda_\infty)$, is reached, where f^3 becomes exactly tangent to the line $y = x$, as shown in Figure 5.26, and a 3-cycle is born. It is at this point that a *tangent bifurcation* ensues. Notice

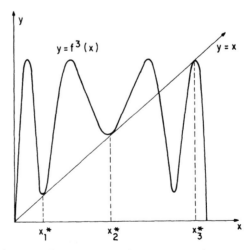

Figure 5.26 Birth of a period-3 orbit: The fixed points x_i^* ($i = 1, 2, 3$) of $f^3(x)$ are exactly tangent to the line $y = x$. The slope of $f^3(x)$ is the same at each of the x_i^*.

that tangency occurs when the slope of f^3 (at the tangent points) is $+1$. (For period doublings, the bifurcations occur when $f^{2n'}(x) = -1$.) For $\lambda > \lambda_*$, each fixed point of the 3-cycle bifurcates into a pair of fixed points—one stable and one unstable. Such bifurcations are also called *saddle-node* bifurcations. This is in contrast to the period-doubling bifurcations, also termed *pitchfork* bifurcations, which involve the transition from an unstable fixed point to a pair of stable fixed points.

5.6.d Other Universality Classes

We emphasize that the "universal" numbers are only universal for a particular class of mappings. The numbers found by Feigenbaum are those relevant to maps with a single quadratic maximum. For the family of maps

$$f(x) = 1 - a|x|^n \qquad (5.5.18)$$

we find period doubling—but for each n, the universal numbers are different:

n:	2	4	6	8
δ (numerical):	5.1224	9.3160	13.3721	17.3987

For area-preserving maps on the plane, that is,

$$x_{n+1} = f(x_n, y_n)$$

$$y_{n+1} = g(x_n, y_n) \qquad (5.5.19)$$

period-doubling bifurcation sequences have also been found. Here the universal exponent is found to be $\delta = 0.7210978\ldots$. All these different universal exponents (and other scaling properties not discussed here) can be calculated by means of the functional renormalization group theory.

SOURCES AND REFERENCES

Texts and General Review Articles

Bergé, P., Y. Pomeau, and C. Vidal, *Order Within Chaos*, Hermann, Paris, 1984.

Chorin, A. J., and J. E. Marsden, *A Mathematical Introduction to Fluid Mechanics*, Springer-Verlag, New York, 1979.

Cvitanović, P., *Universality in Chaos*, Adam Hilger, Bristol, 1984. This volume contains a valuable compilation and extensive bibliography of many fundamental research papers in dissipative dynamical systems.

Dold, A., and B. Eckmann, Eds., *Turbulence Seminar*, Springer-Verlag Lecture Notes in Mathematics, Vol. 615, Springer-Verlag, New York, 1977.

Eckman, J. P., Roads to turbulence in dissipative dynamical systems, *Rev. Mod. Phys.*, **53**, 643 (1981).

Eckman, J. P., and D. Ruelle, Ergodic theory of chaos and strange attractors, *Rev. Mod. Phys.*, **57**, 617 (1985).

Guckenheimer, J., and P. Holmes, *Nonlinear Oscillations, Dynamical Systems and Bifurcations of Vector Fields*, Springer-Verlag, New York, 1983.

Landau, L. D., and E. M. Lifshitz, *Fluid Mechanics*, Pergamon, Oxford, 1959.

Lanford, O. E., Strange attractors and turbulence, in H. L. Swinney and J. P. Gollub, Eds., *Hydrodynamic Instabilities and the Transition to Turbulence*, Springer-Verlag Topics in Applied Physics, Vol. 45, Springer-Verlag, New York, 1981.

Newell, A. C., Chaos and turbulence: is there a connection, *Conference on Mathematics Applied to Fluid Dynamics and Stability* (special proceedings dedicated in memory of Richard C. DiPrima), SIAM, Philadelphia, 1985.

Schuster, H. G., *Deterministic Chaos*, Physik-Verlag, Weinheim, 1984.

Thompson, J. M. T., and H. B. Stewart, *Nonlinear Dynamics and Chaos*, Wiley, Chichester, 1986.

Section 5.1

Aref, H., Integrable, chaotic and turbulent vortex motion in two-dimensional flows, *Ann. Rev. Fluid Mech.*, **15**, 345 (1983).

Section 5.2

Ahlers, G., and R. P. Behringer, Evolution of turbulence from the Rayleigh–Bénard instability, *Phys. Rev. Lett.*, **40**, 712 (1978).

Swinney, H. L., *Suppl. Prog. Theor. Phys.*, **64**, 164 (1978).

Swinney, H. L., and J. P. Gollub, The transition to turbulence, *Phys. Today*, August, 41 (1978).

Section 5.3

Grassberger, P., and I. Procaccia, Measuring the strangeness of strange attractors, *Physica*, **9D**, 189 (1983).

Mandelbrot, B., *The Fractal Geometry of Nature*, Freeman, San Francisco, 1983.

Marsden, J. E., and M. McCracken, *The Hopf Bifurcation and Its Applications*, Springer-Verlag, New York, 1976.

Newhouse, S. E., D. Ruelle, and F. Takens, Occurrence of strange axiom A attractors near quasiperiodic flows on T^m, $m \geq 3$, *Commun. Math. Phys.*, **64**, 35 (1978).

Ruelle, D., and F. Takens, On the nature of turbulence, *Commun. Math. Phys.*, **20**, 167; and **23**, 343 (1971).

Smale, S., Diffeomorphisms with many periodic points, in S. S. Cairns, Ed., *Differential and Combinatorial Topology*, Princeton University Press, Princeton, NJ (1963).

Section 5.4

Crutchfield, J., D. Farmer, N. Packard, R. Shaw, G. Jones, and R. J. Donnelly, Power spectral analysis of a dynamical system, *Phys. Lett.*, **76A**, 1 (1980)

Curry, J., A generalized Lorenz system, *Commun. Math. Phys.*, **60**, 193 (1978).

Doering, C. R., J. D. Gibbon, D. D. Holm, and B. Nicolaenko, Low dimensional behaviour in the complex Ginzburg–Landau equation, *Nonlinearity*, **1**, 279 (1988).

Henon, M., A two dimensional mapping with a strange attractor, *Commun. Math. Phys.*, **50**, 69 (1976).

Lorenz, E. N., Deterministic nonperiodic flow, *J. Atmos. Sci.*, **20**, 130 (1963).

Lozi, R., Un attracteur étrange? du type de Henon, *J. Phys.*, **39**, 9 (1978).

Rössler, O. E, An equation for continuous chaos, *Phys. Lett.*, **57A**, 397 (1976).

Sparrow, C., *The Lorenz Equations*, Springer-Verlag, New York (1982).

Section 5.5

Collet, P., and J. P. Eckman, *Iterated Maps on the Interval as Dynamical Systems*, Progress in Physics, Vol. 1, Birkhäuser, Boston, 1980.

Grebogi, C., E. Ott, and J. A. Yorke, Crises, sudden changes in chaotic attractors, and transient chaos, *Physica*, **7D**, 181 (1983).

Feigenbaum, M. J., Universal behavior in nonlinear systems, *Physica*, **7D**, 16 (1978a).

Feigenbaum, M. J., Quantitative universality for a class of nonlinear transformations, *J. Stat. Phys.*, **19**, 25 (1978b).

Li, T. Y., and J. A. Yorke, Period three implies chaos, *Am. Math. Monthly*, **82**, 985 (1975).

Libchaber, A., S. Fauve, C. Larouche, Two parameter study of routes to chaos, *Physica*, **7D**, 69 (1983).

May, R. M., Simple mathematical models with very complicated dynamics, *Nature*, **261**, 459 (1976).

Mackay, R. S., Renormalization in area-preserving maps, Ph.D. Thesis 1982, Princeton University (University Microfilms Inc., Ann Arbor, MI). An excellent introduction to renormalization in area preserving maps.

6

CHAOS AND INTEGRABILITY IN SEMICLASSICAL MECHANICS

6.1 THE CONNECTION BETWEEN QUANTUM AND CLASSICAL MECHANICS

Although the underlying principles of quantum mechanics, such as Heisenberg's uncertainty principle, have no analogue in classical mechanics, there has always been much interest in determining the nature of the transition from quantum to classical mechanics. For example, as will be discussed below, when the quantal equations of motion, namely, Schrödinger's equation, are solved for particles with large momenta, that is, short de Broglie wavelength, the motion of the associated wave packet is little different from that which can be deduced from solving the corresponding classical equations of motion. (We also mention that this limiting transition is analogous to the transition from wave optics to geometric optics in the limit of short wavelengths.)

6.1.a The Semiclassical Limit for Time-Dependent Problems

Consider a particle of mass m moving in space under the influence of some (smooth) potential $V = V(\mathbf{q})$, where $\mathbf{q} = (q_1, q_2, q_3)$. The classical Hamiltonian is, of course, just

$$H = \frac{1}{2m} \sum_{i=1}^{3} p_i^2 + V(\mathbf{q}) \qquad (6.1.1)$$

and the corresponding time-dependent Schrödinger equation is

$$i\hbar \frac{\partial}{\partial t} \Psi(\mathbf{q}, t) = \hat{\mathcal{H}} \Psi(q, t) \tag{6.1.2}$$

where $\hbar$ is Planck's constant (divided by 2π), $\Psi(\mathbf{q}, t)$ is the wavefunction, and $\hat{\mathcal{H}}$ is the Hamiltonian operator obtained in the usual way by making the substitution $p_i \rightarrow i\hbar(\partial/\partial q_i)$ in (6.1.1), that is,

$$\hat{\mathcal{H}} = \frac{1}{2m} \nabla^2 + V(\mathbf{q}) \tag{6.1.3}$$

where $\nabla^2 = \partial^2/\partial q_1^2 + \partial^2/\partial q_2^2 + \partial^2/\partial q_3^2$ is the Laplacian operator. Making the substitution

$$\Psi(\mathbf{q}, t) = e^{iS(\mathbf{q}, t)/\hbar} \tag{6.1.4}$$

(which we emphasize is, at this stage, an ansatz and not an approximation) in (6.1.2) yields

$$-\frac{\partial S}{\partial t} = \frac{1}{2m} \nabla S \cdot \nabla S + V(\mathbf{q}) - \frac{i\hbar}{2m} \nabla^2 S \tag{6.1.5}$$

If the last term is in some sense negligible, equation (6.1.5) reduces to

$$-\frac{\partial S_0}{\partial t} = \frac{1}{2m} (\nabla S_0)^2 + V(\mathbf{q}) \tag{6.1.6}$$

which is just the classical time-dependent Hamilton–Jacobi equation, and S_0 (now an approximation to the S in (6.1.4)) is identified as the action integral (cf. (2.1.2))

$$S_0(\mathbf{q}, t) = \int_{t_0}^{t} L(\mathbf{q}, \dot{\mathbf{q}}, t') \, dt' \tag{6.1.7}$$

Dropping the last term in (6.1.5) can be thought of, formally, as taking the limit $\hbar \rightarrow 0$. Of course, $\hbar$ is a natural constant and by this limit we really mean the situation in which the quantities having the same dimensions as $\hbar$, namely action, become large relative to $\hbar$. At first sight, the passage from (6.1.2) to (6.1.6) via (6.1.4) might appear straightforward. However, as we shall discuss in this chapter, this limit is very subtle. In fact, this should already be clear from (6.1.4) since the limit $\hbar \rightarrow 0$ leads to ever more rapid oscillations in the wavefunction; that is, the limit is highly *singular* (cf. the brief discussion of singular perturbation theory in Chapter 3). It is completely wrong to think that one can somehow write quantum mechanical quantities as classical quantities plus an expansion of corrections in powers of $\hbar$.

6.1.b The Semiclassical Limit for Time-Independent Problems

In the case of stationary quantum mechanical states, the time and space dependence of the wavefunction is separable and one writes

$$\Psi(\mathbf{q}, t) = \psi(\mathbf{q})e^{iEt/\hbar} \tag{6.1.8}$$

where E is the energy of the stationary state. Now the ansatz (6.1.4) can be written as

$$\Psi(\mathbf{q}, t) = e^{iS(\mathbf{q})/\hbar - iEt/\hbar} \tag{6.1.9}$$

and (6.1.5) becomes

$$\frac{1}{2m}(\nabla S \cdot \nabla S) + V(\mathbf{q}) - E - \frac{i\hbar}{2m}\nabla^2 S = 0 \tag{6.1.10}$$

The $\hbar \to 0$ limit then gives the time-independent Hamilton–Jacobi equation

$$\frac{1}{2m}(\nabla S_0)^2 + V(\mathbf{q}) = E \tag{6.1.11}$$

where

$$S_0(\mathbf{q}) = \int_{\mathbf{q}_0}^{\mathbf{q}} \mathbf{p} \cdot \mathbf{dq} \tag{6.1.12}$$

with $\mathbf{q}_0$ being some initial point on a given trajectory. The formal $\hbar \to 0$ limit is equivalent to requiring that

$$(\nabla S_0)^2 \gg \hbar|\nabla^2 S_0| \tag{6.1.13}$$

and recalling that $\mathbf{p} = \nabla S$ (see (2.3.20a)), this inequality can be written as

$$|\mathbf{p}|^2 = \hbar|\nabla \cdot \mathbf{p}| \tag{6.1.14}$$

For one-dimensional problems, this condition has a simple interpretation in terms of the de Broglie wavelength $\lambda(q) = 2\pi\hbar/p(q)$, namely,

$$\frac{1}{2\pi}\left|\frac{d\lambda}{dq}\right| \ll 1 \tag{6.1.15}$$

which states that the approximation is only valid if the de Broglie wavelength varies slowly over distances of order of itself. It is a standard result (see, for example, the excellent text on quantum mechanics by Landau and Lifshitz (1965)) to further show that this condition breaks down near the classical turning points of the particle motion. In those regimes where (6.1.15) is valid, we clearly have a scheme for obtaining approximate solutions to the Schrödinger equation by using classical quantities. In quantum mechanics this is known as the *semiclassical* (*or quasi-classical*) approximation and is also often referred to as the WKB (Wentzel–Kramers–Brillouin) approximation.

6.2 THE WKB METHOD AND THE BOHR–SOMMERFELD QUANTIZATION CONDITION

The techniques that we are about to describe are not restricted to quantum mechanical problems but have wide applicability to solving (linear) differential equations that depend on a small parameter in a singular manner. For example, consider the simple boundary value problem

$$\epsilon y'' + y = 0, \qquad 0 \le x \le 1 \tag{6.2.1}$$

with boundary conditions $y(0) = 0$, $y(1) = 1$. The exact solution is easily determined to be

$$y(x) = \frac{\sin (x/\sqrt{\epsilon})}{\sin (1/\sqrt{\epsilon})} \tag{6.2.2}$$

(provided that $\epsilon \ne (n\pi)^{-2}$), which breaks down in the limit $\epsilon \to 0$. Clearly, this singular limit will make the solution of more general equations of the form

$$\epsilon y'' + V(x)y = 0$$

where $V(x)$ is some nontrivial function of x, quite difficult. A most useful procedure was developed by Rayleigh, Jefferies, and others (prior to WKB), who demonstrated that the solution could be written in the form of the exponential of a power series in ϵ, namely,

$$y(x) = \exp\left\{\frac{1}{\epsilon} \sum_{n=0}^{\infty} \epsilon^n S_n(x)\right\} \tag{6.2.3}$$

(An excellent discussion of this method is given by Bender and Orszag (1978).)

6.2.a The WKB Expansion

Here we concentrate on the quantum mechanical context in which the small parameter is identified as $\hbar$. Consider for now the one-dimensional, time-independent Schrödinger equation

$$\frac{\hbar^2}{2m}\frac{\partial^2 \psi}{\partial x^2} + (E - V(x))\psi + 0 \qquad (6.2.4)$$

and make the ansatz

$$\psi(x) = e^{iS(x)/\hbar} \qquad (6.2.5)$$

where $S(x)$ is the series

$$S(x) = S_0 + \hbar S_1 + \hbar^2 S_2 + \cdots \qquad (6.2.6)$$

Substituting (6.2.5) into (6.2.4) and equating successive powers of $\hbar$ leads to the hierarchy of equations

$$O(\hbar^0): \quad \left(\frac{\partial S_0}{\partial x}\right)^2 + 2m(E - V(x)) = 0 \qquad (6.2.7a)$$

$$O(\hbar^1): \quad \frac{\partial S_1}{\partial x}\frac{\partial S_0}{\partial x} + \frac{1}{2}\frac{\partial^2 S_0}{\partial x^2} = 0 \qquad (6.2.7b)$$

$$O(\hbar^2): \quad \left(\frac{\partial S_1}{\partial x}\right)^2 + 2\left(\frac{\partial S_0}{dx}\right)\left(\frac{\partial S_2}{\partial x}\right) + \frac{\partial^2 S_1}{\partial x^2} = 0 \qquad (6.2.7c)$$
$$\vdots$$

Recalling that $p = \partial S_0/\partial x$, the first two of these equations are easily solved to give

$$S_0(x) = \pm \int_{x_0}^{x} p(x')\,dx' = \pm \int_{x_0}^{x} \sqrt{2m(E - V(x'))}\,dx' \qquad (6.2.8)$$

and

$$S_1(x) = -\ln\sqrt{p(x)} + \ln c \qquad (6.2.9)$$

where c is a constant of integration and x_0 is some initial point on the trajectory. Expressions for $S_2, S_3, \ldots$ can easily be determined, although they rapidly become quite complicated algebraically. Using (6.2.8) and (6.2.9), the approximate wavefunction to first order in $\hbar$ is just

$$\psi(x) = \frac{A}{\sqrt{p}}\exp\left(+\frac{i}{\hbar}\int_{x_0}^{x} p(x')\,dx'\right) + \frac{B}{\sqrt{p}}\exp\left(-\frac{i}{\hbar}\int_{x_0}^{x} p(x')\,dx'\right) \qquad (6.2.10)$$

where the two roots of (6.2.8) provide the two linearly independent solutions to (6.2.4), which are combined with the arbitrary constants A and B to form the general solution (6.2.10). In the regions $E > V(x)$, $p(x)$ is positive; these are known as the *classically allowed* regions. For $E < V(x)$, the *classically forbidden* regions, $p(x)$ is imaginary. These regions have no meaning classically but quantum mechanically correspond, of course, to regions through which "tunneling" can occur. The solution (6.2.10) clearly breaks down in neighborhood of the classical turning points $E = V(x)$, that is, $p(x) = 0$. In the neighborhood of these regions, the probability density $|\psi(x)|^2$ becomes large. This is exactly in keeping with one's classical intuition that one is most likely to find a particle in the regions where it spends most time, namely, at the turning points (where it moves the slowest). The divergence of the solution (6.2.10) at the turning points is not an insuperable problem, and beautiful mathematical techniques—known as *uniform approximations*—have been developed to find solutions that pass smoothly from the allowed to disallowed regions without divergence. There is an important analogue to these classical divergences in geometric optics. These are the *caustics*, which correspond to the coalescence of rays which leads to points or regions of high intensity such as that observed in focusing (of light). The notion of caustics will play a significant part in our latter discussion of the properties of semiclassical wavefunctions.

6.2.b Bohr–Sommerfeld Quantization

The two exponentials in (6.2.10) can always be combined to give a real solution (in the classically allowed region) of the form

$$\psi(x) = \frac{A}{\sqrt{p}} \sin\left\{\frac{1}{\hbar} \int_{x_0}^{x} p(x')\, dx' + \alpha\right\} \tag{6.2.11}$$

where α is a certain phase factor. It may be shown that in order for the solution (6.2.11) to join smoothly with its (exponentially decaying) form to the left of the classical turning point x_0, the phase α must be $\pi/4$. Thus (6.2.11) becomes

$$\psi(x) = \frac{A}{\sqrt{p}} \sin\left\{\frac{1}{\hbar} \int_{a}^{x} p(x')\, dx' + \frac{\pi}{4}\right\} \tag{6.2.12}$$

where we have relabeled the turning point x_0 as the point a.

Now consider the situation in which the motion is bounded between the two classical turning points a and b, that is, $a \le x \le b$. The wavefunction (6.2.12) could thus also be written as

$$\psi(x) = -\frac{A}{\sqrt{p}} \sin\left\{\frac{1}{\hbar} \int_{x}^{b} p(x')\, dx' + \frac{\pi}{4} + \frac{1}{\hbar} \int_{b}^{a} p(x')\, dx' - \frac{\pi}{2}\right\} \tag{6.2.13}$$

which can be interpreted as taking the wavefunction around a path from x to the (right-hand) turning point b and back to a again. The phase in (6.2.13) illustrates a well-known result in geometric optics—namely, a wave experiences a phase loss of $\pi/2$ as it passes through a caustic (in this case the "round-trip" through the turning point b). The wavefunction (6.2.12) was written with respect to the (left-hand) turning point a—it could just as well have been written with respect to b, that is,

$$\psi(x) = \frac{B}{\sqrt{p}} \sin\left\{\frac{1}{\hbar} \int_x^b p(x') \, dx' + \frac{\pi}{4}\right\} \tag{6.2.14}$$

In order that the wavefunction be single-valued, the forms (6.2.12) and (6.2.14) must be the same, and from (6.2.13) it is easily seen that this will be so, provided that

$$\int_b^a p(x') \, dx' - \frac{\pi}{2} = n\pi, \qquad n = 0, 1, 2, \ldots \tag{6.2.15}$$

and $A = B(-1)^n$. The condition (6.2.15) can also be written as

$$\oint p(x') \, dx' = 2\pi\hbar(n + \tfrac{1}{2}) \tag{6.2.16}$$

where $\oint$ represents the integral around the closed path from a to b and back to a again. The reader will recognize the left-hand side of (6.2.16) as defining the classical action variable for one-dimensional bounded motion. The condition (6.2.16) is known as the *Bohr–Sommerfeld quantization rule*, which was devised in the framework of the "old quantum theory," in which the classical action variables were set equal to integral multiples of $\hbar$. At that time the $\frac{1}{2}$ was introduced as an empirical correction in order to match experimental results (spectroscopically measured energy levels). In the subsequent development of quantum mechanics this correction, the *zero-point energy*, arose naturally as a consequence of the uncertainty principle. Here we have seen (in the semiclassical picture) that it arises as a consequence of phase loss at a caustic.

For one-degree-of-freedom systems, the condition (6.2.16) explicitly determines the quantal eigenvalues E_n, that is,

$$\oint p(x') \, dx' = \oint \sqrt{2m(E_n - V(x'))} \, dx' = 2\pi\hbar(n + \tfrac{1}{2}) \tag{6.2.17}$$

provided that the integral can be inverted. For the case of a simple harmonic oscillator (i.e., $V(x) = \frac{1}{2}\omega^2 x^2$), (6.2.17) easily yields the result $E_n = (n + \frac{1}{2})\omega\hbar$. This is one of the rare cases where the semiclassical quantization condition agrees exactly with the quantum mechanical result. Such cases are known as *correspondence identities*.

6.3 SEMICLASSICAL QUANTIZATION FOR MANY DEGREES OF FREEDOM

In the old quantum theory there was much interest in attempting to generalize the Bohr–Sommerfeld rule to systems of many degrees of freedom (N). Initially it was thought that this could only be achieved if the system was separable (see Section 2.5). In this case each action variable, corresponding to each degree of freedom, was set to a multiple of $\hbar$, that is,

$$I_k = \oint_{\mathscr{C}_k} p_k \, dq_k = n_k \hbar \tag{6.3.1}$$

where $\mathscr{C}_k$ is the closed contour associated with the motion in the kth degree of freedom, and n_k is the corresponding quantum number. Assuming that an explicit transformation to action-angle variables was known, the energy levels were determined by the condition

$$E_{n_1, n_2, \ldots, n_N} = H(I_1 = n_1 \hbar, I_2 = n_2 \hbar, \ldots, I_N = n_N \hbar) \tag{6.3.2}$$

Note that there are as many quantum numbers as there are degrees of freedom and that in the "old" quantum theory there were no zero-point energy corrections.

6.3.a Einstein's Quantization Condition

It was quickly recognized (see Born's *The Mechanics of the Atom* (Born, 1960)) that such a procedure was not unique since if a system was separable in different coordinate systems (e.g., an isotropic 3-D harmonic oscillator is separable in both cartesian and polar coordinates), nonequivalent quantizations would be obtained. This difficulty was resolved in a remarkable paper by Einstein in 1917, in which he demonstrated that the actions must be defined according to Eq. (2.5.13), namely,

$$I_k = \frac{1}{2\pi} \oint_{\mathscr{C}_k} \sum_{l=1}^{N} p_l \, dq_l \tag{6.3.3}$$

that is, in terms of the invariant tori, before (6.3.2) could be used to find the eigenvalues (Einstein, 1917). Furthermore, Einstein appreciated that the $\mathscr{C}_k$ could not always be defined and that if the motion was in some sense ergodic, a "new" type of quantization condition would be required. Soon the "new" quantum theory (i.e., wave mechanics) made these issues redundant, and it was only recently that their significance—in light of our improved understanding of classical mechanics—was appreciated.

In the old quantum theory the quantization conditions (i.e., setting each action equal to an integral multiple of $\hbar$) did not include the "zero-point energy" of $\frac{1}{2}\hbar$ other than as an empirical correction. In the analysis of the one-dimensional WKB wavefunction, we saw that this term arose naturally as a phase loss due to caustics. This can be generalized to multidimensional problems once the corresponding wavefunction is correctly formulated.

6.3.b EBK Quantization

Again we look for a wavefunction of the form $\psi = Ae^{iS/\hbar}$ and to begin with assume a completely integrable, time-independent Hamiltonian system of N degrees of freedom, as described in Chapter 2, with the action function S

$$S(\mathbf{q}, \mathbf{I}) = \int_{\mathbf{q}_0}^{\mathbf{q}} \mathbf{p}(\mathbf{q}', \mathbf{I}) \, d\mathbf{q}' \tag{6.3.4}$$

where $\mathbf{q}_0$ is some (arbitrary) initial point. From (6.3.4) we recall the standard relationship between the conjugate variables, that is,

$$\boldsymbol{\theta} = \nabla_{\mathbf{I}} S(\mathbf{q}, \mathbf{I}) \quad \text{and} \quad \mathbf{p} = \nabla_{\mathbf{q}} S(\mathbf{q}, \mathbf{I}) \tag{6.3.5}$$

On the classical torus with actions $\mathbf{I}$, the classical orbits are distributed uniformly in $\boldsymbol{\theta}$. Thus the associated density of points in the configuration space $\mathbf{q}$ is the projection of the density on the torus onto $\mathbf{q}$ space, that is,

$$\frac{d\boldsymbol{\theta}}{d\mathbf{q}} = \det \left| \frac{\partial^2 S}{\partial q_j \partial I_k} \right|, \qquad j, k = 1, \ldots, N$$

Since the probability density of the wavefunction is $|\psi|^2$, we deduce that the amplitude A is

$$A = \det \left| \frac{\partial^2 S(\mathbf{q}, \mathbf{I})}{\partial q_j \partial I_k} \right|^{1/2} \tag{6.3.6}$$

a result first obtained by Van Vleck in 1928 (Van Vleck, 1928). In the case of a one-degree-of-freedom system, it is easy to see that $A \propto 1/\sqrt{p}$ in keeping with the WKB analysis. Finally, it is most important to recognize that S is a multivalued-valued function of $\mathbf{q}$, which follows from the fact that $\mathbf{p}$ is now multivalued—for example, in the case of one-dimensional bounded motion (see Figure 6.1a), p is a two-valued function of q, that is,

$$p(q, I) = \pm \sqrt{2m(H(I) - V(q))} \tag{6.3.7}$$

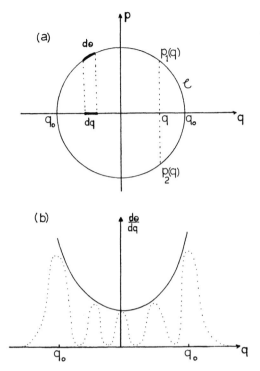

Figure 6.1 (a) Phase plane for one-dimensional bounded motion showing typical constant energy curve $\mathscr{C}$. Momentum p is a two-valued function of q; the two branches $p_1(q)$ and $p_2(q)$ coalesce at classical turning points q_0. Turning points are characterized as those points on $\mathscr{C}$ whose tangents are parallel to p-axis. (b) Projection of $\mathscr{C}$ (given by $d\theta/dq$) onto q-axis, giving smooth envelope of $|\psi(q)|^2$; projection is singular at classical turning points.

Thus a wavefunction of the form $\psi = Ae^{iS/\hbar}$ must be summed over all possible branches of S, that is,

$$\psi(\mathbf{q}) = \sum_r \det\left|\frac{\partial^2 S_r}{\partial q_j \partial I_k}\right|^{1/2} e^{iS_r(\mathbf{q}, \mathbf{I})/\hbar} \qquad (6.3.8)$$

where the sum over r denotes the sum over branches. (The WKB wavefunction (6.2.10) is just the two-branch sum associated with one-dimensional bounded motion.)

In order that the wavefunction (6.3.8) be single-valued, the total phase change on completing one classical "circuit" must be an integral multiple of 2π. For an N-dimensional torus there are N topologically distinct circuits $\mathscr{C}_k$ ($k = 1, \ldots, N$) whose traversal will return one back to the same point. Furthermore, in traversing $\mathscr{C}_k$, one may pass through caustics—each one leading to a phase loss of $\pi/2$. Thus the single-valuedness condition takes the form

$$\frac{1}{\hbar}\oint_{\mathscr{C}_k} \mathbf{p}(\mathbf{q}', \mathbf{I})\, d\mathbf{q}' - \alpha_k \frac{\pi}{2} = 2n_k\pi \qquad (6.3.9)$$

where α_k is the number of caustics traversed. These numbers are usually referred to as the *Maslov indices*. Thus the general multidimensional quantization condition is of the form

$$I_k = \oint_{\mathscr{C}_k} \mathbf{p} \cdot d\mathbf{q} = 2\pi\hbar\left[n_k + \frac{\alpha_k}{4}\right] \qquad (6.3.10)$$

which is usually termed the EBK (Einstein–Brillouin–Keller) quantization rule.

6.3.c Semiclassical Wavepackets*

Interestingly, wavefunctions of the form (6.3.8) need not be confined to the description of stationary states. A wavefunction can be associated with any smooth N-dimensional surface (call it Σ) embedded in the $2N$ phase space (see Figure 6.2). Such a surface in $(\mathbf{p}, \mathbf{q})$ space can be represented as some function of the form $\mathbf{p}(\mathbf{q})$. The only restriction is that this dependence must

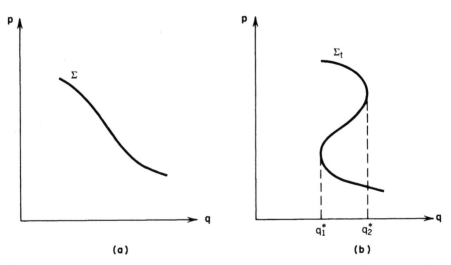

Figure 6.2 (a) Lagrangian manifold Σ in $(\mathbf{p}, \mathbf{q})$ space. Points on Σ are labeled by new canonical variables Q. (b) Evolving Lagrangian manifold Σ_t showing appearance of two caustics q_1^* and q_2^*.

be given in the form of a gradient, that is,

$$\mathbf{p}(\mathbf{q}) = \nabla_{\mathbf{q}} S \qquad (6.3.11)$$

where S is a certain function to be defined shortly. For any smooth function S, the manifold Σ thus constructed is termed a *Lagrangian manifold*. From (6.3.11) we note that Σ has the Lagrangian property

$$\frac{\partial p_i}{\partial q_j} = \frac{\partial p_j}{\partial q_i} \qquad (6.3.12)$$

In the same way that angle variables are used to coordinatize position on tori, we introduce a set of variables $Q_1, \ldots, Q_N$ to label points on the Lagrangian manifold. It then becomes possible to introduce a set of canonical "momenta" P_i ($i = 1, \ldots, N$) conjugate to the Q_i and identify the function S in (6.3.11) as the generating function for the canonical transformation between $(\mathbf{p}, \mathbf{q})$ and $(\mathbf{P}, \mathbf{Q})$ variables, that is,

$$\mathbf{p} = \nabla_{\mathbf{q}} S(\mathbf{q}, \mathbf{P}) \quad \text{and} \quad \mathbf{Q} = \nabla_{\mathbf{P}} S(\mathbf{q}, \mathbf{P}) \qquad (6.3.13)$$

Using the previously given arguments for identifying particle density in $\mathbf{q}$ space (now the projection $|d\mathbf{Q}/d\mathbf{q}|$), the wavefunction associated with the Lagrangian manifold Σ is just

$$\psi(\mathbf{q}) = \det \left| \frac{\partial^2 S}{\partial q_i \partial P_j} \right|^{1/2} e^{(i/\hbar) S(\mathbf{q}, \mathbf{P})} \qquad (6.3.14)$$

where for now we are assuming $\mathbf{p}$ to be a single-valued function of $\mathbf{Q}$. We note that this form of wavefunction is only dependent on our ability to identify a Lagrangian manifold which is *not* restricted to integrable systems.†

Typically, Σ will not be a stationary manifold and will evolve in time. The evolution of Σ (denoted as Σ_t) under the Hamiltonian flow will typically be very complicated. In a 2-D phase space, Σ_t will develop whorls and tendrils; in higher dimensions, the morphologies will only be more complicated. Furthermore, p will typically become an ever more multivalued function of q (see Figure 6.2), and more caustics will appear. Eventually, the proliferation of caustics will lead to a breakdown of the validity of the wavefunction (6.3.14). This very significant point will be discussed in Section 6.6.

---

† The invariant tori of integrable systems are Lagrangian manifolds—but with the important property that they are *stationary*.

6.4 REGULAR AND IRREGULAR SPECTRA: EIGENVALUE-RELATED PROPERTIES

For nonintegrable Hamiltonian systems the KAM theorem tells us that more and more tori are "destroyed" as the (nonintegrable) perturbation is increased. For strongly nonintegrable systems the motion is predominantly chaotic and few tori persist. For most bound states at these energies the EBK rules are clearly no longer applicable; as already mentioned, this problem was appreciated long ago by Einstein. The full significance of this difficulty was, however, only recognized more recently by Percival (1977), who proposed that in the *semiclassical limit* $\hbar \to 0$, the bound-state energy spectrum† might be divided into two parts:

1. A *regular spectrum* corresponding to regimes of integrable (regular) motion in which all states can be quantized according to the EBK rules.
2. An *irregular spectrum* corresponding to regimes of predominantly chaotic (irregular) motion in which EBK is no longer applicable.

The two different classes of spectra may have very different properties, reflecting the differences in the "underlying" classical motion. The notion of an irregular spectrum is particularly exciting since it suggests, in some sense, that there will be a manifestation, in the limit $\hbar \to 0$, of the "underlying" classical chaos in the quantum mechanical properties of a system. This possibility has even led some researchers to invoke notions of "quantum chaos," which are not always associated with the limit $\hbar \to 0$. However, everything that we discuss here is strictly in the context of the semiclassical limit, and any notion of "quantum chaos" implies an association with underlying classical chaos.

Our discussion of the contrasting properties of the regular and irregular spectra is conveniently divided into two parts. The first, covered in this section, is primarily concerned with eigenvalue-related properties; the second, covered in the next section, is primarily concerned with eigenvector(wavefunction)-related properties.

6.4.a Regular and Irregular Bound States

The fundamental distinction between a regular and irregular (semiclassical) state is that the former may be labeled by a complete set of "good" quantum numbers $\mathbf{n} = (n_1, \ldots, n_N)$, where N is the number of degrees of freedom. Thus a state with quantum numbers $\mathbf{n}$ may be associated with that family of trajectories lying on the N-dimensional torus with constant actions given by the EBK rules (6.3.10). Thus there is a one-to-one cor-

---

† The term *spectrum* is here simply taken to mean a set of energy levels.

respondence between a regular state and a classical torus. By contrast, for a state of the irregular spectrum, there is not only no meaningful assignment of "good" quantum numbers but also no clear association between such a state and a given region of phase space, other than it should "occupy" a volume of order $(2\pi\hbar)^N$. Various workers have, nonetheless, suggested that by using certain classical perturbation techniques in the chaotic regimes "approximate" tori can still be constructed (ultimately, of course, such perturbation series must diverge in these regimes) and used for EBK quantization purposes.† In the model systems tested, the energy levels thus obtained agreed quite well with the exact quantal calculations. These results touch on a very important point—namely, that the systems studied did not contain many bound states and therefore corresponded to a situation in which $\hbar$ is relatively "large." Clearly, if the regions of phase space occupied by chaotic motion are much smaller than $O((2\pi\hbar)^N)$, that is, the "size" of a quantum state, they will have little significance quantum mechanically. Thus these results suggest that $\hbar$ plays some sort of "smoothing" role that can patch up otherwise destroyed tori. Clearly, however, once $\hbar$ becomes sufficiently small, this approach is no longer applicable and one is faced with the fundamental issue of trying to find a semiclassical quantization condition for chaotic systems. As yet, no "direct" scheme has been found, although an "indirect" method using the classical periodic orbits is available. This important technique will be introduced in the context of "quantum maps" in Section 6.7.

This lack of "good" quantum numbers for the irregular states has led Percival (1977) to conjecture that the regular and irregular spectra will be distinguishable by means of their transition probabilities. Transitions between states of the regular spectrum will be characterized by strong selection rules; that is, one expects to see a spectrum (in the sense of a spectroscopic observable) of just a few intense lines corresponding to strongly coupled states. On the other hand, states of the irregular (energy) spectrum are expected to be coupled with similar intensity to all those states of a similar energy that correspond to the same "chaotic" regions of phase space; that is, one expects to see a spectrum of many weak lines. There are, as yet, no reported experimental observations that unambiguously indicate the existence of an irregular spectrum.

6.4.b Power Spectra and the Correspondence Principle

The correspondence between classical power spectra (as described in Chapter 4) and quantal (transition) spectra is easily understood in the regular regime. Consider the two EBK states $E_n = H(\mathbf{I} = \mathbf{n}\hbar)$ and $E_m = H(\mathbf{I} = \mathbf{m}\hbar)$, where for convenience we drop the Maslov terms. For E_m sufficiently close to E_n we may expand the former about the latter in a

†See Jaffé and Reinhardt (1979).

Taylor series to obtain, to first order,

$$E_m = H(n\hbar) + \hbar(m-n) \cdot [\nabla_I H(I)]_{I=n\hbar} + \cdots \tag{6.4.1}$$

The quantal transition frequency ω_{mn} is therefore given by

$$\omega_{mn} = \frac{(E_m - E_n)}{\hbar} \simeq (m-n) \cdot \omega(n\hbar) \tag{6.4.2}$$

where we use (2.5.15b). Thus the power spectrum of a classical trajectory belonging to the torus with action $I = n\hbar$ will have lines that correspond (approximately) to the $n \to m$ quantal transition. In the limit $\hbar \to 0$ or $|n| \gg |n-m|$ the classical and quantal frequencies become equal. Furthermore, the squared moduli of the classical Fourier coefficients correspond to the quantal transition probabilities. In practice (i.e., for finite $\hbar$), the best agreement between classical and quantal spectra is often obtained by comparing the quantal spectrum for $n \to m$ transition with the classical spectrum of a trajectory lying on the torus with the "mean action" $I = \hbar(m+n)/2$ rather than the "initial state action" $I = n\hbar$.

In contrast to the regular regime, the power spectra of irregular trajectories are immensely complicated and display, essentially, an infinity of lines. It is not at all clear at this stage how one might compare such spectra with the quantal spectrum. It may well be that a meaningful comparison is only possible when both spectra are averaged over some range of trajectories and states, respectively. Much more consideration of a "correspondence principle" for the irregular regime is now required.

6.4.c Sensitivity to Perturbation

Percival (1977) has also predicted that regular and irregular states will be distinguishable by their behavior under perturbation. Irregular states will be very sensitive to an external or slowly varying perturbation—in some sense reflecting the sensitivity to initial conditions of chaotic classical orbits—whereas regular states will be relatively stable. This conjecture was first tested by Pomphrey (1974), who investigated the eigenstates of the Henon–Heiles-type Hamiltonian

$$H = \tfrac{1}{2}(p_x^2 + p_y^2 + x^2 + y^2) + \lambda(x^2 y - \tfrac{1}{3}y^3) \tag{6.4.3}$$

using a value of $\lambda = 0.088$. The quantity calculated was the "second difference," $\Delta^2 E_i$, where

$$\Delta^2 E_i = E_i(\lambda + \Delta\lambda) + 2E_i(\lambda) - E_i(\lambda - \Delta\lambda) \tag{6.4.4}$$

which gives a measure of the sensitivity of the ith eigenvalue to small changes ($\Delta\lambda$) in the perturbation. A number of very large second differences were found for states with energies in the predominantly chaotic regime. This would appear to confirm Percival's prediction, and other studies have backed up these findings. In particular, we mention one by Noid et al. (1980), which again worked with Hamiltonian (6.4.3) but now with $\lambda = 0.1118$. Owing to the symmetry of the potential, the eigenvalues can have either A (nondegenerate) or E (doubly degenerate) symmetry. Furthermore, each state can be assigned a principal quantum number and an approximate "angular-momentum" quantum number. It was found that high-angular-momentum states all had small $\Delta^2 E_i$, whereas low-angular-momentum states had consistently larger $\Delta^2 E_i$. (This phenomenon has also been noted by McDonald and Kaufman (1979, 1988) in their study of the eigenstates of the "stadium" described below.) This behavior is consistent with the underlying classical dynamics. The high-angular-momentum states could all be associated with tori (i.e., stable motion) and hence quantized by EBK, even in the predominantly chaotic regime. On the other hand, the low-angular-momentum states, when they could be computed semiclassically, were found to be associated with those tori that were the first to be destroyed at higher energy. This study also revealed another interesting feature, namely, the possibility of both level "crossings" and "avoided crossings." At high energies and where symmetry permitted, a number of states were found to cross as a function of λ. If this effect were not taken into account, spuriously large values of $\Delta^2 E_i$ would have been calculated.

6.4.d   Level-Spacing Distributions

Berry (1983) has emphasized the importance of examining energy spectra on different scales of $\hbar$ in order to fully reveal the differences between regular and irregular spectra. On the finest scales, one is identifying individual eigenvalues. This can be achieved for regular states by means of the EBK rules but, as discussed above, is not yet possible for irregular states. By contrast, on the largest scales, one is only able to characterize the spectrum in terms of an average *density of states* $\bar{\rho}(E)$ given by the Thomas–Fermi formula

$$\bar{\rho}(E) = \frac{1}{(2\pi\hbar)^N} \int \int \delta(E - H(\mathbf{p}, \mathbf{q})) \, \mathbf{dp} \, \mathbf{dq} \qquad (6.4.5)$$

which is just the Liouville measure of classical phase space at energy E divided by the statistical volume $(2\pi\hbar)^N$ "occupied" by a quantum state. Clearly, at this coarsest scale there is no means of distinguishing between the effects of regular and irregular classical motion. At intermediate scales of resolution, the various clusterings and distributions of the energy levels reveal much about the underlying motion.

At these intermediate scales, one of the most actively investigated aspects of the energy spectrum has been the statistics of level-spacing distributions. These are the properties of the spectrum on scales of the order of the mean level spacing, which, from (6.4.5), is seen to be of order $\hbar^N$, that is, $(\bar{\rho})^{-1}$. An overview of this topic has been given by Berry in the 1987 Bakerian Lecture of the Royal Society (Berry, 1987). The most significant quantity is the *nearest-neighbor level-spacing distribution* characterized by the probability function $P(s)\, ds$, which gives the probability that the spacing of a pair of neighboring (in energy) levels lies between s and $s + ds$. This quantity was much used in the study of the statistical properties of nuclear energy levels which have a very high density of states. By assuming that the nuclear energy levels could be modeled by the eigenvalues of random matrices whose elements are drawn from a Gaussian ensemble, Wigner (1957) made the remarkable conjecture that

$$P(s) = s(\pi/2)e^{-(\pi/4)s^2} \tag{6.4.6}$$

This result was subsequently shown to be in almost perfect agreement with the exact solution to this model (the derivation of which proved to be a formidable task in mathematical physics; see Mehta (1967)). Experimental data on nuclear energy levels amply confirmed Wigner's prediction. Important points to note are that $P(s) \to 0$ as $s \to 0$, which implies some form of level "repulsion" or "avoided crossing," and that (nuclear) energy levels can belong to different symmetry classes, whereas the Wigner distribution (6.4.6) only applies to each such class. Thus if the distributions from different classes are mixed together, a Poisson distribution, that is,

$$P(s) = e^{-s} \tag{6.4.7}$$

results which corresponds to a completely random, uncorrelated organization of levels.

To study the level-spacing statistics of the eigenvalues of a *single* Hamiltonian, one can create an "ensemble" by taking the limit $\hbar \to 0$ such that the density of states tends to infinity at a given energy E. In this way one can take meaningful statistical samples of the level spacings (suitably "normalized" by the mean spacing) in a small energy band about E.

For completely integrable systems (i.e., ones for which all states can be quantized by EBK), it has been proved (Berry and Tabor 1977) that $P(s)$ takes the Poissonian form (6.4.7), provided that the Hamiltonian is nondegenerate, that is,

$$\det\left|\frac{\partial^2 H}{\partial I_i \partial I_j}\right| \neq 0 \tag{6.4.8}$$

Thus the most probable level spacing is zero, implying a strong clustering of levels. That this is so should not be too surprising. The regular states have a full complement of good quantum numbers and can therefore form strongly correlated *sequences* in quantum number, that is, $(n_1 + 1, n_2, \ldots, n_N)$, $(n_1 + 2, n_2, \ldots, n_N)$, $(n_1 + 3, n_2, \ldots, n_N)$, ... It is these sequences that can lead to Percival's strong selection rules. However, from the point of view of energy spacings, all these sequences will be jumbled up and little correlation is to be expected. In the case of degenerate Hamiltonians, such as systems of harmonic oscillators, different distributions are obtained that depend on subtle number-theoretic properties of the fundamental frequencies.

For nonintegrable systems there has been much speculation as to the form of $P(s)$, with the anticipation that it should be more like the Wigner distribution, thereby, in some sense, reflecting the "randomness" of the irregular spectrum. Many numerical studies have been performed, all of which indicate a transition from Poissonian behavior in integrable regimes to some Wigner-like distribution, which tends to zero as $s \to 0$, in chaotic regimes. Various workers have given arguments to support the claim that

$$\lim_{s \to 0} P(s) \sim s^\gamma \qquad (6.4.9)$$

where γ is some exponent. Berry (1981) has argued that $\gamma = m - 1$, where m is the number of parameters of the system that must be varied in order to produce degeneracies of energy levels. Although the idea of a connection between the semiclassical irregular spectrum and the eigenvalue spectra associated with various ensembles of matrices is very tantalizing, a rigorous demonstration of this has yet to be made. It is also not clear whether level-spacing distributions can be obtained with sufficient accuracy experimentally to enable one to make an unambiguous identification of a regular or irregular spectrum.

6.4.e Spectral Rigidity*

Another statistical characterization of level spacings, which can distinguish between integrable and chaotic systems, is that of *spectral rigidity*. An energy spectrum can be described in terms of the "spectral staircase" function

$$\mathcal{N}(E) = \sum_n \Theta(E - E_n) \qquad (6.4.10)$$

where Θ is the unit step function and the E_n are a sequence of ordered (in energy) eigenvalues. Thus for increasing E, $\mathcal{N}(E)$ exhibits a step of unit height at each eigenvalue E_n. The derivative of $\mathcal{N}(E)$ is the density of states, that is,

$$\rho(E) = \frac{d\mathcal{N}}{dE}(E) = \sum_n \delta(E - E_n) \qquad (6.4.11)$$

whose phase-space average, $\bar{\rho}(E)$, is just the Thomas–Fermi result (6.4.5). The average of $\mathcal{N}(E)$, that is,

$$\bar{\mathcal{N}}(E) = \frac{1}{(2\pi\hbar)^N} \int \int \Theta(E - H(\mathbf{p}, \mathbf{q})) \, \mathbf{dp} \, \mathbf{dq} \qquad (6.4.12)$$

is the total phase-space volume up to energy E divided by $(2\pi\hbar)^N$ and gives the total number of states below energy E.

The spectral rigidity $\Delta(L)$ is defined as the local average of the mean square deviation of the staircase function $\mathcal{N}(E)$ from the best-fitting straight line over a range of energy corresponding to L mean level spacings (recall from (6.4.5) that the mean level spacing is $(\bar{\rho})^{-1}$), that is,

$$\Delta(L) = \left\langle \min_{A,B} \frac{\bar{\rho}(E)}{L} \int_{-L/2\bar{\rho}(E)}^{L/2\bar{\rho}(E)} d\epsilon [\mathcal{N}(E + \epsilon) - A - B\epsilon]^2 \right\rangle \qquad (6.4.13)$$

A semiclassical analysis of (6.4.13) by Berry (1985) has shown that $\Delta(L)$ can be expressed as a sum of contributions from the closed (i.e., periodic) classical orbits of the system, with the very long orbits playing the most significant role. (A more detailed discussion of the role of periodic orbits in describing the semiclassical limit of the energy spectrum is given in Section 6.7.) His results show that over a certain range of L, (i) $\Delta(L) = L/15$ if the system is classically integrable, (ii) $\Delta(L) = \ln L/\pi^2 + E$ if the system is classically chaotic with time-reversal symmetry, and (iii) $\Delta(L) = \ln L/2\pi^2 + D$ if the system is classically chaotic and does not have time-reversal symmetry. Here D and E are certain constants. For very large values of L, $\Delta(L)$ behaves in a nonuniversal manner. For more details of these important results, the reader is referred to Berry's original paper (Berry, 1985).

6.5 REGULAR AND IRREGULAR SPECTRA: EIGENVECTOR-RELATED PROPERTIES

In this section we examine various properties of the wavefunctions of regular and irregular states.

6.5.a Regular Bound-State Wavefunctions

For regular states the semiclassical wavefunction takes the form (6.3.8). Thus for a state with quantum numbers $\mathbf{n}$, we write $\psi_{\mathbf{n}}(\mathbf{q})$ (except for certain phase factors) as

$$\psi_{\mathbf{n}}(\mathbf{q}) = \sum_r \det \left| \frac{\partial^2 S_r(\mathbf{q}, \mathbf{I_n})}{\partial q_j \partial I_k} \right|^{1/2} e^{iS_r(\mathbf{q}, \mathbf{I_n})/\hbar} \qquad (6.5.1)$$

where I_n identifies the phase-space torus with actions $I = (n + \alpha/4)\hbar$. The probability density $|\psi_n|^2$ evaluated from (6.5.1) will include oscillatory cross-terms corresponding to interference between the different branches of S. These terms can be eliminated by a local averaging over some width Δ which vanishes more slowly than $\hbar$ as $\hbar \to 0$. This leads to a "coarse graining" of the form

$$\bar{f}(q) = \frac{1}{\Delta} \int_{q-\Delta/2}^{q+\Delta/2} f(q') \, dq' \qquad (6.5.2)$$

for a given function $f(q)$, where

$$\lim_{\hbar \to 0} \Delta = 0 \quad \text{and} \quad \lim_{\hbar \to 0} \frac{\Delta}{\hbar} = 0$$

Thus the coarse-grained probability density is

$$\overline{|\psi_n(\mathbf{q})|^2} = \sum_r \left| \frac{\partial^2 S_r}{\partial q_j \partial I_k} \right| \qquad (6.5.3)$$

This has the geometrical interpretation of corresponding to the projection of the torus associated with the nth state onto the coordinate plane (see Section 6.3). As a simple example, consider one-dimensional bounded motion for which

$$\overline{|\psi(q)|^2} = \left| \frac{\partial^2 S}{\partial q \partial I} \right| = \left| \frac{d\theta}{dq} \right| \qquad (6.5.4)$$

This coarse-grained form of $|\psi(q)|^2$, sketched in Figure 6.1b, is singular at the classical turning points; these singularities correspond, as discussed in the previous section, to caustics. However, this form of $\overline{|\psi|^2}$ provides, in the limit $\hbar \to 0$, the envelope of the oscillations of the true quantal probability density (in the classically allowed region).

So far, though, everything we have said applies only to the wavefunctions of the regular states. Defining a semiclassical form of the wavefunction for irregular states is much more difficult. The problem is that for irregular trajectories, p is no longer a finitely multivalued function of q; instead, it has infinitely many branches. (Alternatively, one may say that in the chaotic regime, no global solution to the Hamilton–Jacobi equation exists.)

6.5.b The Wigner Function

To be able to compare the semiclassical forms of the regular and irregular states, it has been suggested that the use of the Wigner function may provide a convenient alternative to the study of the wavefunctions themselves. The Wigner function provides a quantal analogue to classical phase-space density. It takes the form

$$W(\mathbf{p}, \mathbf{q}) = \frac{1}{(\pi \hbar)^N} \int d\mathbf{x} \; e^{-2i\mathbf{p} \cdot \mathbf{x}/\hbar} \psi(\mathbf{q} + \mathbf{x}) \psi^*(\mathbf{q} - \mathbf{x}) \qquad (6.5.5)$$

Examination of $W(\mathbf{p}, \mathbf{q})$ in a particular phase plane (p_i, q_i) provides one with a quantal analogue to the Poincaré surface of section. The Wigner function has many interesting properties, but for our purposes the most important is that its projection onto the coordinate plane gives the quantal probability density, that is,

$$\int W(\mathbf{p}, \mathbf{q}) \, d\mathbf{p} = |\psi(\mathbf{q})|^2 \qquad (6.5.6)$$

In the case of regular states the semiclassical form of the wavefunction $\psi_m(\mathbf{q})$ can be used to evaluate the associated pure state Wigner function $W_m(\mathbf{p}, \mathbf{q})$. It may then be shown (Berry, 1977a) that in the classical limit, $\hbar = 0$, $W_m(\mathbf{p}, \mathbf{q})$ reduces to

$$\bar{W}_m(\mathbf{p}, \mathbf{q}) = \frac{1}{(2\pi)^N} \delta(\mathbf{I}(\mathbf{p}, \mathbf{q}) - \mathbf{I}_m) \qquad (6.5.7)$$

that is, the Wigner function "collapses" onto a delta function on the classical torus associated with the **m**th state. As described above, projection of this torus onto the coordinate plane gives the limiting form of the probability density. The results of these projections depend on the "orientation" of the torus in phase space. This can lead to a variety of different caustic structures. There are subtle differences between the results obtained for separable and nonseparable integrable systems.

For finite $\hbar$ (i.e., in the semiclassical limit), the Wigner function displays a regular structure of "diffraction fringes." Projection of this form of $W_m(\mathbf{p}, \mathbf{q})$ onto the coordinate plane yields the correct oscillatory form of $|\psi_m(\mathbf{q})|^2$. Comparison of the classical and semiclassical limiting forms of $W_m(\mathbf{p}, \mathbf{q})$ shows that for regular states, the role of $\hbar$ is to add a regular structure (the quantum oscillations) onto a smooth classical background. In Figure 6.3 we show a Wigner surface of section, for a regular state of a Henon–Heiles-type system, calculated by Hutchinson and Wyatt (1980).

An investigation of the Wigner function for irregular states might again appear to be frustrated by our lack of knowledge of a semiclassical form for the wavefunction. However, some progress can be made by adopting a slightly different point of view. For regular states, the classical limit of the Wigner function is the phase-space manifold (the torus) with which the state

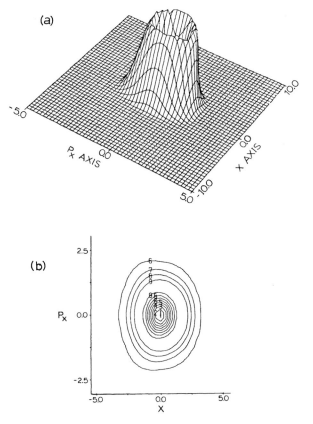

Figure 6.3 Wigner distribution (p_x, x)-plane of 10th eigenstate of the Hamiltonian $H = \frac{1}{2}(p_x^2 + p_y^2 + 0.49x^2 + 1.69y^2) - 0.10(xy^2 - x^3)$. (a) Perspective and (b) contours showing smooth concentric pattern of phase-space density. (Reproduced, with permission, from Hutchinson and Wyatt (1980).)

is associated. In the extreme chaotic regime, an irregular state is probably associated with a large portion of the corresponding energy shell. This being so, a reasonable conjecture for the classical limit of the Wigner function is that it be the (normalized) microcanonical distribution (see Section 6.5.f), that is,

$$W(\mathbf{p}, \mathbf{q}) = \frac{\delta(E - H(\mathbf{p}, \mathbf{q}))}{\int d\mathbf{p} \int d\mathbf{q}\ \delta(E - H(\mathbf{p}, \mathbf{q}))} \qquad (6.5.8)$$

For finite $\hbar$, although the exact form of W is not known, one anticipates that a surface of section of W would display a random splatter of phase-space density. This is analogous to the Poincaré surface of section observed for irregular trajectories.

Working with the above form of W, one can investigate the limiting form of the associated $|\psi(\mathbf{q})|^2$ by use of (6.5.6). This has been done by Berry (1977b), who shows that for systems of two or more degrees of freedom, $|\psi(\mathbf{q})|^2$ vanishes at the classical boundaries. This "anticaustic" structure is in sharp contrast to the caustic structure found for regular states.

6.5.c Spatial Correlations in Wave Functions

The Wigner function can also be used to provide information about the wavefunctions themselves rather than just about their squared moduli. A spatial autocorrelation function for the state $\psi(\mathbf{q})$ can be defined as

$$C(\mathbf{x}, \mathbf{q}) = \frac{\overline{\psi(\mathbf{q}+\mathbf{x})\psi^*(\mathbf{q}-\mathbf{x})}}{|\psi(\mathbf{q})|^2} \tag{6.5.9}$$

where the overbars denote, as before, a local averaging. There is a simple relationship between $C(\mathbf{x}, \mathbf{q})$ and $W(\mathbf{p}, \mathbf{q})$, namely,

$$C(\mathbf{x}, \mathbf{q}) = \frac{\int d\mathbf{p}\, \bar{W}(\mathbf{p}, \mathbf{q}) e^{2i\mathbf{p}\cdot\mathbf{x}/\hbar}}{|\psi(\mathbf{q})|^2} \tag{6.5.10}$$

where $\bar{W}(\mathbf{p}, \mathbf{q})$ is the coarse-grained Wigner function, that is, its classical limiting form. The behavior of $C(\mathbf{x}, \mathbf{q})$ has been investigated for both regular and irregular states. It is concluded that for regular states, $C(\mathbf{x}, \mathbf{q})$ is anisotropic, whereas for irregular states, it is isotropic—taking the form of a Bessel function for certain forms of potential. Overall, regular states are expected to exhibit strong, anisotropic interference oscillations with just a few scales of oscillations. In the case of irregular states, ψ should exhibit more moderate, spatially isotropic oscillations with a continuous spectrum of wavevectors $(\mathbf{p}/\hbar)$, that is, oscillations on all scales. This would imply that $\psi(\mathbf{q})$ is a Gaussian random function of $\mathbf{q}$.

The role played by $\hbar$ in the chaotic regime is very different from that played in the regular regime. Chaotic classical dynamics displays structure down to arbitrarily fine scales. Here $\hbar$ "smooths away" this fine structure, and irregular states display structure only down to scales of order $\hbar$. Again we emphasize that the notion of an "irregular state" is a semiclassical one, namely, an $\hbar \to 0$ phenomenon. However strong the nonintegrable perturbation, if $\hbar$ is not sufficiently "small," one should not expect to observe irregular states. Indeed, the interplay between the limits $\hbar \to 0$ and $\epsilon \to 0$, where ϵ is the (nonintegrable) perturbation parameter, is highly nontrivial and can lead to a variety of different regimes of regular and irregular spectra.

6.5.d Some Numerical Results

With the above remarks in mind, we now turn to some numerical investigations of wavefunctions in the regular and irregular regimes. Noid et al. (1979) have computed the exact quantal probability density for states of a Henon–Heiles-type system (with the fundamental frequencies in the ratio 2:1 rather than 1:1). In Figure 6.4 we show their $|\psi(\mathbf{q})|^2$ computed for a regular state. A classical trajectory belonging to the associated (EBK) torus is also shown. The probability density is clearly confined to the region enclosed by the "boxlike" caustic structure of the associated trajectory.

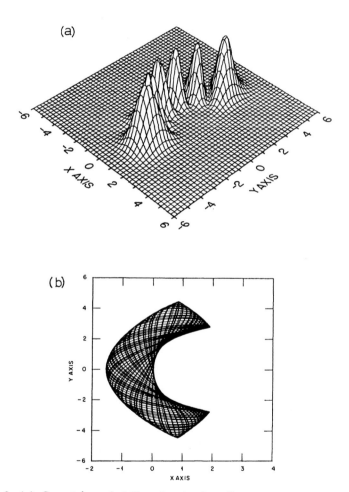

Figure 6.4 (a) Quantal probability density $|\psi(\mathbf{q})|^2$ of "regular state" of the Hamiltonian $H = \frac{1}{2}(p_x^2 + p_y^2 + 1.96x^2 + 0.49y^2) - 0.08(xy^2 - 0.08y^3)$ with eigenvalue $E = 4.265$. (b) Trajectory of orbit belonging to associated EBK torus. (Reproduced, with permission, from Noid et al. (1979).)

From the regularity of the oscillations of $|\psi(\mathbf{q})|^2$, it is clear that $\psi(\mathbf{q})$ will have a regular pattern of nodal lines. In Figure 6.5 we show $|\psi(\mathbf{q})|^2$ computed for a state at a higher energy and for which EBK quantization was not possible. A typical irregular trajectory at that energy is also shown. For this state, $|\psi(\mathbf{q})|^2$ appears to have spread, like the trajectory, over most of the classically allowed configuration space. The whole structure of the wavefunction is now much less regular than before.

Another interesting study is that made of the wavefunctions of the "stadium" by McDonald and Kaufman (1979). However, it should be noted that for this system the potential is infinite at the classical boundaries and hence $\psi(\mathbf{q})$ must go to zero there (i.e., the "anticaustic" conjecture cannot be tested). For aspect ratio $\gamma = 0$ the "stadium" reduces to a circle. In

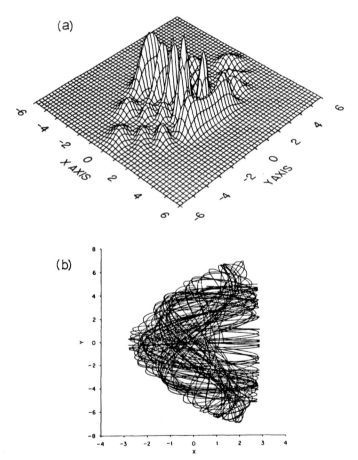

Figure 6.5 (a) Quantal probability density $|\psi(\mathbf{q})|^2$ of an "irregular state" at energy $E = 8.0$ of Hamiltonian used in Figure 6.4 (b) Trajectory of typical irregular orbit at same energy. (Reproduced, with permission, from Noid et al. (1979).)

(a)

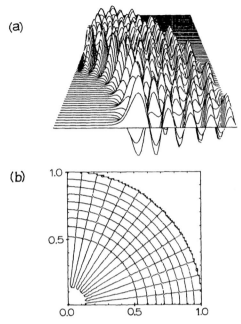

(b)

Figure 6.6 (a) Perspective of one quadrant of wavefunction amplitude $\psi(\mathbf{q})$ for state with eigenvalue $k = 65.38142$ of stadium with aspect ratio $\gamma = 0$ (i.e., circle). (b) Nodal structure of this state. Apparent noncrossing of nodal lines is due to computer graphics. (Reproduced, with permission, from McDonald and Kaufman (1979).)

(a)

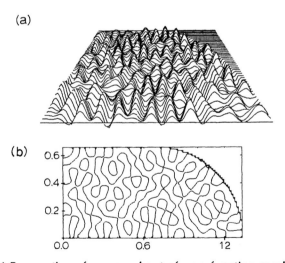

(b)

Figure 6.7 (a) Perspective of one quadrant of wavefunction amplitude $\psi(\mathbf{q})$ for state with eigenvalue $k = 65.036$ of stadium with $\gamma = 1$. (b) Nodal structure of this state—in this case there are actually no crossings of nodal lines. (Reproduced, with permission, from McDonald and Kaufman (1979).)

Figure 6.6 we show McDonald and Kaufman's computations of a wavefunction for this case. The amplitude shows regular, strongly directional oscillations and rises to a maximum around an inner circle that corresponds to an underlying caustic. The nodal structure displays a regular pattern of intersecting nodal lines. In Figure 6.7 we show the results obtained for a state of almost the same energy but now in a stadium with $\gamma = 1$. The difference is striking. The amplitude is now uniformly distributed throughout the whole of the classically allowed region, and the nodal structure is very irregular with apparently no crossings of nodal lines.

6.5.e Nodal Patterns

There has been some discussion about the relationship between the changes in nodal structure and the onset of chaotic motion. The noncrossing of nodal lines has been carefully discussed by Pechukas (1972) in the context of nonseparable systems, that is, systems whose Hamiltonian cannot be separated in orthogonal coordinates. It is now clear that one cannot distinguish between a change of nodal pattern as a genuine manifestation of some form of underlying chaos and that due to an increasing "nonseparability" (which, of course, does not imply nonintegrability) of the system in the coordinate space in which the wavefunction is plotted. Indeed, it is possible to construct integrable systems whose wavefunctions have arbitrary nodal complexity.

6.5.f Localization Theorems*

In both of the examples cited in Section 6.5.d, the changes in the wavefunctions are undoubtedly related, to some extent, to the changes in the underlying classical dynamics. However, in both cases it is not easy to make a direct, one-to-one correlation between the changes in the classical dynamics and those in the quantum mechanics. In the regular case we can, since the semiclassical form of the wavefunction is known. In the irregular case, lack of knowledge of the semiclassical mechanics makes a direct comparison much more difficult. To date, the only rigorous results available are theorems to the effect that sequences of quantum states localize, in the limit $\hbar \to 0$, to those regions of classical phase space that support an invariant measure (Shnirelman, 1974; Helton and Tabor, 1985). For sequences of regular states, the invariant torus is the obvious invariant set. For irregular systems, one possible invariant set is just the energy shell, that is, the Liouville measure $\delta(E - H)$. However, other invariant sets may still be possible candidates.

6.5.g Microwave Ionization Experiments

The study of regular and irregular spectra is fascinating in its own right, even if there has been little success in verifying any of the predictions experimentally. However, recently, one class of experiments has provided a direct insight into the connection between quantum mechanics and classical chaos. These concern the microwave ionization of the hydrogen atom. Here, microwave radiation can be used to strip off the electrons in highly excited states with quantum numbers of order 80 plus, that is, well within the regime where semiclassical approximations are valid. From the classical point of view, the radiated atom can be modeled as a driven oscillator, and the dynamics can ultimately be reduced to that of the Chirikov map. The onset of high ionization yields seems to be well correlated with the onset of widespread chaos classically. Beautiful, controlled experiments have been performed, and the results can be compared with classical, semiclassical, and quantal calculations, both theoretical and numerical. Much of this work has been recently reviewed by Jensen (1985).

6.6 QUANTUM MAPS: EVOLUTION OF WAVEPACKETS

A problem in studying the semiclassical mechanics of nonintegrable systems is to find model systems for which detailed computations (analytical and numerical) of both the classical and quantum mechanics are feasible. For conservative systems the lowest dimensionality that can display classical chaos is two degrees of freedom. Although for such systems quantities such as Lyapunov exponents and surfaces of section are now relatively easy to compute, other desired properties such as the enumeration of all the closed orbits are virtually unachievable.† Similarly, accurate quantum mechanical computations of energy levels and eigenfunctions for small $\hbar$ (i.e., high density of states) is still, computationally, very expensive and prone to increasing instability as $\hbar \to 0$. These difficulties suggest the desirability of studying simpler, but generic, dynamical systems. As we discussed in Chapter 4, area-preserving mappings are a suitable candidate since they exhibit all the generic properties of nonintegrable systems. Furthermore, being one-dimensional (i.e., on the plane), the associated classical (and, as we shall see, quantal) properties are relatively easy to compute. By studying the semiclassical mechanics of area-preserving maps, we will be able to draw on many of the results given in Chapter 4—although for continuity we will repeat some of the relevant equations here.

6.6.a The Classical Map

First recall one-dimensional, time-dependent Hamiltonians of the form

$$H(p, q, t) = \begin{cases} \dfrac{1}{\gamma} V(q), & 0 < t < \gamma T \\[2mm] \dfrac{p^2}{2\mu(1 - \gamma)}, & \gamma T < t < T \end{cases} \tag{6.6.1}$$

---

† "Analyzing a general potential system with two degrees of freedom is beyond the capability of modern science" (Arnold 1978, page 22).

where μ is the particle mass and $0 < \gamma < 1$, which corresponds to a sequence of periodically alternating kinetic and potential motions. Integrating the associated Hamilton's equations over any one period $t = nT$ to $t = (n + 1)T$ yields the area-preserving map of the phase plane onto itself, that is,

$$p_{n+1} = p_n - TV'(q_n) \qquad (6.6.2a)$$

$$q_{n+1} = q_n + \frac{T}{\mu} p_{n+1} \qquad (6.6.2b)$$

where prime denotes differentiation with respect to q. At this point it is convenient to recall Percival's discrete Lagrangian (Section 4.2.d)

$$L(q_{n+1}, q_n) = \frac{\mu}{2} \left(\frac{q_{n+1} - q_n}{T} \right)^2 - V(q_n) \qquad (6.6.3)$$

the associated action function

$$W(q_{n+1}, q_n) = TL(q_{n+1}, q_n) = \frac{\mu}{2} \frac{(q_{n+1} - q_n)^2}{T} - TV(q_n) \qquad (6.6.4)$$

and the "generating" relations

$$p_n = -\frac{\partial W}{\partial q_n}(q_{n+1}, q_n), \qquad p_{n+1} = \frac{\partial W}{\partial q_{n+1}}(q_{n+1}, q_n) \qquad (6.6.5)$$

6.6.b The Quantum Map

Having identified the Hamiltonian (6.6.1) associated with the mapping (6.6.2), the "quantum mechanics" of the mapping can now be formulated. The idea is to find the corresponding quantal operator $\hat{U}$ that maps the state† of the system $|n\rangle$ at "time" n to its state $|n + 1\rangle$ at "time" $n + 1$, that is,

$$|n + 1\rangle = \hat{U}|n\rangle \qquad (6.6.6)$$

This can be transformed to the position representation, namely, $\psi_n(q) = \langle q | n \rangle$, as follows

$$\psi_{n+1}(q) = \langle q | n + 1 \rangle = \langle q | \hat{U} | n \rangle$$

$$= \int \langle q | \hat{U} | q' \rangle \langle q' | n \rangle \, dq'$$

$$= \int \langle q | \hat{U} | q' \rangle \psi_n(q') \, dq' \qquad (6.6.7)$$

†Here we are using standard "bra," "ket" notation.

where we have used the standard identity $\int |q\rangle\langle q|\, dq = 1$. Again we emphasize that the subscripts n, $n+1$ refer to "time," corresponding to iterates of the map (6.6.2), and not quantum number. The Hamiltonian operator $\hat{\mathcal{H}}$ corresponding to (6.6.1) is obtained by the usual rules and we write

$$\hat{\mathcal{H}}(\hat{p}, \hat{q}, t) = \begin{cases} \dfrac{1}{\gamma}\, V(\hat{q}), & 0 < t < \gamma T \\[2mm] \dfrac{\hat{p}^2}{2\mu(1-\gamma)}, & \gamma t < t < T \end{cases} \tag{6.6.8}$$

where $\hat{q}$ and $\hat{p}$ are the position and momentum operators, respectively (in position space $\hat{q} = q$ and $\hat{p} = -i\hbar\, \partial/\partial q$). The evolution operator $\hat{U} = e^{-i\hat{\mathcal{H}}t/\hbar}$ is evaluated over a period T and is easily seen to take the form

$$\hat{U} = e^{-V(\hat{q})T/\hbar}\, e^{-\hat{p}^2 T/2\mu\hbar} \tag{6.6.9}$$

which, like the classical map, does not depend on γ (just integrate the time-dependent Schrödinger equation (6.1.2) from $t=0$ to $t=T$.) Note that since the kinetic and potential energy operators act separately in (6.6.8), $\hat{U}$ conveniently factorizes into the form (6.6.9). In the position representation the matrix elements $\langle q|\hat{U}|q'\rangle$ are straightforward to evaluate and one finds that

$$\langle q|\hat{U}|q'\rangle = \left(\frac{\mu}{2\pi\hbar T}\right)^{1/2} e^{i(\mu(q-q')^2/2T - V(q)T)/\hbar - i\pi/4} \tag{6.6.10}$$

Using (6.6.4) and setting $q = q_{n+1}$ and $q' = q_n$, we note that (6.6.10) can also be cast in the form

$$\langle q_{n+1}|\hat{U}|q_n\rangle = \left(\frac{i}{2\pi\hbar}\right)^{1/2} \left[\frac{\partial^2 W(q_{n+1}, q_n)}{\partial q\, \partial q'}\right]^{1/2} e^{iW(q_{n+1}, q_n)/\hbar} \tag{6.6.11}$$

that is, the propagator (Green's function) from the "state" $|q_n\rangle$ to $|q_{n+1}\rangle$ expressed as an amplitude times a phase—the latter just being the classical action along a path from q_n to q_{n+1}. The matrix element (6.6.10) is used in (6.6.7) to obtain the integral evolution equation

$$\psi_{n+1}(q) = \left(\frac{\mu}{2\pi\hbar T}\right)^{1/2} e^{-i\pi/4 - iV(q)T/\hbar} \int_{-\infty}^{\infty} dq'\, \psi_n(q')\, e^{i\mu(q-q')^2/2T\hbar} \tag{6.6.12}$$

It is not difficult to verify that in the limit $T \to 0$ the classical map (6.6.2) reduces to Hamilton's equations $\dot{q} = \partial\bar{H}/\partial p$, $\dot{p} = -\partial\bar{H}/\partial q$ for the continuous evolution of the system governed by the Hamiltonian $\bar{H}$, which corresponds to the time average of (6.6.1), that is,

$$\bar{H} = \tfrac{1}{2}p^2 + V(q) \tag{6.6.13}$$

Similarly, in the limit $T \to 0$ the "quantum map" reduces to the standard time-dependent Schrödinger equation (6.1.2) with Hamiltonian operator corresponding to (6.6.13). In fact, the time step T can be thought of as a perturbation parameter: In the limit $T \to 0$ the motion reduces to an integrable one-degree-of-freedom Hamiltonian and the (p, q) phase plane is just covered by invariant curves. For $T > 0$ the dynamics becomes that of the discrete mapping (6.6.2), and, as T is increased, more and more of the invariant curves are destroyed and the phase plane exhibits the generic structure of regular and chaotic motions.

6.6.c The Evolution of Classical and Quantal States

The above formulation can now be used to investigate the evolution of quantum states in regimes of chaotic motion in a way that permits a one-to-one comparison between the classical and quantal behavior. Firstly, consider the conservative one-degree-of-freedom Hamiltonian (6.6.13). The quantal stationary states, $\psi^{(m)}$, of this system can be computed (typically numerically) to a high degree of accuracy. Furthermore, in the semi-classical limit, each of these states can be associated with an invariant Lagrangian manifold, $\mathscr{C}_0$, in the (p, q) phase space through the Bohr–Sommerfeld quantization rule, that is,

$$I = \oint_{\mathscr{C}_0} p \, dq = (m + \tfrac{1}{2})\hbar \tag{6.6.14}$$

where we associate the eigenvalue $E_m = H(I = (m + \tfrac{1}{2})\hbar)$ with the stationary state $\psi^{(m)}$. Now consider "switching on," at $t = 0$, the switching perturbation T. The classical motion is now governed by the Hamiltonian (6.6.1). The invariant curves $\mathscr{C}_0$ (of $\bar{H}$), identified by (6.6.14) are no longer invariant curves of (6.6.1), and they will evolve into complicated morphologies depending on the phase-space structure of (6.6.1); these morphologies are precisely the whorls and tendrils discussed in Chapter 4. Similarly, the stationary states $\psi^{(m)}$ (of $\bar{H}$) are no longer stationary states of $\hat{\mathscr{H}}$ (6.6.8), and they will now evolve according to the integral equation (6.6.12).

The evolving Lagrangian manifolds can be compared (in the limit $\hbar \to 0$) with their associated evolving quantum states by computing (at each time step n) $|\psi_n^{(m)}(q)|^2$ and the projection of $\mathscr{C}_n$ onto the q-axis—the latter quantity corresponding to the coarse-grained $\overline{|\psi(q)|^2}$ (see Section 6.4). We illustrate this by citing some results taken from Berry et al. (1979). This study investigated the mapping

$$p_{n+1} = p_n - q_{n+1}^3 \qquad\qquad (6.6.15a)$$

$$q_{n+1} = q_n + p_n \qquad\qquad (6.6.15b)$$

which has the averaged Hamiltonian $\bar{H} = p^2/2 + q^4/4$, that is, a simple quartic oscillator. The phase plane of (6.6.15) is shown in Figure 6.8a. The largest curve superimposed on this picture in Figure 6.8b is an invariant curve of $\bar{H}$—with the units chosen, it corresponds, through (6.6.14), to the 18th bound state of the corresponding quantal system. In Figure 6.9 we show how this curve evolves under the classical map; after only five iterations, $\mathscr{C}_5$ already displays remarkable complexity, reminiscent of the passive scalar patterns discussed in Section 4.8. In $\mathscr{C}_4$ one can clearly see

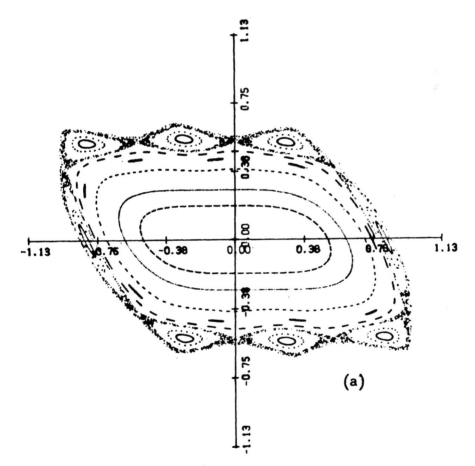

Figure 6.8 (a) Typical phase plane of the area-preserving map (6.6.15). (b) Same phase plane as in (a) but with three invariant curves of Hamiltonian (6.6.13) superimposed. (Reproduced, with permission, from Berry et al. (1979).)

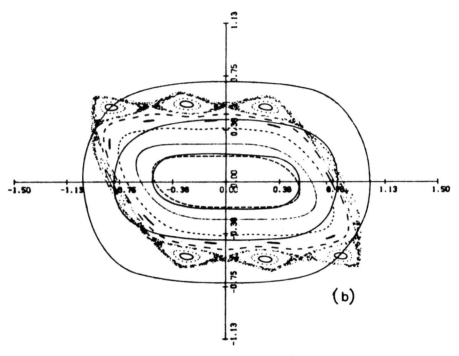

Figure 6.8 (*Continued*)

the small whorls associated with the island chain shown in Figure 6.8 and the long thin tendrils associated with the hyperbolic regions through which $\mathscr{C}$ passes. The overall spiral shape already apparent in $\mathscr{C}_1$ corresponds to a "giant" whorl associated with rotation about the central fixed point $q = p = 0$. In Figure 6.10a we show the projections of the $\mathscr{C}_n$. After only two iterations there is a striking proliferation of caustics; this seems to be a characteristic feature of the chaotic regime. Now compare with the quantum map pictures of $|\psi(q)|^2$ (Figure 6.11a). At $n = 2$ there is a striking change from a regular structure featuring only one scale of oscillations to a structure featuring multiple scales of oscillations. The projections of the associated evolving curves provide the smooth envelope for $|\psi|^2$ for $n = 0$ and $n = 1$ only. Thereafter, the proliferating caustic structure and $|\psi|^2$ appear to bear little relation; this is hardly surprising since the caustics are now clustering on scales smaller than the characteristic de Broglie wavelength. Clearly, the quantal wavefunctions cannot resolve classical features (in phase space) on scales less than $O(\hbar)$. In order to make a comparison between the classical projections and the $|\psi|^2$ for $n \geq 2$, both must be smoothed over a scale of $O(\hbar)$. This is shown in the sequences in Figure 6.10b and Figure 6.11b; quite good agreement between the two is observed.

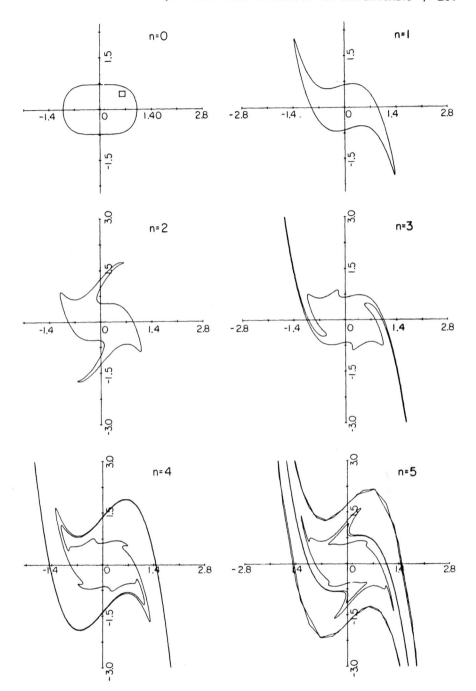

Figure 6.9 Classical maps $\mathscr{C}_n$ of initial family of trajectories $\mathscr{C}_0$, which is the outermost curve shown in Figure 6.8(*b*). Small square marked in $\mathscr{C}_0$ has "area" $\hbar$. (Reproduced, with permission, from Berry et al. (1979).)

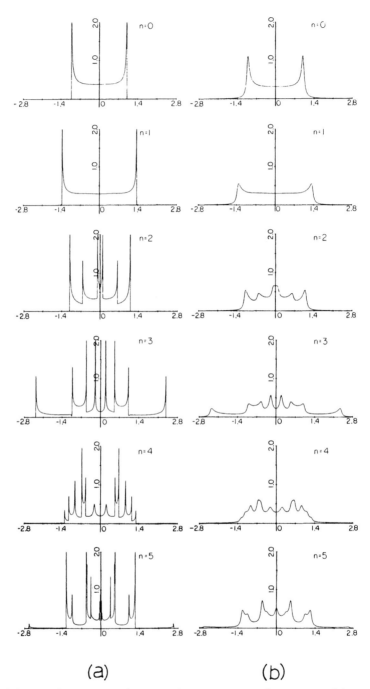

(a)   (b)

Figure 6.10 (a) Projection of maps $\mathscr{C}_n$ onto q-axis showing proliferation of caustics after $n = 2$. (b) Projections smoothed over width $\Delta q = 0.05$. (Reproduced, with permission, from Berry et al. (1979).)

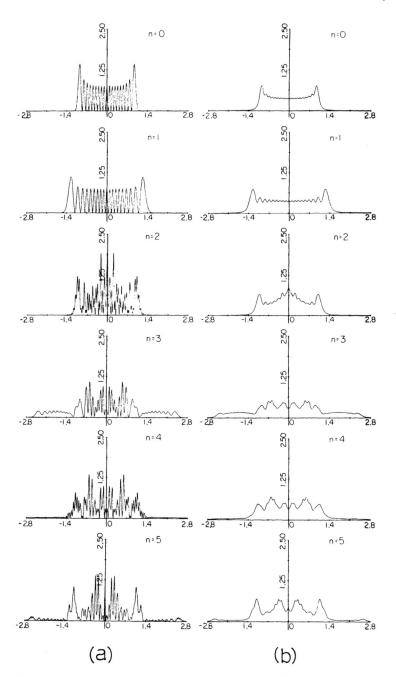

(a)   (b)

Figure 6.11 (a) Quantum maps showing evolution of $|\psi(q)|^2$. The initial state ($n = 0$) is 18th bound state of $\bar{H} = p^2/2 + q^4/4$ (with associated WKB curve $\mathscr{C}_0$). (b) Quantum maps smoothed over same width as used in Figure 6.10b. (Reproduced, with permission, from Berry et al. (1979).)

These results show $\hbar$ playing two different roles, depending on the classical regime. Initially $\hbar$ is, in effect, adding quantum structure (the oscillations in ψ) onto the smooth classical background (the initial curves $\mathscr{C}_0$ and $\mathscr{C}_1$). However, as the classical structure evolves complexities (whorls and tendrils) on scales of less than $O(\hbar)$, $\hbar$ begins to play a smoothing role in that it cannot "resolve" these fine-scale structures. More detailed discussions are given in the paper by Berry et al. (1979). Other discussions of quantum maps have been given by Casati et al. (1981) and Berman and Zavlaskii (1978).

6.7 QUANTUM MAPS: CLOSED-ORBIT QUANTIZATION*

As mentioned in Section 6.4, there is no "direct," EBK-like quantization condition for nonintegrable Hamiltonians. However, an alternative semi-classical technique, pioneered by Gutzwiller (1971), has been developed which uses the closed classical orbits of the system. For general Hamiltonian systems, its formulation is rather complicated. However, quantum maps provide a simpler framework for introducing the basic ideas. To begin, we review the required quantum mechanical background.

6.7.a Quantum Mechanical Background

A particle of mass m in a potential field $V(\mathbf{q})$ ($\mathbf{q} = (q_1, q_2, q_3)$) satisfies the Schrödinger equation

$$\left[\frac{\hbar^2}{2m}\nabla^2 + E_n - V(\mathbf{q})\right]\psi_n(\mathbf{q}) = 0 \tag{6.7.1}$$

where the $\psi_n(\mathbf{q})$ form a complete set of eigenfunctions (labeled by some vector of quantum numbers collectively denoted as $\mathbf{n}$) with corresponding eigenvalues E_n. The quantum mechanical density of states is defined as

$$\rho(E) = \sum_n \delta(E - E_n) \tag{6.7.2}$$

that is, a delta-function spike at each eigenvalue. The Green's function (propagator) for Eq. (6.7.1) satisfies the equation

$$\left[\frac{\hbar^2}{2m}\nabla^2 + E - V(\mathbf{q})\right]G(\mathbf{q}, \mathbf{q}') = \delta(\mathbf{q} - \mathbf{q}') \tag{6.7.3}$$

and can be written in the "bilinear form"

$$G(\mathbf{q}, \mathbf{q}') = \sum_n \frac{\psi_n(\mathbf{q})\psi_n^*(\mathbf{q}')}{E - E_n} \tag{6.7.4}$$

On taking the "trace" of the Green's function (i.e., setting $\mathbf{q}'$ equal to $\mathbf{q}$ and integrating over all $\mathbf{q}$), one obtains

$$\int d\mathbf{q} \, G(\mathbf{q}, \mathbf{q}) = \sum_n \frac{1}{E - E_n} \tag{6.7.5}$$

Using the formal relationship

$$\frac{1}{E - E_n + i\epsilon} = \mathscr{P}\frac{1}{E - E_n} - i\epsilon\pi\delta(E - E_n) \tag{6.7.6}$$

(where $\mathscr{P}$ denotes Cauchy principal value), writing

$$G^{(+)}(\mathbf{q}, \mathbf{q}') = \sum_n \frac{\psi_n(\mathbf{q})\psi_n^*(\mathbf{q}')}{E - E_n + i\epsilon} \tag{6.7.7}$$

and taking the limit $\epsilon \to 0$ gives

$$\rho(E) = -\frac{1}{\pi}\text{Im}\int d\mathbf{q} \, G^{(+)}(\mathbf{q}, \mathbf{q}) \tag{6.7.8}$$

Later we shall see that in the semiclassical limit it is possible to represent $G(\mathbf{q}, \mathbf{q}')$ in terms of the *classical paths* connection $\mathbf{q}$ and $\mathbf{q}'$ and that the trace operation picks out precisely the closed orbits of the system.

6.7.b The Quasi-Energy Spectrum

For Hamiltonians such as (6.6.1) with periodic time dependence of the form

$$H(p, q, t + T) = H(p, q, T) \tag{6.7.9}$$

we have seen (cf. (6.6.6)) that its quantal state at "time" N is obtained by N applications of the associated unitary operator $\hat{U}$, that is,

$$|\psi(t + NT)\rangle = \hat{U}^N|\psi(t)\rangle \tag{6.7.10}$$

The operator $\hat{U}$ can possess a spectrum of eigenstates $|n\rangle$ such that

$$\hat{U}^N|n\rangle = e^{-iN\alpha_n/\hbar}|n\rangle \tag{6.7.11}$$

where the eigenvalues α_n, which lie on the unit circle, are termed the *quasi-energies*. Using (6.7.11), we can then write

$$\text{Tr}(\hat{U}^N) = \sum_n \langle n|\hat{U}^N|n\rangle = \sum_n e^{-i(N\alpha_n/\hbar)} \tag{6.7.12}$$

It is convenient to introduce the scaled eigenvalues

$$\epsilon_n = \frac{\alpha_n}{T} \tag{6.7.13}$$

and then define the quasi-energy density of states as the Fourier sum

$$\rho(E) = \frac{1}{2\pi\hbar} \sum_{N=-\infty}^{\infty} e^{iNET/\hbar} \operatorname{Tr}(\hat{U}^N) = \frac{1}{2\pi\hbar} \sum_{N=-\infty}^{\infty} \sum_n e^{i(NT/\hbar)(E-\epsilon_n)} \tag{6.7.14}$$

The function $\rho(E)$ can also be represented as a sum of delta functions (cf. (6.7.2)) by invoking the identity†

$$\sum_{n=-\infty}^{\infty} \delta(x-n) = \sum_{M=-\infty}^{\infty} e^{2\pi iMx} \tag{6.7.15}$$

from which we see that

$$\rho(E) = \sum_{M=-\infty}^{\infty} \sum_n \delta(M\omega\hbar - (E-\epsilon_n)) \tag{6.7.16}$$

where $\omega = 2\pi/T$ is the natural period of the associated Hamiltonian (6.7.9). Note that associated with each quasi-energy level ϵ_n there is an infinite set of states, that is, $E = \epsilon_n + M\omega\hbar$, $M = 0, \pm1, \pm2, \ldots$.

6.7.c The Quantum Map Propagator

The starting point is the "one-step" propagator from $|q_n\rangle$ to $|q_{n+1}\rangle$ given in (6.6.11). To form the "n-step" propagator (i.e., to take the system from the state $|q_0\rangle$ to $|q_n\rangle$), one must evaluate the $(n-1)$-dimensional integral

$$\langle q_n|\hat{U}|q_0\rangle = \int dq_1 \langle q_0|\hat{U}|q_1\rangle\langle q_1|\hat{U}|q_2\rangle \int dq_2 \langle q_2|\hat{U}|q_3\rangle \cdots \int dq_{n-1} \langle q_{n-1}|\hat{U}|q_n\rangle \tag{6.7.17}$$

---

†This identity is a special case of a very useful mathematical trick, much used in semiclassical problems, known as the *Poisson sum formula*. Its general form is

$$\sum_{l=0}^{\infty} g(l) = \sum_{M=-\infty}^{\infty} \int_0^{\infty} g(x) e^{2\pi iMx} dx$$

which enables one to represent an infinite sum of functions by an infinite sum of related Fourier transforms. It can be derived by operating on both sides of (6.7.15) with $\int_0^{\infty} dx\, g(x)$. Note that the crude and commonly used approximation of replacing a sum $\sum_l$ by the integral $\int dl$ just corresponds to using the term $M = 0$ in the Poisson sum formula.

In the limit $\hbar \to 0$ this integral can be evaluated by the all-important *method of stationary phase*. This technique is of central importance in semiclassical problems, and the reader is urged to refer to Appendix 6.1 for an introductory account. To see how this technique is used to evaluate (6.7.17), we first of all consider the "two-step" propagator

$$\langle q_2 | \hat{U} | q_0 \rangle = \left(\frac{i}{2\pi\hbar} \right) \int dq_1 \left[\frac{\partial^2 W(q_2, q_1)}{\partial q_2 \partial q_1} \frac{\partial^2 W(q_1, q_0)}{\partial q_1 \partial q_0} \right]^{1/2}$$
$$\times \exp \frac{i}{\hbar} [W(q_2, q_1) + W(q_1, q_0)] \qquad (6.7.18)$$

In the limit $\hbar \to 0$, the exponential term is highly oscillatory except at its stationary phase point determined by the condition

$$\frac{\partial}{\partial q_1} [W(q_2, q_1) + W(q_1, q_0)] = -p_1(q_2, q_1) + p_1(q_1, q_0) = 0 \quad (6.7.19)$$

where we have used the "generating" relations (6.6.5). This condition will be satisfied if q_1 is the point on the classical path connecting q_2 and q_0. We denote this point as $q_1 = q_1^c$. The next step in the stationary phase approximation is to evaluate the amplitude factor (see Appendix 6.1)

$$A_{20} = \left[\frac{\dfrac{\partial^2}{\partial q_2 \partial q_1} W(q_2, q_1) \dfrac{\partial^2}{\partial q_1 \partial q_0} W(q_1, q_0)}{-\dfrac{\partial}{\partial q_1} [p_1(q_2, q_1) - p_1(q_1, q_0)]} \right]_{q_1 = q_1^c} \qquad (6.7.20)$$

Differentiating $p_1(q_2, q_1) - p_1(q_1, q_0) = 0$ with respect to q_0, one obtains, via the chain rule,

$$-\frac{\partial}{\partial q_1} [p_1(q_2, q_1) - p_1(q_1, q_0)] = \frac{\partial}{\partial q_0} p_1(q_1, q_0) \Big/ \frac{\partial q_1}{\partial q_0} = \frac{\partial^2}{\partial q_0 \partial q_1} W(q_1, q_0) \Big/ \frac{\partial q_1}{\partial q_0}$$

and hence

$$A_{20} = \left[-\frac{\partial^2}{\partial q_2 \partial q_1} W(q_2, q_1) \cdot \frac{\partial q_1}{\partial q_0} \right]_{q_1 = q_1^c} = \left[-\frac{\partial^2}{\partial q_2 \partial q_0} W(q_2, q_0) \right]_{q_1 = q_1^c}$$
$$(6.7.21)$$

where we have used the classical "combination" rule (i.e., the additivity of the action along a classical path)

$$W(q_2, q_1^c) + W(q_1^c, q_0) = W(q_2, q_0) \qquad (6.7.22)$$

Using these results, the semiclassical approximation to (6.7.18) is thus

$$\langle q_2 | \hat{U} | q_0 \rangle = \left(\frac{i}{2\pi\hbar} \right)^{1/2} \left[\frac{\partial^2}{\partial q_2 \partial q_0} W(q_2, q_0) \right]^{1/2} e^{(i/\hbar) W(q_2, q_0)} \qquad (6.7.23)$$

where $W(q_2, q_0)$ is the classical action along the (classical) path from q_0 to q_2. Repeated application of the stationary phase method to (6.7.17) gives the semiclassical approximation to the n-step propagator, namely,

$$\langle q_n | \hat{U} | q_0 \rangle = \left(\frac{i}{2\pi\hbar} \right)^{1/2} \left[\frac{\partial^2}{\partial q_n \partial q_0} W(q_n, q_0) \right]^{1/2} e^{(i/\hbar) W(q_n, q_0)} \qquad (6.7.24)$$

where we are assuming that the classical path does not pass through any caustics—hence the absence of any extra phase factors.

6.7.d Tracing the Propagator

In order to discuss the tracing of the propagator, it is convenient to introduce the notation

$$K(q_n, q_0, \tau) \equiv \langle q_n | \hat{U} | q_0 \rangle \qquad (6.7.25)$$

where $\tau = nT$ is the time to evolve from q_0 to q_n. The trace, that is,

$$\text{Tr}(K) = \int dq \, K(q, q, \tau) \qquad (6.7.26)$$

consists of contributions from paths that begin and end at the same point. These contributions are of two types:

  (i) Paths of zero length, that is, $q = q$ with $\tau = 0$, and
  (ii) Paths of length-n steps, that is, $q = q_n$, $q = q_0$ with $\tau = nT$.

It is only the second type of path that gives a nontrivial contribution to the trace. Using the semiclassical form of K, that is, (6.7.24), we have to evaluate

$$\text{Tr}(K) = \left(\frac{i}{2\pi\hbar} \right)^{1/2} \int dq \left[\frac{\partial^2}{\partial q_n \partial q_0} W(q_n, q_0) \right]^{1/2} e^{(i/\hbar) W(q_n, q_0)} \qquad (6.7.27)$$

where it is understood that $q_n = q_0 = q$. Again using the method of stationary phase, the stationary phase points come from those paths that satisfy the condition

$$\left[\frac{\partial}{\partial q_n} W(q_n, q_0) + \frac{\partial}{\partial q_0} W(q_n, q_0)\right]_{q_n = q_0 = q} = [p_n(q_n, q_0) - p_0(q_n, q_0)]_{q_n = q_0 = q}$$

$$= p_n(q, q) - p_0(q, q) = 0 \qquad (6.7.28)$$

These paths correspond to the *closed orbits* of the system since they begin and end at the same position in both q and (by (6.7.28)) p space. Provided that each closed orbit is isolated (we will return to this point later), the calculation can procede. Evaluation of the amplitude factor is highly nontrivial. It takes the form

$$\left[\frac{\partial^2}{\partial q_n \partial q_0} W(q_n, q_0) \middle/ \left(\frac{\partial^2}{\partial q_n^2} W(q_n, q_0) + 2\frac{\partial^2}{\partial q_n \partial q_0} W(q_n, q_0)\right.\right.$$
$$\left.\left. + \frac{\partial^2}{\partial q_0^2} W(q_n, q_0)\right)\right]_{q_n = q_0 = q}$$

which can then be shown (see Tabor (1983) for details) to be directly proportional to the residue R (defined in Section 4.6), of the orbit. Overall, one finds that the contribution to $\mathrm{Tr}(K)$ from each closed orbit takes the form

$$\mathrm{Tr}(K) = \frac{1}{2}\left(\frac{1}{R}\right)^{1/2} e^{(i/\hbar)\bar{W}(\tau)} \qquad (6.7.29)$$

where $\bar{W}(\tau)$ denotes the action around the orbit.

The closed orbits fall into three categories:

(i) $0 < R < 1$—the orbits are stable and manifested in the phase plane of the mapping as elliptic fixed points. The residue can be expressed in the form

$$R = \sin^2\left(\frac{v}{2}\right) \qquad (6.7.30)$$

where v is the so-called *stability angle.*

(ii) $R < 0$—the orbits are unstable, corresponding to hyperbolic fixed points. Now $v = iu$ and

$$R = -\sin^2\left(\frac{u}{2}\right) \qquad (6.7.31)$$

(iii) $R > 1$—the orbits are again unstable but now corresponding to hyperbolic-with-reflection fixed points. In this case, $v = \pi + iu$ and

$$R = \cosh^2\left(\frac{u}{2}\right) \tag{6.7.32}$$

Returning to the stationary phase result (6.7.29), it must also be borne in mind that there is a contribution from repeated traversals of each orbit. Thus the final result for a given stable orbit takes the form

$$\text{Tr}(K) = \sum_{N=1}^{\infty} \frac{1}{\sin(Nv/2)} \, e^{(i/\hbar)N\bar{W}} \tag{6.7.33}$$

where N corresponds to the number of traversals and where the factor $\frac{1}{2}$ in (6.7.29) disappears when we take account of the contributions from both the "forward" and "backward" traversals. For hyperbolic fixed points, the corresponding result is

$$\text{Tr}(K) = -i \sum_{N=1}^{\infty} \frac{1}{\sinh(Nu/2)} \, e^{(i/\hbar)N\bar{W}} \tag{6.7.34}$$

and for hyperbolic-with-reflection fixed points, we obtain

$$\text{Tr}(K) = \sum_{N=1}^{\infty} \frac{1}{\cosh(Nu/2)} \, e^{(i/\hbar)N\bar{W}} \tag{6.7.35}$$

In each of the formulae (6.7.33), (6.7.34), and (6.7.35), the exponent $\bar{W}$ corresponds to the action of just one circuit of the given orbit.

In the case of stable orbits, (6.7.33) suffers a serious defect in that it will diverge every time $Nv/2$ is a multiple of π. This is due to the appearance of a caustic in the classical propagator and represents a breakdown in the method of stationary phase used to evaluate (6.7.29). The expression for hyperbolic orbits (6.7.34) also suffers from a divergence problem—but only for $u = 0$, which rarely occurs in practice. Resolution of this issue is nontrivial, and various "first-order" approximations to (6.7.33) and (6.7.34) have been proposed to overcome these divergences. Ultimately, this defect must be remedied by suitable modifications of the stationary phase approximation (i.e., uniform approximations), which takes account of coalescing stationary phase points.

These results can be assembled to provide a semiclassical representation of the density of states (6.7.14) in terms of contributions from the closed orbits. Overall, one obtains

$$\rho(E) = \rho^0 + \sum_{j} \rho_j^{\text{I}}(E) + \rho_j^{\text{II}}(E) + \rho_j^{\text{III}}(E) \tag{6.7.36}$$

where ρ^0 is the mean density of states $\rho^0 = T/2\pi\hbar$ (this is the contribution from paths of zero length); the sum over j represents the sum of contributions from each topologically distinct orbit. The three terms under the summation sign in (6.7.36) represent the contributions from the three different orbital stability types, that is,

$$\rho_j^{\mathrm{I}}(E) = \frac{T}{2\pi\hbar} \sum_{N=1}^{\infty} \frac{1}{\sin(Nv_j/2)} \sin\left[\frac{N}{\hbar}(ET + \bar{W}_j)\right] \qquad (6.7.37)$$

$$\rho_j^{\mathrm{II}}(E) = \frac{-T}{2\pi\hbar} \sum_{N=1}^{\infty} \frac{1}{\sinh(Nu_j/2)} \cos\left[\frac{N}{\hbar}(ET + \bar{W}_j)\right] \qquad (6.7.38)$$

$$\rho_j^{(\mathrm{III})}(E) = \frac{T}{2\pi\hbar} \sum_{N=1}^{\infty} \frac{1}{\cosh(Nu_j/2)} \sin\left[\frac{N}{\hbar}(ET + \bar{W}_j)\right] \qquad (6.7.39)$$

6.7.e Discussion of Closed-Orbit Theory

From the practical point of view, closed-orbit quantization is not the most useful technique since it requires the enumeration of all closed orbits and their associated properties—an enormous task for mappings, let alone generic Hamiltonians of two or more degrees of freedom. However, a very important feature of the method is the insight it gives into the *clustering* of energy levels. Each closed orbit gives an oscillatory contribution to $\rho(E)$, whose "wavelength," ΔE, in energy space is determined by the relationship $\Delta E(d\bar{W}/dE) = 2\pi\hbar$, where $\bar{W}$ is the total action around the orbit. (This total action includes the possibility of repeated circuits of a given orbit.) Since $d\bar{W}/dE$ is just the period of the orbit, $T(E)$, we see that $\Delta E = 2\pi\hbar/T(E)$. Thus each closed orbit gives an oscillatory contribution to $\rho(E)$ on a scale of $O(\hbar)$, which is far greater (for more than one degree of freedom) than the mean level spacing, which is $O(\hbar^N)$.

This can be seen from a different point of view by considering the "smoothed density of states." This is defined as the density of states obtained when $\rho(E)$ (6.7.2) is smoothed by a continuous weighting function—typically of Lorentzian form, that is,

$$\rho_\gamma(E) = \int_{-\infty}^{\infty} dE' \rho(E') \frac{\gamma}{\pi} \frac{1}{(E - E')^2 + \gamma^2} \qquad (6.7.40)$$

where γ is a smoothing width. This width corresponds to adding an imaginary part, $i\gamma$, to the energy in (6.7.5) since

$$\rho_\gamma(E) = -\frac{\mathrm{Im}}{\pi} \sum_{\mathbf{n}} \frac{1}{E - E_{\mathbf{n}} + i\gamma} = \frac{\gamma}{\pi} \sum_{\mathbf{n}} \frac{1}{(E - E_{\mathbf{n}})^2 + \gamma^2} \qquad (6.7.41)$$

Thus each delta function in $\rho(E)$ is replaced by a Lorentzian peak in $\rho_\gamma(E)$. As γ gets larger, the Lorentzians overlap and $\rho_\gamma(E)$ will only be able to resolve certain clusterings of energy levels in $\rho(E)$. The connection with the closed-orbit representation of $\rho(E)$ is easily seen since, to first order, the effect of adding an imaginary part to the energy in the phase factor is just

$$\bar{W}(E + i\gamma) \simeq \bar{W}(E) + i\gamma \frac{d\bar{W}}{dE} = \bar{W}(E) + i\gamma T(E) \qquad (6.7.42)$$

Thus the contribution of each closed orbit to $\rho_\gamma(E)$ is damped, for a fixed γ, by an amount proportional to the period of the orbit. Thus long orbits, or repeated circuits of a given orbit, only make exponentially small contributions to $\rho_\gamma(E)$. In this way, just a few (short) periodic orbits can give an approximate representation of $\rho(E)$ with the oscillations in $\rho_\gamma(E)$ representing various clusterings of the energy levels.

By contrast, in the case of the spectral rigidity function (discussed in Section 6.4.e), which measures certain long-range correlations in the energy spectrum, the semiclassical analysis shows that this quantity is primarily determined by the very long closed orbits of the system.

The closed-orbit theory formulated here is for nonintegrable systems in that it assumes all the closed orbits to be isolated. However, in the case of integrable systems, this is no longer so and they form one parameter families lying on tori. In these cases also, the spectrum can be represented in terms of the closed orbits using a theory developed by Balian and Bloch (1974) and Berry and Tabor (1976). (For some nonintegrable systems, certain closed orbits form "small," continuous families—these too can be handled by the above-mentioned techniques)

APPENDIX 6.1:   THE METHOD OF STATIONARY PHASE

The method of stationary phase concerns itself with the evaluation of oscillatory integrals of the form

$$I = \int_{x_1}^{x_2} g(x) \, e^{(i/\hbar)f(x)} \, dx \qquad (6.A.1)$$

where x_1 and x_2 are the limits of integration, and $\hbar$ is a small parameter—Planck's constant in semiclassical applications. In the limit $\hbar \to 0$ the integrand will oscillate rapidly, and most of these oscillations will cancel each other out (i.e., there is destructive interference). The dominant contribution to the integral will come from those points where the phase is *stationary*, that is, those points x_i for which

$$f'(x_i) = 0 \qquad (6.A.2)$$

where prime denotes differentiation with respect to x. The points x_i are termed *stationary phase points*, and for now we will assume that they are (1) suitably separated from each other and (2) suitable far away from the end points.

Near each isolated stationary phase point, $f(x)$ is mapped onto the quadratic form

$$f(x) = f(x_i) + at^2 \tag{6.A.3}$$

where t is a variable similar to x and where

$$a = +1 \quad \text{if} \quad f''(x_i) > 0$$

and

$$a = -1 \quad \text{if} \quad f''(x_i) < 0$$

Provided that $x_1 \ll x_i \ll x_2$, the limits of integration are pushed to $\pm\infty$ and the integral (6.A.1) becomes

$$I = e^{(i/\hbar)f(x_i)} \int_{-\infty}^{\infty} g(x(t)) \, e^{(i/\hbar)at^2} \frac{dx}{dt} \, dt \tag{6.A.4}$$

The next stage is to expand the slowly varying part of the integrand $(dx(t)/dt)g(x(t))$ in powers of t about the stationary phase point $t = 0$. (Note, from (6.A.3), the correspondence between the stationary phase point x_i in (6.A.1) and $t = 0$ in (6.A.4).) To a first approximation, we retain only the leading term, that is,

$$\frac{dx(t)}{dt} g(x(t)) \simeq g(x_i) \left(\frac{dx}{dt}\right)_{t=0} \tag{6.A.5}$$

since $g(x(0)) = g(x_i)$. The derivative dx/dt can be obtained by differentiating the mapping relationship (6.A.3). A first differentiation gives

$$\frac{dx}{dt} \frac{df}{dx} = 2at \tag{6.A.6}$$

but since $(df/dx)_{x=x_i}$ is zero we must differentiate again, yielding

$$\left(\frac{dx}{dt}\right)^2 \frac{d^2 f}{dx^2} + \frac{d^2 x}{dt^2} \frac{df}{dx} = 2a \tag{6.A.7}$$

The second term on the left-hand side of (6.A.7) is again zero, so we obtain

$$\left(\frac{dx}{dt}\right)_{t=0} = \sqrt{\frac{2a}{(\partial^2 f/\partial x^2)_{x=x_i}}} = \sqrt{\frac{2}{|f''(x_i)|}} \tag{6.A.8}$$

Thus the integral (6.A.4) becomes

$$I = \sqrt{\frac{2}{|f''(x_i)|}} \, g(x_i) \, e^{(i/\hbar)f(x_i)} \int_{-\infty}^{+\infty} e^{(i/\hbar^2)at^2} \, dt \tag{6.A.9}$$

The remaining integral is just a simple Gaussian integral, that is,

$$\int_{-\infty}^{\infty} e^{iat^2/\hbar} \, dt = (\pi\hbar)^{1/2} \, e^{ia\pi/4} \tag{6.A.10}$$

and we obtain the final "stationary phase approximation"

$$I = \sqrt{\frac{2\pi\hbar}{|f''(x_i)|}} \, g(x_i) \exp\!\left((i/\hbar)f(x_i) + \frac{i \, \mathrm{sgn}(f''(x_i))\,\pi}{4}\right) \tag{6.A.11}$$

where

$$\mathrm{sgn}(x) = \begin{cases} +1, & x > 0 \\ -1, & x < 0 \end{cases}$$

For a set of isolated stationary phase points x_i $(i = 1, 2, \ldots, n)$, each one makes a contribution to (6.A.1) of the form (6.A.11), and the final result is just the sum of all such contributions.

A most important modification of the stationary phase approximation concerns the evaluation of (6.A.1) in the case of nonisolated stationary phase points. This is the topic of *uniform approximations*. Consider the phase function f to now be a function of some parameter μ, that is, $f = f(x; \mu)$ with the (local) behavior, as a function of μ, as sketched in Figure 6.12. For $\mu \ll \mu_c$, f exhibits two well-separated stationary points, x_1 and x_2, which can each make separate contributions to (6.A.1) of the form (6.A.11). As μ approaches μ_c, x_1 and x_2 become closer and merge at $x_1 = x_2 = x_c$ when $\mu = \mu_c$. At this point both $f'(x_c)$ and $f''(x_c)$ are zero and (6.A.11) breaks down. Beyond μ_c there are no (real) stationary phase points. Clearly, the simple mapping (6.A.3) is no longer valid and a more general mapping, which "models" this behavior, is required. This mapping takes the form

$$f(x) = (\tfrac{1}{3})t^3 - a(\mu)t + b(\mu) \tag{6.A.12}$$

where t is similar to $x - x_c$, $a(\mu)$ is similar to $\mu_c - \mu$, and $b(\mu)$ is related to the isolated stationary phase points x_1 and x_2. The details of the ensuing analysis is beyond the scope of this text. The overall result is an approximation in which the Gaussian integral (6.A.10) is replaced by an Airy function such that the solution behaves uniformly over the full range of μ,

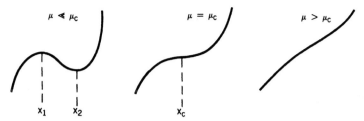

Figure 6.12 Separate stationary phase points x_1 and x_2, for $\mu \ll \mu_c$, merging at $x_1 = x_2 = x_c$, for $\mu = \mu_c$. For $\mu > \mu_c$ there are no real stationary phase points.

that is, goes through $\mu = \mu_c$ without divergences and, in the limit $\mu \ll \mu_c$, reduces to the stationary phase result for the isolated points x_1 and x_2.

For situations with more coalescing stationary phase points, in one or more dimensions, still more general mappings than (6.A.12) are required. It turns out that these can be systematically identified through the use of Thom's theorem on the singularities of gradient mappings—more popularly known as *Catastrophe Theory*.

SOURCES AND REFERENCES

Texts and General Review Articles

Berry, M. V., Semiclassical mechanics of regular and irregular motion, in G. Ioos, R. H. G. Helleman, and R. Stora, Eds., *Chaotic Behavior of Deterministic Systems*, Les Houches Lectures, Vol. 18, North-Holland, Amsterdam, 1983.

Eckhardt, B., Quantum mechanics of classically nonintegrable systems, *Phys. Reports C*, **163**, 205 (1988).

Landau, L. D., and E. M. Lifshitz, *Quantum Mechanics*, 2nd ed., Pergamon, Oxford, 1965.

Percival, I. C., Semiclassical theory of bound states, *Adv. Chem. Phys.*, **36**, 1 (1977).

Section 6.2

Bender, C. M., and S. A. Orszag, *Advanced Mathematical Methods for Scientists and Engineers*, McGraw-Hill, New York, 1978.

Norcliffe, A., and I. C. Percival, Correspondence identities: I, *J. Phys. B*, **1**, 774 (1968); **1**, 784 (1968); **2**, 578 (1969).

Section 6.3

Berry, M. V., and N. L. Balazs, Evolution of semiclassical states in phase space, *J. Phys. A*, **12**, 625 (1979).

Born, M., *The Mechanics of the Atom*, Ungar, New York, 1960.

Einstein, A., Zum Quantensatz von Sommerfeld und Epstein, *Verh. Deutsch. Phys. Ges.*, **19**, 82 (1917).

Van Vleck, J. H., The correspondence principle in the statistical interpretation of quantum mechanics, *Proc. Math. Acad. Sci. USA* **14**, 178 (1928).

Section 6.4

Berry, M. V., Quantizing a classically ergodic system: Sinai's billiard and the KKR method, *Ann. Phys. NY*, **131**, 163 (1981).

Berry, M. V., "Semiclassical theory of spectral rigidity," *Proc. R. Soc. London*, **A400**, 229 (1985).

Berry, M. V., *Quantum Chaology* (Bakerian Lecture of the Royal Society), *Proc. R. Soc. London*, **A413**, 183 (1987).

Berry, M. V., and M. Tabor, Level clustering in the regular spectrum, *Proc. R. Soc. London*, **A356**, 375 (1977).

Bohigas, O., and M. J. Giannoni, Chaotic motion and random-matrix theory, in J. S. Dehesa, J. M. G. Gomez, and A. Polls, Eds., *Mathematical and Computational Methods in Nuclear Physics*, Lecture Notes in Physics, Vol. 209, Springer-Verlag, New York, 1984.

Jaffé, C., and W. P. Reinhardt, Time dependent methods in classical mechanics: calculation of invariant tori and semiclassical energy levels via classical Van Vleck transformations, *J. Chem. Phys.*, **71**, 1962 (1979).

McDonald, S. W., and A. N. Kaufman, Spectrum and eigenfunctions for a Hamiltonian with stochastic trajectories, *Phys. Rev. Lett.*, **42**, 1189 (1979). Additional photographs were kindly furnished by the authors from "Ray and wave optics of integrable and stochastic systems," Lawrence Berkeley Lab. Report No. 9465.

McDonald, S. W., and A. N. Kaufman, Wave chaos in the stadium: Statistical properties of short-wave solutions of the Helmholtz equation, *Phys. Rev. A*, **37**, 3067 (1988).

Mehta, M. L., *Random Matrices and the Statistical Theory of Energy Levels*, Academic, New York, 1967.

Noid, D. W., M. L. Koszykowski, M. Tabor, and R. A. Marcus, Properties of vibrational energy levels in the quasi-periodic and chaotic regimes, *J. Chem. Phys.*, **72**, 6169 (1980).

Pomphrey, N., Numerical identification of regular and irregular spectra, *J. Phys. B*, **7**, 1909 (1974).

Porter, C. E., *Statistical Theories of Spectra: Fluctuations*, Academic, New York, 1965.

Wigner, E. P., Gatlinberg Conf. on Neutron Physics, Oak Ridge Natl. Lab. Rept no. ORNL-2309, p. 57 (1957). This report is reproduced in the volume by Porter.

Section 6.5

Berry, M. V., Semi-classical mechanics in phase space: A study of Wigner's function, *Philos. Trans. R. Soc. London*, **287**, 237 (1977a).

Berry, M. V., Regular and irregular semiclassical wavefunctions, *J. Phys. A*, **10**, 2083 (1977b).

Helton, J. W., and M. Tabor, On the classical support of quantum mechanical wavefunctions, *Physica*, **14D**, 409 (1985).

Hutchinson, J. S., and R. E. Wyatt, Quantum ergodicity and the Wigner distribution, *Chem. Phys. Lett.*, **72**, 378 (1980).

Jensen, R. V., in G. Casati, Ed., *Chaotic Behavior in Quantum Systems*, NATO ASI Series, Series B: Physics, Vol. 120, Plenum, New York, 1985.

McDonald, S. W., and A. N. Kaufman, Spectrum and eigenfunctions for a Hamiltonian with stochastic trajectories, *Phys. Rev. Lett.*, **42**, 1189 (1979).

Noid, D. W., M. L. Koszykowski, and R. A. Marcus, Semiclassical calculation of bound states in multidimensional systems with Fermi resonance, *J. Chem. Phys.*, **71** 2864 (1979).

Ozorio de Almeida, A. M., and J. H. Hannay, Geometry of two dimensional tori in phase space: projections, sections and the Wigner function, *Ann. Phys. NY*, **138**, 115 (1982).

Pechukas, P., Semiclassical approximation of multidimensional bound states, *J. Chem. Phys.*, **57**, 5577 (1972).

Shnirelman, A. I., Ergodic properties of eigenfunctions, *Usp. Mat. Nauk.*, **29**, 181 (1974).

Section 6.6

Arnold, V. I., *Mathematical Methods of Classical Mechanics*, Springer-Verlag, New York, 1978.

Berman, G. P., and G. M. Zavlaskii, Condition of stochasticity in quantum nonlinear systems, *Physica*, **91A**, 450 (1978).

Berry, M. V., N. L. Balazs, M. Tabor, and A. Voros, Quantum maps, *Ann. Phys. NY*, **122**, 26 (1979).

Casati, G., B. V. Chirikov, F. M. Israelev and J. Ford, in G. Casati and J. Ford, Eds., *Stochastic Behavior in Classical and Quantum Hamiltonian Systems*, Lecture Notes in Physics, Vol. 93, Springer-Verlag, New York, (1981).

Korsch, H. J., and M. V. Berry, Evolution of Wigner's phase space density under a nonintegrable quantum map, *Physica*, **3D**, 627 (1981).

Section 6.7

Balian, R., and C. Bloch, Distribution of eigenfrequencies, *Ann. Phys. NY*, **60**, 401 (1970). Also see the sequels: *Ann. Phys. NY*, **64**, 271 (1971); **69**, 76 (1972); **85**, 514 (1974).

Berry, M. V., and M. Tabor, Closed orbits and the regular bound spectrum, *Proc. R. Soc. London*, **A349**, 101 (1976).

Gutzwiller, M. C., Bound states of an atom, *J. Math. Phys.*, **8**, 1979 (1967). Also see the sequels: *J. Math. Phys.*, **10**, 100 (1969); **11**, 1791 (1970); **12**, 343 (1971).

Tabor, M., A semiclassical quantization of area preserving maps, *Physica*, **6D**, 195 (1983).

7

NONLINEAR EVOLUTION EQUATIONS AND SOLITONS

7.1 HISTORICAL BACKGROUND

A principal theme of the preceding chapters has been that nonlinear systems of just a few degrees of freedom can display very complex behavior. A natural question to ask is, what happens to all this dynamics in the limit of the number of degrees of freedom becoming infinite? In this limit, the degrees of freedom, or modes, are treated as a continuum with a continuous label x rather than a discrete index $i = 1, \ldots, N$. Thus the description of a system in terms of a finite number of ordinary differential equations (o.d.e.'s), with time as the only independent variable, goes over to a partial differential equation (p.d.e.) with both x and t as the independent variables. If only a few nonlinear o.d.e.'s can display complex behavior, it might be thought that a continuum of them (i.e., a nonlinear p.d.e.) could only display more complicated behavior. In many cases this is indeed so, and nonlinear p.d.e.'s will display chaos in both time and space. However, there is also an important class of nonlinear p.d.e.'s whose behavior is remarkably regular, that is, they are, in effect, integrable. The properties of these systems and the behavior of their solutions has given rise to what is generally considered to be one of the most significant advances in post-war mathematical physics.

7.1.a Russell's Observations

The story begins over 150 years ago with the now famous observations made by the Scottish engineer John Scott Russell while riding (on horseback) near the Union Canal outside Edinburgh. His report (Russell, 1844) reads as follows:

> I was observing the motion of a boat which was rapidly drawn along a narrow channel by a pair of horses, when the boat suddenly stopped—not so the mass of the water in the channel which it had put in motion; it accumulated round the prow of the vessel in a state of violent agitation, then suddenly leaving it behind, rolled forward with great velocity, assuming the form of a large solitary elevation, a rounded, smooth and well-defined heap of water, which continued its course along the channel apparently without change of form or diminution of speed. I followed it on horseback and overtook it still rolling on at a rate of some eight or nine miles an hour, preserving its original figure some thirty feet long and a foot to a foot and a half in height. Its height gradually diminished and after a chase of two miles I lost it in the windings of the channel. Such in the month of August 1834 was my first chance interview with that singular and beautiful phenomenon which I have called the Wave of Translation . . . .

Russell carried out many experiments on solitary waves and concluded that the persistence of their form was genuine and that the speed of propagation in a channel of uniform depth was

$$c = \sqrt{g(h + \eta)} \qquad (7.1.1)$$

where η is the amplitude of the wave, h is the depth of the (undisturbed) channel, and g is the gravitational constant.

Russell's results were controversial since it was not believed at that time that such a wave could be stable. The Astronomer Royal, Sir John Herschel, dismissed it as "merely half of a common wave that has been cut off." There was also a dispute with Airy, who had developed a shallow-water wave theory in which such waves were not stable. The controversy was resolved in 1895 by Korteweg and de Vries (1895), who derived an equation governing weakly nonlinear shallow-water waves of the form

$$\frac{\partial \eta}{\partial t} = \frac{3}{2} \sqrt{\frac{g}{h}} \left(\eta \frac{\partial \eta}{\partial x} + \frac{2}{3} \frac{\partial \eta}{\partial x} + \frac{1}{3} \sigma \frac{\partial^3 \eta}{\partial x^3} \right) \qquad (7.1.2)$$

in which $\sigma = h^3/3 - Th/g\rho$, where T is the surface tension of the liquid of density ρ. This equation was found to have solitary wave solutions of permanent form. After Korteweg and de Vries's work, the problem disappeared and it was not until the early 1960s that Eq. (7.1.2) reappeared in certain plasma physics problems.

At this stage we note that introduction of the scaled variables

$$t' = \frac{1}{2}\sqrt{\frac{g}{h\sigma}}\, t, \qquad x' = \frac{-x}{\sqrt{\sigma}}, \qquad u = -\frac{1}{2}\eta - \frac{1}{3}\alpha$$

reduces (7.1.2) to the form

$$u_t - 6uu_x + u_{xxx} = 0 \qquad (7.1.3)$$

7.1.b The FUP Experiment

A motivation for studying the Korteweg–de Vries equation (hereafter referred to as the KdV equation) was provided by the work of Fermi, Ulam, and Pasta (FUP) in 1955 (Fermi et al., 1955). Recall (from Chapter 3) that here the physical question was one of energy distribution in a chain of nonlinear oscillators. The speculation was that as the number of oscillators tended to infinity (the "statistical limit"), the energy would distribute itself uniformly among all the modes, implying ergodicity on the energy shell. Their model consisted of a one-dimensional nonlinear chain, of equal masses, with nearest neighbors connected by a force law of the form $F(\Delta) = k(\Delta + \alpha\Delta^2)$. This gave the following system of coupled nonlinear o.d.e.'s:

$$m\ddot{y}_i = k(y_{i+1} + y_{i-1} - 2y_i) + k\alpha[(y_{i+1} - y_i)^2 - (y_i - y_{i-1})^2] \qquad (7.1.4)$$

where $y_i = y_i(t)$ $(i = 1, 2, \ldots, N-1)$ and $y_0 = y_n = 0$. Initial conditions were typically chosen to be $y_i(0) = \sin(i\pi/N)$, $\dot{y}_i(0) = 0$. Working with $N = 64$ the system of equations (7.1.4) was integrated numerically on the Los Alamos MANIAC computer. (It is worth noting that this was one of the first peacetime uses of the computer for scientific research.†) Their results showed that the bulk of the energy tended to cycle periodically through the initially populated modes and that there was little energy sharing—an unexpected result at the time.

7.1.c Discovery of the Soliton

The scene now changes to Princeton, 1965, for the work of Kruskal and Zabusky. They were interested in the continuum limit of the Fermi–Ulam–Pasta chain, which they derived in the following way (Zabusky and Kruskal (1965)). Setting the distance between the springs to be h and introducing

---

†In 1977, at the First International Conference on Stochastic Behavior in Classical and Quantum Systems held in Como, Italy, Dr. Pasta reminisced about those computations. The program was, of course, punched on cards. A "DO" loop was executed by the operator feeding in the deck of cards over and over again until the loop was completed!

the variables $t' = \omega t$, $\omega = \sqrt{k/m}$, and $x' = x/h$ (where $x = ih$), they showed that by expanding the $y_{i\pm1}$ in Taylor series to fourth order in h, (7.1.4) becomes (dropping primes)

$$y_{tt} = y_{xx} + \epsilon y_x y_{xx} + \frac{h^2}{12} y_{xxxx} + O(\epsilon h^2, h^4) \tag{7.1.5}$$

where $\epsilon = 2\alpha h$. The next stage is to look for an asymptotic solution of the form

$$y \sim \phi(x, T)$$

where $T = \epsilon t/2$ and $X = x - t$, that is, a right moving wave. Noting that $y_t = -\phi_X + \frac{1}{2}\epsilon\phi_T$, one obtains

$$\phi_{TX} + \phi_X \phi_{XX} + \delta^2 \phi_{XXXX} = 0 \tag{7.1.6}$$

where $\delta = h^2/12$. Finally, setting $u = \phi_X$ yields

$$u_T + uu_X + \delta^2 u_{XXX} = 0 \tag{7.1.7}$$

which is, within trivial scalings, just the reduced form of the KdV equation (7.1.3)!

Zabusky and Kruskal (1965) studied the KdV equation (7.1.7) numerically, imposing the periodic boundary conditions $u(L, t) = u(0, t)$, $u_x(L, t) = u_x(0, t)$, and $u_{xx}(L, t) = u_{xx}(0, t)$. (We immediately comment that this choice of periodic boundary conditions was for numerical convenience and does not affect the fundamental result.) Working with an initial condition of the form $u(x, 0) = \cos(2\pi x/L)$, $0 \le x \le L$, they found that the solution broke up into a train of (eight) solitary waves of successively larger amplitude. The larger waves traveled faster than the smaller ones, and, remarkably, the former traveled "through" the latter and emerged from the "collisions" apparently unscathed! This behavior is akin to the superposition principle for linear waves, although in this case the waves are highly nonlinear. The term *soliton* was introduced by Zabusky and Kruskal (1965) to describe these remarkably stable nonlinear solutions. Their numerical results were followed by the development of a remarkable new solution technique by Kruskal and co-workers (Miura et al., 1968) which led to the development of a whole new area of mathematical physics—which might loosely be termed *soliton mathematics*. To begin with, though, we first investigate some of the more elementary properties of the KdV equation.

7.2   BASIC PROPERTIES OF THE KdV EQUATION

There are two basic forces at work in the KdV equation—which from now on we consider in the form (7.1.3). These are (1) the nonlinearity, that is, the term uu_x, which tends to "sharpen" the wave up, and (2) the dispersion, due to the term u_{xxx}, which tends to "spread" the wave out.

7.2.a   Effects of Nonlinearity and Dispersion

For a smooth initial condition, such as the one considered by Zabusky and Kruskal (1965), the term u_{xxx} is relatively small compared to the nonlinear term and one can consider the initial evolution to be governed by

$$u_t - 6uu_x = 0 \qquad (7.2.1)$$

This is a standard, quasi-linear, first-order p.d.e which is capable of developing shock solutions. Briefly, this can be seen as follows.† Consider the solution to (7.2.1) to be of the form

$$u(s) = u(x(s), t(s)) \qquad (7.2.2)$$

where s parameterizes certain paths, termed *characteristics*, in the (x, t)-plane. Then, from the differential equation

$$\frac{du}{ds} = \frac{dx}{ds} u_x + \frac{dt}{ds} u_t \qquad (7.2.3)$$

we deduce

$$\frac{dt}{ds} = 1 \qquad (7.2.4a)$$

$$\frac{dx}{ds} = -6u \qquad (7.2.4b)$$

$$\frac{du}{ds} = 0 \qquad (7.2.4c)$$

The set of equations (7.2.4) are easily integrated (ignoring constants of

---

†A fuller discussion can be found in any standard text on partial differential equations, such as Carrier and Pearson (1976).

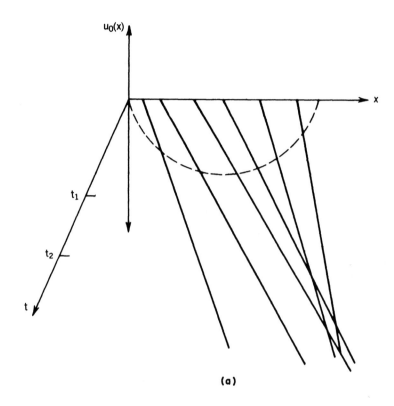

$u_0(x)$

x

t_1

t_2

t

(a)

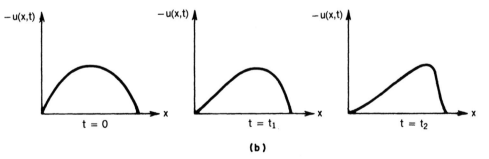

$-u(x,t)$

x

$t = 0$

$-u(x,t)$

x

$t = t_1$

$-u(x,t)$

x

$t = t_2$

(b)

Figure 7.1 (a) Straight-line characteristics associated with Eq. (7.2.1) for smooth initial condition $u_0(x)$ (dotted line). (b) Sequence showing sharpening of $u(x, t)$ as solution evolves toward crossing characteristics.

integration) to give

$$t(s) = s \tag{7.2.5a}$$

$$x(s) = -6su_0(x) \tag{7.2.5b}$$

$$u(s) = u_0(x) \tag{7.2.5c}$$

where $u_0(x) = u(x, 0)$ is the initial condition for (7.2.1). Thus along the characteristics defined by (7.2.5a) and (7.2.5b), the solution $u(s)$ is constant, that is, maintaining the initial amplitude given at $s = 0$. However, the characteristics—which are straight line paths—have a slope proportional to $u_0(x)$, and, depending on the precise shape of this initial data, the characteristics can cross. It is not difficult to see (see Figure 7.1) that this will lead to a steepening of the wave and hence to a wave-breaking-like phenomenon known as *shock formation*.

As the wave steepens, the term u_{xxx} in (7.1.3) will become significant and we must now consider how the evolution of the wave will be affected by the linear part of the equation

$$u_t + u_{xxx} = 0 \tag{7.2.6}$$

Such an equation always admits a solution of the form

$$u(x, t) = e^{i(kx - \omega t)} \tag{7.2.7}$$

and, by direct substitution into (7.2.6), one obtains the "dispersion relation"

$$\omega(k) = -k^3 \tag{7.2.8}$$

Thus longer wavenumbers travel with faster phase velocities, given by $c = \omega(k)/k = -k^2$, and the wave (7.2.7) will spread out as it evolves. Thus, in some sense, this dispersive effect will balance the nonlinear effects and lead to the formation of the stable solitary waves.

7.2.b A Traveling Wave Solution

A simple form of solitary wave solution can be obtained as follows. One assumes a right traveling wave solution of the form

$$u(x, t) = f(x - ct) \equiv f(z) \tag{7.2.9}$$

where $z = x - ct$, and, by direct substitution into (7.1.3), one obtains the ordinary differential equation (prime denotes differentiation with respect to z)

$$f''' - 6ff'' - cf' = 0 \tag{7.2.10}$$

A first integration with respect to z yields

$$f'' = 3f^2 + cf + d \qquad (7.2.11)$$

where d is a constant of integration, which is the o.d.e. for Weierstrass elliptic functions described in Chapter 1. A second integration yields

$$(\tfrac{1}{2}f')^2 = f^3 + \tfrac{1}{2}cf^2 + df + e \qquad (7.2.12)$$

where e is a second constant of integration. Equation (7.2.12) can now be integrated by quadratures to yield the elliptic integral

$$z - z_0 = \int \frac{df}{\sqrt{2(f^3 + \tfrac{1}{2}cf^2 + df + e)}} \qquad (7.2.13)$$

If (7.1.3) is defined on the infinite domain and one takes the boundary conditions f, f', $f'' \to 0$ as $z \to \pm\infty$, it is easy to deduce from (7.2.11) and (7.2.12) that both the constants of integration d and e are zero. In this case the quadrature (7.2.13) reduces to

$$z - z_0 = \int \frac{df}{f\sqrt{2f + c}} \qquad (7.2.14)$$

which is easily integrated and inverted to yield

$$f(z) = -\tfrac{1}{2}c \,\mathrm{sech}^2(\tfrac{1}{2}\sqrt{c}(z - z_0)) \qquad (7.2.15)$$

Owing to the minus sign, the solution is of the form of a negative amplitude traveling wave (the sign would be positive if (7.1.3) were $u_t + 6uu_x + u_{xxx} = 0$), with the amplitude proportional to wave speed; that is, larger waves travel faster. The sech^2 form of (7.2.15) gives the wave its localized "lump"-like structure as first seen by Russell (1844). In the numerical experiments of Zabusky and Kruskal (1965), each member of the family of traveling waves was also observed to have a sech^2-like shape. However, the reason for the appearance of such a family and their stability cannot be explained by a simple traveling wave analysis. A much deeper theory is required.

7.2.c Similarity Solutions

Another type of solution to (7.1.3) can also be found by what is termed a *similarity transformation*. If the variables x, t, u are scaled according to $x \to k^\alpha x$, $t \to k^\beta t$, $u \to k^\gamma u$, direct substitution into the KdV equation shows that it is invariant to these scalings if $\beta = 3\alpha$ and $\gamma = -2\alpha$. Any choice of α

may be made, for example, $\alpha = 1$. Thus (7.1.3) is invariant to the scalings

$$x \to kx, \qquad t \to k^3 t, \qquad u \to k^{-2} u$$

Furthermore, the combinations of variables

$$x/t^{1/3}, \qquad ut^{2/3}$$

are also scale invariant. These results suggest a change of variable, namely,

$$z = x/(3t)^{1/3} \tag{7.2.16a}$$

$$u(x, t) = -(3t)^{-2/3} f(z) \tag{7.2.16b}$$

where the factor 3 in the scalings has been chosen for convenience. Noting that

$$\frac{\partial}{\partial t} = \frac{\partial z}{\partial t} \frac{\partial}{\partial z} = -\frac{z}{3t} \frac{\partial}{\partial z}$$

and

$$\frac{\partial}{\partial x} = \frac{\partial z}{\partial x} \frac{\partial}{\partial z} = \frac{1}{(3t)^{1/3}} \frac{\partial}{\partial z}$$

the KdV equation transforms to

$$f''' + (6f - z)f' - 2f = 0 \tag{7.2.17}$$

Another type of similarity reduction is obtained by setting

$$z = x + 3t^2 \tag{7.2.18a}$$

and

$$u(x, t) = t + f(z) \tag{7.2.18b}$$

This leads (after one integration) to

$$f'' = 3f^2 - t + c \tag{7.2.19}$$

where c is a constant of integration. This equation is a special ordinary differential equation known as the *first Painlevé transcendent*. Equation (7.2.17) is related to another of these special equations, namely, the *second Painlevé transcendent*. The significance of the appearance of these special ordinary differential equations on making similarity reductions of the KdV equation will be discussed in Chapter 8.

7.2.d Conservation Laws

If we think of a p.d.e., such as the KdV equation, to be a dynamical system with an infinite number of degrees of freedom, it is natural to ask, in keeping with our preceding discussions, if the equations have any integrals of motion. For p.d.e.'s the notion of integrals of motion is replaced by the notion of *conservation laws*. These are relations of the form

$$T_t + X_x = 0 \qquad (7.2.20)$$

where T and X are certain functions of the solution to the p.d.e., u, and its derivatives. T is termed the *density*, and $-X$ is called the *flux*. If T and X are connected by a gradient relationship (i.e., $T = F_x$) and hence from (7.2.20), $X = -F_t$, the conservation law is trivial since

$$(F_x)_t + (-F_x)_t = 0 \qquad (7.2.21)$$

If, for systems defined on the infinite interval $(-\infty \le x \le \infty)$, the flux X decays to zero as $x \to \infty$, integrating both sides of (7.2.20) with respect to x yields

$$\frac{\partial}{\partial t} \int_{-\infty}^{\infty} T\, dx = \int_{-\infty}^{\infty} X_x\, dx = \left. X \right|_{-\infty}^{\infty} = 0 \qquad (7.2.22)$$

This has the consequence that

$$\int_{-\infty}^{\infty} T\, dx = \text{constant} \qquad (7.2.23)$$

Thus we can think of these quantities as the p.d.e. analogues to the integrals of motion for o.d.e.'s.

For the KdV equation (7.1.3) the equation itself is in conservation law form, that is,

$$(u)_t + (-3u^2 + u_{xx})_x = 0$$

From this conservation law we see that

$$\int_{-\infty}^{\infty} u\, dx = \text{constant} \qquad (7.2.24)$$

which represents the conservation of mass. Multiplying the KdV equation through by u, it is not difficult to obtain the second conservation law

$$(u^2)_t + (-2u^3 - \tfrac{1}{2}u_x^2 + uu_{xx})_x = 0$$

In this case we see that

$$\int_{-\infty}^{\infty} u^2 \, dx = \text{constant} \tag{7.2.25}$$

which represents the conservation of momentum. A certain amount of experimentation yields a third conservation law of the form

$$(u^3 + \tfrac{1}{2}u_x^2)_t = (\tfrac{9}{2}u^4 - 3u^2 u_{xx} + 6uu_x^2 - u_x u_{xxx} + \tfrac{1}{2}u_{xx}^2)_x \tag{7.2.26}$$

and hence

$$\int_{-\infty}^{\infty} (u^3 + \tfrac{1}{2}u_x^2) \, dx = \text{constant} \tag{7.2.27}$$

Later on we shall describe how the integral (7.2.27) actually represents the Hamiltonian for the KdV equation.

Having found three conservation laws, one naturally asks if there are any more and, indeed, if there could be an infinite number corresponding to the infinite number of degrees of freedom? This latter result would imply, in some sense, a type of complete "integrability." Kruskal and co-workers (Miura et al., 1968) found, at first, by little more than brute-force pencil-and-paper computations, nine conservation laws. A heroic effort by Miura (1968) produced a tenth, which, at that time, strongly suggested that there were indeed an infinite number of conserved quantities.

7.2.e The Miura Transformation

An important part of Miura's investigation of conservation laws for the KdV equation was the simultaneous study of a closely related equation, called the modified KdV (mKdV) equation, which takes the form

$$u_t + 6u^2 u_x + u_{xxx} = 0 \tag{7.2.28}$$

This equation can be derived, in a similar manner to the KdV equation, from the Fermi–Ulam–Pasta lattice if the nonlinearity is taken to be cubic rather than quadratic. Miura found a set of conservation laws for the mKdV equation parallel to the set for the KdV equation. His crucial observation was that the two sets of conservation laws were connected by the "Miura transformation"

$$u = v_x + v^2 \tag{7.2.29}$$

where u and v denote solutions to the KdV and mKdV equations, respectively. Furthermore, if we make the notation

$$P(u) = u_t - 6uu_x + u_{xxx} = 0 \tag{7.2.30a}$$

and

$$K(v) = v_t + 6v^2 v_x + v_{xxx} = 0 \qquad (7.2.30\text{b})$$

then one finds, using (7.2.29), that

$$P(u) = \left(2v + \frac{\partial}{\partial x}\right) K(v) \qquad (7.2.31)$$

These results subsequently motivated Miura et al. (1968) to introduce a slightly different transformation of the form

$$u = w + \epsilon w_x + \epsilon^2 w^2 \qquad (7.2.32)$$

where ϵ is some (small) parameter. With this transformation, one finds that

$$P(u) = \left(1 + \epsilon \frac{\partial}{\partial x} + 2\epsilon^2 w\right) Q(w) \qquad (7.2.33)$$

where

$$Q(w) = w_t - 6(w + \epsilon^2 w^2) w_x + w_{xxx} = 0 \qquad (7.2.34)$$

is known as the *Gardner equation*. Note that $Q(w)$ can also be written in conservation law form, that is,

$$(w)_t + (-3w^2 - 2\epsilon^2 w^3 + w_{xx})_x = 0 \qquad (7.2.35)$$

The idea is to expand w in a small ϵ power series, that is,

$$w = \sum_{j=0}^{\infty} \epsilon^j w_j \qquad (7.2.36)$$

The individual w_j are easily found by solving (7.2.32) recursively, giving

$$w_0 = u \qquad (7.2.37\text{a})$$

$$w_1 = -u_x \qquad (7.2.37\text{b})$$

$$w_2 = u_{xx} - u^2 \qquad (7.2.37\text{c})$$

and so on. The conservation laws are found by substituting (7.2.36) into (7.2.35) and equating powers of ϵ. This result follows from the fact that

(7.2.36) is itself in conservation form. The first few laws are

$$O(\epsilon^0): \quad (w_0)_t = (3w_0^2 - w_{0_{xx}})_x \tag{7.2.38a}$$

$$O(\epsilon^1): \quad (w_1)_t = (6w_0 w_1 - w_{1_{xx}})_x \tag{7.2.38b}$$

$$O(\epsilon^2): \quad (w_2)_t = (3w_1^2 + 6w_0 w_2 + 2w_0^3 - w_{2_{xx}})_x \tag{7.2.38c}$$

The reader will be able to verify that (7.2.38a) and (7.2.38c) correspond to the conservation laws (7.2.24) and (7.2.25), respectively. (Derivation of the latter requires use of (7.2.38a) to rearrange the left-hand side of (7.2.38c).) The law found at $O(\epsilon^1)$ is just the differential of the law found at $O(\epsilon^0)$. This is a general result: The conservation laws found at the odd powers of ϵ are just derivatives of those found at the preceding even powers.

7.2.f Galilean Invariance

Gardner's transformation provides an algorithm to compute an infinity of conserved densities for the KdV equation. As discussed in Chapter 2 the existence of an integral—here a conserved density—implies the existence of some special symmetry or invariance. That the KdV equation should possess an infinity of such symmetries suggests that it must have some very special properties.

One basic invariance possessed by the KdV equation is Galilean (or translation) invariance. If one makes the change of variables, namely,

$$t' = t, \qquad x' = x - ct, \qquad u'(x', t') = u(x, t) + \tfrac{1}{6}c$$

which corresponds to transforming to a frame of reference moving to the right, the KdV equation becomes

$$u'_{t'} - 6u' u'_{x'} + u'_{x'x'x'} = 0$$

that is, it is invariant to such a transformation. By contrast, the mKdV equation is easily seen not to be Galilean invariant. However, if the above change of variables is applied to the Miura transformation (7.2.29), one obtains the Gardner transformation (7.2.32) on setting $c = \tfrac{3}{2}\epsilon^2$.

7.3 THE INVERSE SCATTERING TRANSFORM: BASIC PRINCIPLES

So far the basic facts we have learned about the KdV equation are: (1) it exhibits (numerically) solitons, (2) it possesses a variety of special solutions, (3) it is Galilean invariant, and (4) it possesses an infinite number of conservation laws which are connected to those of the mKdV equation through the Miura transformation

$$v_x + v^2 = u \tag{7.3.1}$$

This was, approximately, the information at the disposal of Gardner, Greene, Kruskal, and Miura (GGKM) in 1967 (Gardner et al., 1967). Their observation was that (7.3.1) is a Riccati equation for v which can be linearized (see Chapter 1) by making the substitution

$$v = \frac{\psi_x}{\psi} \tag{7.3.2}$$

to yield

$$\psi_{xx} = u(x, t)\psi \tag{7.3.3}$$

Furthermore, since the KdV equation is Galilean invariant, u can be replaced by $u - \lambda$, where λ is any (at this stage) constant. On making this shift, (7.3.3) becomes

$$\psi_{xx} - (u(x, t) - \lambda)\psi = 0 \tag{7.3.4}$$

which is just the one-dimensional time-independent Schrödinger equation for a "potential" $u(x, t)$ with eigenvalues λ.

7.3.a The Connection with Quantum Mechanics

GGKM then made a remarkable intuitive leap by proposing that the time evolution of $u(x, t)$, according to the KdV equation, could be studied through the properties of the quantum mechanical problem (7.3.4). Their idea was as follows. Given an initial condition $u = u(x, 0)$, solve the "*direct scattering*" problem, that is, treat $u(x, 0)$ as a potential in the Schrödinger equation (7.3.4) and find all the associated eigenvalues and eigenfunctions. As u evolves, or deforms as a function of t, these associated quantum mechanical properties—termed the *scattering data*—will also evolve. At this point it is most important to emphasize that the variable t in $u(x, t)$ should be thought of as some *deformation parameter* in the KdV equation and should in no way be confused with the time variable that apppears in the traditional time-dependent Schrödinger equation. The idea of GGKM was that the evolution of the scattering data, initially obtained from $u(x, 0)$, might somehow be obtained without having to solve the KdV equation directly. If this could be achieved, the scattering data, thereby obtained at some later value of t, could be used to "reconstruct" the "potential" $u(x, t)$.

This latter step involves solving the quantum mechanical *inverse scattering* problem, that is, going from the scattering data to the potential, in contrast to the direct scattering problem of going from the potential to the scattering data. This indirect route to solving the KdV equation is sketched below:

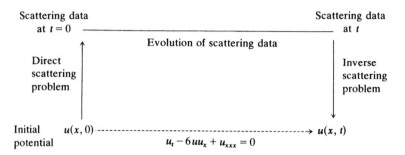

7.3.b Analogy with Fourier Transforms

Such a scheme is not as far fetched as it might at first sound and is, in fact, closely analogous to the use of Fourier transforms in solving linear evolution equations. Consider such an equation, on the interval $-\infty \le x \le \infty$, of the form

$$u_t = \mathscr{L}\left(\frac{\partial}{\partial x}\right)u \tag{7.3.5}$$

where $\mathscr{L}(\partial/\partial x)$ is a polynomial in $\partial/\partial x$, that is, a linear operator. A simple example would be $\mathscr{L} = \partial^2/\partial x^2$, in which case (7.3.5) is just the standard diffusion equation. Now define the Fourier transform of $u(x, t)$:

$$\bar{u}(k, t) = \int_{-\infty}^{\infty} u(x, t)e^{ikx}\, dx \tag{7.3.6}$$

Also define the "inverse" Fourier transform:

$$u(x, t) = \frac{1}{2\pi}\int_{-\infty}^{\infty} \bar{u}(k, t)e^{-ikx}\, dk \tag{7.3.7}$$

Fourier transforming the p.d.e. (7.3.5) gives the evolution equation for $\bar{u}(k, t)$, that is,

$$\frac{d\bar{u}}{dt} = \mathscr{L}(ik)\bar{u}$$

This linear equation has the simple solution

$$\bar{u}(k, t) = \bar{u}(k, 0)e^{\mathscr{L}(ik)t} \tag{7.3.8}$$

where the initial Fourier transform data $\bar{u}(k, 0)$ is determined from the given initial data $u(x, 0)$, that is,

$$\bar{u}(k, 0) = \int_{-\infty}^{\infty} u(x, 0) e^{ikx} \, dx \qquad (7.3.9)$$

The evolution of the "Fourier data" $\bar{u}(k, t)$ is governed by the trivial relation (7.3.8) and can be inverted at any subsequent value of t to give the desired $u(x, t)$, namely,

$$u(x, t) = \frac{1}{2\pi} \int_{-\infty}^{\infty} \bar{u}(k, 0) e^{\mathcal{L}(ik)t} e^{-ikx} \, dk \qquad (7.3.10)$$

Thus the solution route is analogous to that proposed by GGKM, that is,

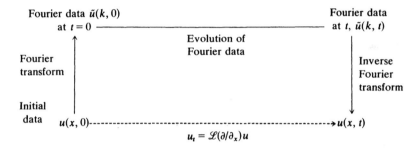

Of course the GGKM method is more complicated since the evolution is nonlinear. To understand how it works, we must first discuss in more detail the quantum mechanical direct and inverse scattering problems.

7.3.c The Direct Scattering Problem

Depending on its precise shape, a given potential $u_0(x) = u(x, 0)$ can support bound states. The Schrödinger equation (7.3.4) will then admit a corresponding set of discrete eigenvalues, $\lambda_n = -k_n^2$ $(n = 1, \ldots, N)$ corresponding to "negative energy" bound states with associated eigenfunctions $\psi_n(x)$, that is,

$$\psi_{n,xx} = (u_0(x) + k_n^2) \psi_n = 0 \qquad (7.3.11)$$

Bound-state eigenfunctions are required to be square integrable and normalized to unity, that is,

$$\int_{-\infty}^{\infty} |\psi_n(x)|^2 \, dx = 1 \qquad (7.3.12)$$

The normalizing constant, c_n, that ensures (7.3.12) is defined by

$$\lim_{x \to \infty} e^{k_n x} \psi_n(x) = c_n \tag{7.3.13}$$

This follows from the assumption that $u_0(x)$ decays to zero sufficiently rapidly as $x \to \pm\infty$ such that (7.3.11) reduces to $\psi_{n,xx} - k_n^2 \psi_n = 0$. One may also define, equivalently, the c_n by

$$c_n = \left[\int_{-\infty}^{\infty} |\psi_n(x)|^2 \, dx \right]^{-1/2} \tag{7.3.14}$$

The set of eigenvalues λ_n $(n = 1, \ldots, N)$ is termed the bound-state spectrum.

At positive energy the Schrödinger equation for $u_0(x)$ exhibits a continuous spectrum and we set $\lambda = k^2$. As is well known, quantal wavefunctions can exhibit reflection above a potential barrier. Thus the asymptotic form of $\psi(x)$ in the limit $x \to \infty$ is

$$\lim_{x \to \infty} \psi(x) = e^{-ikx} + b(k)e^{+ikx} \tag{7.3.15}$$

where the first term on the right-hand side represents an incoming wave and the second term represents the reflected wave with *reflection coefficient* $b(k)$. In the limit $x \to -\infty$, we have

$$\lim_{x \to -\infty} \psi(x) = a(k)e^{-ikx} \tag{7.3.16}$$

which represents the transmitted wave with *transmission coefficient* $a(k)$.

7.3.d The Inverse Scattering Problem

The term *scattering data* is used to mean, for a given potential, the set of all bound-state eigenvalues, λ_n, normalizing constants, c_n, and the continuum functions $a(k)$ and $b(k)$. A remarkable result of Gelfand, Levitan, and Marchenko in the 1950s (quite independent of anything to do with solitons!) was to show how the scattering data could be used to uniquely find the associated potential function $u_0(x)$ (Gelfand and Levitan, 1955; Marchenko, 1955). Assuming that $u_0(x)$ satisfies the boundedness condition

$$\int_{-\infty}^{\infty} (1 + |x|) |u_0(x)| \, dx < \infty \tag{7.3.17}$$

one defines the following quantity

$$B(\zeta) = \sum_{n=1}^{N} c_n^2 e^{-k_n \zeta} + \frac{1}{2\pi} \int_{-\infty}^{\infty} b(k) e^{ik\zeta} \, dk \tag{7.3.18}$$

which can be thought of as a sort of Fourier transform of the scattering data. The next step is to solve the following linear integral equation

$$K(x, y) + B(x + y) + \int_{-\infty}^{\infty} B(x + z)K(z, y)\, dz = 0 \qquad (7.3.19)$$

for the function $K(x, y)$. If $K(x, y)$ can be found, then it is possible to show that the potential $u_0(x)$ giving rise to the scattering data used in (7.3.18) is given by

$$u_0(x) = -2(d/dx)K(x, x) \qquad (7.3.20)$$

A truly remarkable result!

In order to use all this to solve the KdV equation, we have to:

(i) find out how the scattering data "evolves" as $u_0(x)$ is "deformed" into $u(x, t)$ and

(ii) be able to solve the Gelfand–Levitan–Marchenko equation (Eq. (7.3.19)).

As it turns out, (i) can be solved relatively easily. The problem, in keeping with the conservation of difficulty principle, is (ii). Unfortunately, (7.3.19) can only be solved exactly for rather special cases, but these include the case required to explain the appearance of solitons. These results are discussed in the next section.

7.4 THE INVERSE SCATTERING TRANSFORM: THE KdV EQUATION

To begin with, we again emphasize that the variable t appearing in the KdV equation

$$u_t - 6uu_x + u_{xxx} = 0 \qquad (7.4.1)$$

should not be thought of as "real" time but rather as a deformation parameter. Thus in the Schrödinger equation we can assign a t dependence to the eigenvalue λ without confusion, that is,

$$\psi_{xx} - (u(x, t) - \lambda(t))\psi = 0 \qquad (7.4.2)$$

where $\psi = \psi(x, t)$.

7.4.a The Isospectral Deformation

Using (7.4.2) to express u as a function of ψ, that is,

$$u = \frac{\psi_{xx}}{\psi} + \lambda \qquad (7.4.3)$$

it follows that

$$u_t = \frac{\psi_{xxt}}{\psi} - \frac{\psi_{xx}\psi_t}{\psi^2} + \lambda_t \qquad (7.4.4)$$

with analogous expressions obtainable for uu_x and u_{xxx}. For these latter terms it is convenient to eliminate third and higher derivatives (with respect to x) of ψ by repeated use of (7.4.2). In this manner the KdV equation (7.4.1) can be reexpressed as

$$\lambda_t \psi^2 + (\psi M_x - \psi_x M)_x = 0 \qquad (7.4.5)$$

where

$$M = \psi_t - 2(u + 2\lambda)\psi_x + u_x\psi \qquad (7.4.6)$$

For bound-state eigenfunctions the ψ are square integrable, so integrating both sides of (7.4.5) from $-\infty$ to $+\infty$ yields

$$\lambda_t \int_{-\infty}^{\infty} \psi^2 \, dx = -\int_{-\infty}^{\infty} (\psi M_x - \psi_x M)_x \, dx$$

$$= -|\psi M_x - \psi_x M|_{-\infty}^{\infty}$$

$$= 0 \qquad (7.4.7)$$

Since $\int_{-\infty}^{\infty} \psi^2 \, dx$ is just a nonzero constant, (7.4.7) implies that

$$\lambda_t = 0 \qquad (7.4.8)$$

This is an immensely significant result since it tells us that for a potential $u(x, t)$, deformed according to the KdV equation, the bound-state eigenvalues $\lambda_n(t)$ $(n = 1, \ldots, N)$ remain *unchanged*! This is an example of what is termed an *isospectral deformation*. In the continuum (i.e., $\lambda > 0$), there is a solution to the Schrödinger equation for every value of λ. Thus we can simply argue that at every positive energy, λ is fixed and hence $\lambda_t = 0$. Either way, (7.4.5) gives

$$M_{xx}\psi - M\psi_{xx} = 0 \qquad (7.4.9)$$

and by using (7.4.2) for ψ_{xx} we write this as a second-order differential equation for M, that is,

$$M_{xx} - (u - \lambda)M = 0 \qquad (7.4.10)$$

The general solution to (7.4.10) is of the standard form

$$M = A\psi + B\varphi \tag{7.4.11}$$

where ψ and φ are the two linearly independent solutions. Obviously one of these solutions is just the eigenfunction ψ (just compare (7.4.2) and (7.4.10)). It is a standard result to show that the second solution φ can be computed from

$$\varphi = \psi \int^x \frac{dx'}{\psi^2} \tag{7.4.12}$$

which is easily verified by checking that the Wronskian $\varphi_x\psi - \psi_x\varphi = 1$. However, it is not difficult to show from the asymptotic properties of (7.4.10) that $B = 0$ for both the bound states and the continuum. (In the limit $x \to \pm\infty$, (7.4.10) becomes $M_{xx} + \lambda M = 0$; and using the asymptotic forms of ψ in (7.4.11), this can only be satisfied for nontrivial φ if $B = 0$.) Thus, overall, we have

$$M = \psi_t - 2(u + 2\lambda)\psi_x + u_x\psi = A\psi \tag{7.4.13}$$

For bound states, we can go further and also show that $A = 0$. Multiply both sides of (7.4.13) by ψ to obtain

$$\psi\psi_t - 2(u + 2\lambda)\psi^2 + u_x\psi^2 = A\psi^2 \tag{7.4.14}$$

and rewrite this as

$$\tfrac{1}{2}(\psi^2)_t + (u\psi^2 - 2\psi_x^2 - 4\lambda\psi^2)_x = A\psi^2 \tag{7.4.15}$$

For square integrable bound-state eigenfunctions, we can now integrate over x to obtain

$$\frac{1}{2}\left(\int_{-\infty}^{\infty} \psi^2 \, dx\right)_t + \left|u\psi^2 - 2\psi_x^2 - 4\lambda\psi^2\right|_{-\infty}^{\infty} = A \int_{-\infty}^{\infty} \psi^2 \, dx \tag{7.4.16}$$

The second term on the left-hand side of (7.4.16) is zero—as is the first term by the constancy of the normalization integral. Thus $A = 0$ and we have

$$\psi_t - 2(u + 2\lambda)\psi_x + u_x\psi = 0 \tag{7.4.17}$$

7.4.b Evolution of the Scattering Data

Equation (7.4.17) can now be used to derive the "evolution" equation for the normalization constants $c_n(t)$. In the limit $x \to \infty$, for suitably decaying u and u_x, (7.4.17) reduces to

$$\psi_{n,t} + 4k_n^2\psi_n = 0 \tag{7.4.18}$$

where we have set $\lambda = -k_n^2$ for the eigenfunction ψ_n. By definition,

$$\lim_{x\to\infty} \psi_n(x, t) = c_n(t)e^{-k_n x} \tag{7.4.19}$$

so by direct substitution of (7.4.19) into (7.4.18) we obtain the first-order ordinary differential equation for c_n, namely,

$$\frac{dc_n}{dt} = 4k_n^3 c_n \tag{7.4.20}$$

This has the simple solution

$$c_n(t) = c_n(0)e^{4k_n^3 t} \tag{7.4.21}$$

where the initial value $c_n(0)$ is the bound-state normalization constant for the nth eigenfunction of $u_0(x) = u(x, 0)$.

In the case of the continuum, we still have to work with (7.4.13), but setting $\lambda = k^2$ and again taking the limit $x \to \infty$, (7.4.13) reduces to

$$\psi_t - 4k^2\psi_n = A\psi \tag{7.4.22}$$

Recalling the asymptotic form

$$\lim_{x\to\infty} \psi(x, t) = e^{-ikx} + b(k, t)e^{ikx} \tag{7.4.23}$$

direct substitution into (7.4.22) and the choice $A = 4ik^3$ yields the evolution equation for $b(k, t)$, namely,

$$\frac{db}{dt} = 8ik^3 b \tag{7.4.24}$$

This has the solution

$$b(k, t) = b(k, 0)e^{8ik^3 t} \tag{7.4.25}$$

where $b(k, 0)$ is the reflection coefficient for $u_0(x)$. Using the same arguments for the limit $x \to -\infty$, it is easy to show that

$$\frac{da}{dt} = 0 \tag{7.4.26}$$

and hence

$$a(k, t) = a(k, 0) \tag{7.4.27}$$

The reader will now see that the desired miracle has occurred—namely, that under deformation according to the KdV equation the scattering data $\lambda_n(0)$, $c_n(0)$, $a(k, 0)$, $b(k, 0)$, associated with the initial potential $u_0(x)$, evolve according to simple linear equations. Since the deformation is isospectral, we have $\lambda_n(t) = \lambda_n(0)$. Using (7.4.21) and (7.4.25), we can construct the corresponding quantity

$$B(\zeta; t) = \sum_{n=1}^{\infty} c_n^2(t)e^{-k_n\zeta} + \frac{1}{2\pi}\int_{-\infty}^{\infty} b(k, t)e^{ik\zeta}\, dk \qquad (7.4.28)$$

and, on solving (if we are lucky!) the Gelfand–Levitan–Marchenko equation for $K(x, y; t)$, we obtain $u(x, t)$ from

$$u(x, t) = -2(d/dx)K(x, x; t) \qquad (7.4.29)$$

This whole procedure, called the *inverse scattering transform*, or IST, is, in effect, an indirect linearization of the KdV equation with a strikingly close analogy to the Fourier transform method. Indeed, IST is often referred to as a *nonlinear Fourier transformation*.

As was hinted earlier on, the real problem lies with solving the integral equation for $K(x, y; t)$. It turns out, to date, that it can only be solved in closed form for those scattering problems which are *reflectionless*, that is, those for which $b(k, t) = b(k, 0) = 0$.

7.4.c A Two-Soliton Solution

A standard illustration of the IST for the KdV equation is provided by the study of potentials of the form $u_0(x) = - V \operatorname{sech}^2 x$, where V is a constant. Here we work with the particular case

$$u(x, 0) = - 6 \operatorname{sech}^2 x \qquad (7.4.30)$$

The associated Schrödinger equation

$$\psi_{xx} + (6 \operatorname{sech}^2 x + \lambda)\psi = 0 \qquad (7.4.31)$$

can be solved exactly. (Details are given in the excellent account by Drazin (1983)). There are just two bound-state eigenfunctions:

$\psi_1 = \tfrac{1}{4}\operatorname{sech}^2(x)$,   with $\lambda_1 = - k_1^2 = - 4$ and $c_1^2(0) = 12$

$\psi_2 = \tfrac{1}{2}\tanh(x)\operatorname{sech}(x)$,   with $\lambda_2 = - k_2^2 = - 1$ and $c_2^2(0) = 6$

Fortunately, all sech^2 potentials and reflectionless, so $b(k) = 0$. Thus

$$B(\zeta, t) = \sum_{n=1}^{2} c_n^2(t) e^{-k_n \zeta}$$

$$= \sum_{n=1}^{2} c_n^2(0) e^{8k_n^3 t - k_n \zeta}$$

$$= 12 e^{64t - 2\zeta} + 6 e^{8t - \zeta} \qquad (7.4.32)$$

For reflectionless potentials the Gelfand–Levitan–Marchenko equation can be solved by assuming the kernel to be of the form

$$K(x, y; t) = \sum_{n=1}^{N} p_n(x, t) e^{-k_n y} \qquad (7.4.33)$$

Using this separation for the problem at hand, one may show that

$$K(x, y; t) = \frac{-3(2e^{28t+x-2y} + 2e^{36t-x-2y} - e^{36t-2x-y} + e^{-24t+2x-y})}{(3\cosh(x - 28t) + \cosh(3x - 36t))} \qquad (7.4.34)$$

Then, using (7.4.29), the solution can (eventually) be expressed as

$$u(x, t) = -\frac{12(3 + 4\cosh(2x - 8t) + \cosh(4x - 64t))}{(3\cosh(x - 28t) + \cosh(3x - 36t))^2} \qquad (7.4.35)$$

Note that $u(x, 0) = -6\,\mathrm{sech}^2 x$. To understand the properties of this solution we follow the analysis given by Drazin (1983). This involves introducing the variables $x_1 = x - 4k_1^2 t = x - 16t$ and $x_2 = x - 4k_2^2 t = x - 4t$. Expressing the arguments of the cosh terms in (7.4.35) in terms of x_1 we obtain

$$u(x, t) = -\frac{12(3 + 4\cosh(2x_1 + 24t) + \cosh(4x_1))}{(3\cosh(x_1 - 12t) + \cosh(3x_1 + 12t))^2} \qquad (7.4.36)$$

Now, for fixed x_1, take the limit $t \to \infty$ and discard the exponentially decaying portions of the cosh terms, that is,

$$\lim_{t \to \infty} u(x, t) \simeq \frac{-96 e^{2x_1 + 24t}}{(3e^{-x_1 + 12t} + e^{3x_1 + 12t})^2}$$

$$= \frac{-32}{((1/\sqrt{3})e^{2x_1} + \sqrt{3}e^{-2x_1})^2}$$

$$= \frac{-32}{(e^{2x_1 - \ln\sqrt{3}} + e^{-2x_1 + \ln\sqrt{3}})^2}$$

$$= -8\,\mathrm{sech}^2(2x_1 + \delta) \qquad (7.4.37)$$

where $\delta = \ln\sqrt{3}$. Similarly, by expressing (7.4.35) in terms of x_2, we obtain

$$u(x, t) = \frac{-12(3 + 4\cosh(2x_2) + \cosh(4x_2 - 48t))}{(3\cosh(x_2 - 24t) + \cosh(3x_2 - 24t))^2}$$

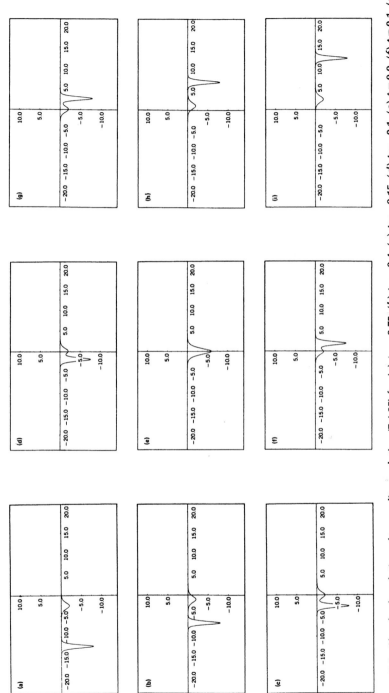

Figure 7.2 Sketch of evolution of two-soliton solution (7.4.35) for (a) $t = -0.75$, (b) $t = -0.4$, (c) $t = -0.15$, (d) $t = -0.1$, (e) $t = 0.0$, (f) $t = 0.1$, (g) $t = 0.15$, (h) $t = 0.4$, and (i) $t = 0.75$ showing the two separate solitons of depths 8 and 2, respectively, merging at $t = 0$ to give the initial potential $u(x, 0) = -6 \operatorname{sech}^2 x$ and then separating again with the deeper wave overtaking the smaller one. The "mass" $\int_{-\infty}^{\infty} u(x, t)\, dx$ is conserved throughout. (Computation by Mr. E. Dresselhaus (private communication).)

and, on taking the limit $t \to \infty$, we have

$$
\begin{aligned}
\lim_{t \to \infty} u(x, t) &\simeq \frac{-24e^{-4x_2+48t}}{(3e^{-x_2+24t} + e^{-3x_2+24t})^2} \\
&= \frac{-8}{(\sqrt{3}e^{x_2} + (1/\sqrt{3})e^{-x_2})^2} \\
&= -2 \operatorname{sech}^2(x_2 + \delta)
\end{aligned}
\tag{7.4.38}
$$

In the same spirit, one can, in fact, show that in the limit $t \to \pm\infty$ for either fixed x_1 or x_2, $u(x, t)$ behaves as

$$
\lim_{t \to \pm\infty} u(x, t) = -2 \operatorname{sech}^2(x_2 \pm \delta) - 8 \operatorname{sech}^2(2x_1 \mp \delta)
\tag{7.4.39}
$$

From the above analysis we can now understand the behavior observed by Zabusky and Kruskal (1965). The solution represents the interaction of two solitary traveling waves (cf discussion of traveling wave solutions in the previous section). At $t = -\infty$ the deeper wave lies to left of the shallow one (see Figure 7.2). As $t \to 0$, the deeper wave catches up with the shallower one and at $t = 0$ they merge into the original potential $u(x, 0) = -6 \operatorname{sech}^2 x$. Remarkably, as $t \to \infty$ they separate again, with the deeper wave moving ahead of the shallower one. At $t = +\infty$ the solution is again just the sum of two separate solitary waves, with the only consequence of the collision being the small phase shift δ.

7.4.d   More General Solutions

The result obtained for this "two-soliton" potential is fairly easily generalized to any potential of the form $u(x, 0) = - V \operatorname{sech}^2 x$. For such a potential supporting N bound states with eigenvalues $\lambda_n = - k_n^2$ $(n = 1, \ldots, N)$, the asymptotic form of solution is

$$
\lim_{t \to +\infty} u(x, t) = \sum_{n=1}^{N} -2\lambda_n^2 \operatorname{sech}^2(k_n(x - 4k_n^2 t - \delta_n))
\tag{7.4.40}
$$

where the phase shift δ_n is given by

$$
\delta_n = \frac{1}{2k_n} \ln \left\{ \frac{c_n^2(0)}{2k_n} \prod_{m=1}^{N-1} \left(\frac{k_n - k_m}{k_n + k_m} \right)^2 \right\}
\tag{7.4.41}
$$

Such a solution is called an *N-soliton solution*. As $t \to \infty$ the initial condition takes up into a train of solitary traveling waves, with the deepest at the front and the shallowest at the rear. As t goes from $-\infty$ to $+\infty$, the only effect of the interactions is just the phase shift δ_n.

Our identification of the solitons has come about through considering the limit $t \to \pm\infty$. In fact, it is possible to show that the N-soliton solutions can be represented exactly in the general form

$$u(x, t) = \sum_{n=1}^{N} -4k_n \psi_n^2(x, t) \qquad (7.4.42)$$

where the ψ_n are the bound-state eigenfunctions with eigenvalues $\lambda_n = -k_n^2$.

For potentials which are not reflectionless, the continuum portion of the quantal spectrum renders the Gelfand–Levitan–Marchenko equation intractable to exact solution. However, the same basic picture holds, and in the limit $t \to \infty$ the initial condition separates into a procession of isolated traveling waves. The continuum has the effect of introducing an oscillatory portion to the solution which dies out dispersively as $t \to \infty$. This phenomenon is sometimes termed *radiation*. Little is known about the exact nature of this part of the solution, although some asymptotic estimates are available. A discussion of some of these results can be found, for example, in Ablowitz and Segur (1981) or Drazin (1983). (The latter also gives a simple example.)

7.4.e The Lax Pair*

The reader will have noticed that the quantum mechanical problem boils down (for the bound state spectrum) to the pair of linear equations

$$\psi_{xx} = (u - \lambda)\psi \qquad (7.4.43a)$$

and

$$\psi_t = 2(u + 2\lambda)\psi_x - u_x\psi \qquad (7.4.43b)$$

In order for these equations to be consistent with each other, they must satisfy the "integrability condition"

$$\psi_{xxt} = \psi_{txx} \qquad (7.4.44)$$

Differentiating (7.4.43a) with respect to t and using (7.4.43b) yields

$$\psi_{xxt} = (u_t - uu_x + \lambda u_x - \lambda_t)\psi + 2(u + 2\lambda)(u - \lambda)\psi_x \qquad (7.4.45)$$

where, for now, we are not assuming $\lambda_t = 0$. Similarly, differentiating (7.4.43b) with respect to x finally gives

$$\psi_{txx} = (5uu_x + \lambda u_x - u_{xxx})\psi + 2(u + 2\lambda)(u - \lambda)\psi_x \qquad (7.4.46)$$

In order to satisfy condition (7.4.44), we immediately see that the following two conditions must be imposed, namely,

$$u_t - 6uu_x + u_{xxx} = 0 \tag{7.4.47}$$

and

$$\lambda_t = 0 \tag{7.4.48}$$

The pair of equations (7.4.43) is termed a *Lax pair* after Lax, who showed (immediately following GGKM) that the KdV equation and other closely related nonlinear evolution equations are equivalent to the isospectral integrability condition for pairs of linear operators.

7.5 OTHER SOLITON SYSTEMS

For a while it was thought that the IST technique developed by GGKM was only applicable to the KdV equation. However, within a few years a whole host of other, physically important, nonlinear evolution equations were found to have soliton solutions and suitable ISTs developed. For these systems the quantal problem does not, typically, involve the time-independent Schrödinger equation any more and a different, but related, eigenvalue problem has to be solved. Here we just introduce a few of these equations and some of their simple solutions and briefly describe the basic IST procedure. Fuller accounts can be found in the cited texts.

7.5.a The Modified KdV Equation

An evolution equation that we have already mentioned is the mKdV equation, namely,

$$v_t + 6v^2 v_x + v_{xxx} = 0 \tag{7.5.1}$$

The first step is to look for traveling wave solutions of the form

$$v(x, t) = f(z) \tag{7.5.2}$$

where $z = x - ct$. Direct substitution into (7.5.2), followed by two integrations, yields the quadrature

$$z - z_0 = \int \frac{df}{\sqrt{cf^2 - f^4 + \frac{1}{2}df + e}} \tag{7.5.3}$$

where d and e are the first two integration constants. The general solution to this problem is in terms of Jacobi elliptic functions, but for the choice of boundary conditions $f, f' \to 0$ as $z \to \pm\infty$ the quadrature reduces to

$$z - z_0 = \int \frac{df}{f\sqrt{c - f^2}} \qquad (7.5.4)$$

This is easily integrated and inverted to give the solitary wave solution

$$f(z) = -\sqrt{c} \operatorname{sech}(\sqrt{c}(z - z_0)) \qquad (7.5.5)$$

Note that the solution is a sech rather than the analogously obtained sech^2 solution (7.2.15) for the KdV equation.

Another simple solution can be obtained by a similarlity transformation. Following the arguments given in Section 7.2, it is not difficult to show that (7.5.1) is invariant to the scalings $x \to kx$, $t \to k^3 t$, $u \to k^{-1}u$. This suggests the change of variables,

$$z = x/t^{1/3}, \qquad u(x, t) = t^{-1/3}f(z) \qquad (7.5.6)$$

which gives

$$f'' + 6ff' - \tfrac{1}{3}zf' - \tfrac{1}{3}f = 0 \qquad (7.5.7)$$

This can be integrated once to yield

$$f'' + 2f^3 - \tfrac{1}{3}zf + c = 0 \qquad (7.5.8)$$

which is the special ordinary differential equation known as the second Painlevé transcendent.

7.5.b The Sine–Gordon Equation

A very important nonlinear p.d.e. is the Sine–Gordon equation

$$u_{tt} - u_{xx} + \sin u = 0 \qquad (7.5.9)$$

This equation, as well as various solution techniques, was already used in the last century, where it appeared in various problems of differential geometry. A more contemporary use for it is in relativistic field theory. It is often convenient to study (7.5.9) in the variables

$$\xi = \tfrac{1}{2}(x - t), \qquad \eta = \tfrac{1}{2}(x + t) \qquad (7.5.10)$$

which transforms it to

$$u_{\xi\eta} = \sin u \qquad (7.5.11)$$

The periodicity of the sine introduces some interesting properties. If (7.5.9) is linearized (i.e., expanded to first order) about the solution $\psi = 0$, one obtains

$$u_{tt} - u_{xx} + u = 0 \qquad (7.5.12)$$

for which the dispersion relations are easily seen to be

$$\omega = \sqrt{k^2 + 1} \qquad (7.5.13)$$

This is real for all real k, implying that $\psi = 0$ is a stable equilibrium point. This should hardly be too surprising since if the space-dependent part of (7.5.9) is dropped, one is just left with the simple pendulum equation $u_{tt} + \sin u = 0$. On the other hand, if (7.5.9) is expanded about the solution $u = \pi$, the result is

$$u_{tt} - u_{xx} - u = 0 \qquad (7.5.14)$$

which has the dispersion relations

$$\omega = \sqrt{k^2 - 1} \qquad (7.5.15)$$

This demonstrates that the solution $u = \pi$ is unstable for $0 < k < 1$. This is again consistent with the properties of the space-independent problem.

A traveling wave analysis (i.e., setting $u(x, t) = f(z)$), yields after one integration the quadrature

$$z - z_0 = (c^2 - 1)^{1/2} \int \frac{df}{\sqrt{2(d - 2\sin^2(\frac{1}{2}f))}} \qquad (7.5.16)$$

where d is the first constant of integration. For the particular choice $d = 0$, (7.5.16) is easily solved to give

$$z - z_0 = \pm \sqrt{1 - c^2} \, \ln(\pm \tan(\tfrac{1}{4}f)) \qquad (7.5.17)$$

and hence

$$f(z) = \pm 4 \tan^{-1}(e^{\pm(z - z_0)/\sqrt{1 - c^2}}) \qquad (7.5.18)$$

This solution takes on different shapes, depending on the choice of signs. For the case of both signs being positive, $f(z)$ rises, from left to right, from zero to a height of 2π (see Figure 7.3). Such a solution is called a *kink*. The solutions whose amplitude decays from 2π down to zero are termed *antikinks*. Although, at first sight, these solutions look quite different from solitons, their derivative has the characteristic sech shape, that is,

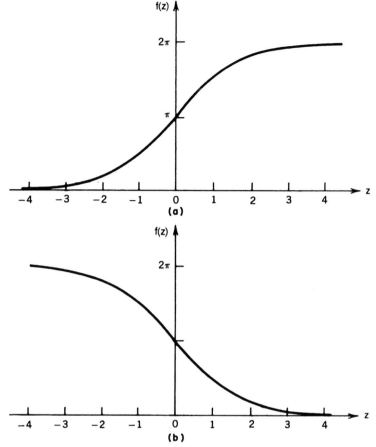

Figure 7.3 (a) Kink solution $f(z) = 4 \tan^{-1}[\exp(z)]$ and (b) anti-kink solution $f(z) = 4 \tan^{-1}[-\exp(z)]$.

$$u_x(x, t) = f'(z) = \frac{2}{\sqrt{1 - c^2}} \operatorname{sech}((z - z_0)/\sqrt{1 - c^2}) \qquad (7.5.19)$$

Kinks (and antikinks) show all the collisional properties of solitons; that is, they emerge unscathed from collision, suffering only a phase shift. A two-kink solution that demonstrates this property was derived, using standard separation of variables techniques, by Perring and Skyrme (1962). (This predates the work of Kruskal and co-workers, but at that time the full significance of their result was not appreciated.) The solution (see Drazin (1983) for details) takes the form

$$u(x, t) = 4 \tan^{-1} \left[\frac{c \sinh(x/\sqrt{1 - c^2})}{\cosh(ct/\sqrt{1 - c^2})} \right] \qquad (7.5.20)$$

The limits $t \to \pm\infty$ yield

$$\lim_{t \to -\infty} u(x, t) = 4 \tan^{-1}[e^{(x+ct-\delta)/\sqrt{1-c^2}} - e^{-(x-ct+\delta)/\sqrt{1-c^2}}] \qquad (7.5.21)$$

and

$$\lim_{t \to +\infty} u(x, t) = 4 \tan^{-1}[-e^{-(x+ct+\delta)/\sqrt{1-c^2}} + e^{(x-ct-\delta)/\sqrt{1-c^2}}] \qquad (7.5.22)$$

where we have introduced the phase shift

$$\delta = \sqrt{1 - c^2} \ln\left(\frac{1}{c}\right) \qquad (7.5.23)$$

The behavior of this solution is sketched in Figure 7.4.

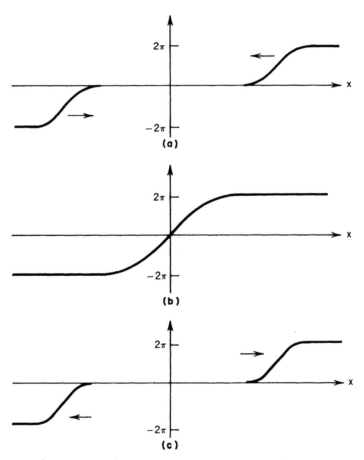

Figure 7.4 Evolution of Perring–Skyrme solution (7.5.20) for (a) $t < 0$, (b) $t = 0$, and (c) $t > 0$.

Another special solution to the Sine–Gordon equation can be obtained by a similarity transformation. Working with (7.5.11), one notes that it is invariant to the scalings $\xi \to k\xi$ and $\eta \to \eta/k$. Making the substitution $u(\xi, \eta) = f(z)$, where $z = \xi\eta$, yields

$$zf'' + f' = \sin(f) \tag{7.5.24}$$

The change of variable $g = e^{if}$ then gives

$$g'' - \frac{(g')^2}{g} + \frac{(2g' - g^2 + 1)}{2z} = 0 \tag{7.5.25}$$

which is a special case of the *third Painlevé transcendent*.

7.5.c The NLS Equation

Another significant nonlinear p.d.e. is the nonlinear Schrödinger equation (NLS), which takes the form

$$iu_t = u_{xx} + 2u|u|^2 \tag{7.5.26}$$

where u represents the amplitude of an almost monochromatic wave train. Note that this is a complex equation. Thus $|u|^2$ gives the (real) amplitude of the wavetrain envelope, which is also found to be of sech^2 form.

7.5.d A General IST Scheme*

As was mentioned at the beginning of the section, the IST for the above equations is different from that used for the KdV equation. The relevant schemes were developed by Zakharov and Shabat (1979) and Ablowitz et al. (1974). Here the associated eigenvalue problem is the two-component system of equations

$$v_{1x} = -i\zeta v_1 + qv_2 \tag{7.5.27a}$$

$$v_{2x} = i\zeta v_2 + rv_1 \tag{7.5.27b}$$

where $q = q(x, t)$ and $r = r(x, t)$ are the potentials and ζ is the eigenvalue. Note that if $r = -1$, (7.5.27) reduces to the Schrödinger equation $v_{2xx} + (q + \zeta^2)v = 0$. The associated time-dependent part of the problem takes the general form

$$v_{1t} = Av_1 + Bv_2 \tag{7.5.28a}$$

$$v_{2t} = Cv_1 - Av_2 \tag{7.5.28b}$$

where A, B, and C are various functions of q and r and the spectral parameter ζ. The derivation of the precise forms of A, B, and C for the particular equations considered here is not, in fact, all that difficult. However, we omit the details and refer the reader to the account given in Ablowitz and Segur (1981).

For the Sine–Gordon equation, one sets $q = -r = -u_x/2$ and obtains the scattering problem

$$\begin{bmatrix} v_1 \\ v_2 \end{bmatrix}_x = \begin{bmatrix} -i\zeta & -\frac{1}{2}u_x \\ \frac{1}{2}u_x & i\zeta \end{bmatrix} \begin{bmatrix} v_1 \\ v_2 \end{bmatrix} \tag{7.5.29}$$

and

$$\begin{bmatrix} v_1 \\ v_2 \end{bmatrix}_t = \begin{bmatrix} \dfrac{i}{4\zeta}\cos u & \dfrac{i}{4\zeta}\sin u \\ \dfrac{i}{4\zeta}\sin u & -\dfrac{i}{4\zeta}\cos u \end{bmatrix} \begin{bmatrix} v_1 \\ v_2 \end{bmatrix} \tag{7.5.30}$$

The reader may easily verify that these equations are the Lax pair for the Sine–Gordon equation since the "integrability condition"

$$\begin{bmatrix} v_1 \\ v_2 \end{bmatrix}_{xt} = \begin{bmatrix} v_1 \\ v_2 \end{bmatrix}_{tx} \tag{7.5.31}$$

will only be satisfied if (i) $\zeta_t = 0$ (i.e., the deformation is isospectral) and (ii) $u_{xt} = \sin u$.

For the NLS equation the scattering problem is found to be

$$\begin{bmatrix} v_1 \\ v_2 \end{bmatrix}_x = \begin{bmatrix} -i\zeta & u \\ \pm u^* & i\zeta \end{bmatrix} \begin{bmatrix} v_1 \\ v_2 \end{bmatrix} \tag{7.5.32}$$

where the asterisk denotes complex conjugate, and

$$\begin{bmatrix} v_1 \\ v_2 \end{bmatrix}_t = \begin{bmatrix} 2i\zeta^2 \pm iuu^* & 2u\zeta + iu_x \\ \mp 2u^*\zeta \pm iu_x^* & -2i\zeta^2 \mp iuu^* \end{bmatrix} \begin{bmatrix} v_1 \\ v_2 \end{bmatrix} \tag{7.5.33}$$

The integrability condition for this system is

$$iu_t = u_{xx} \pm 2u^2u^* \tag{7.5.34}$$

It turns out that the case with the minus sign cannot exhibit soliton solutions, whereas the positive-sign case can.

For the mKdV equation the scattering problem is

$$\begin{bmatrix} v_1 \\ v_2 \end{bmatrix}_x = \begin{bmatrix} -i\zeta & u \\ \mp u & i\zeta \end{bmatrix} \begin{bmatrix} v_1 \\ v_2 \end{bmatrix} \tag{7.5.35}$$

and

$$\begin{bmatrix} v_1 \\ v_2 \end{bmatrix}_t = \begin{bmatrix} -4i\zeta^3 \pm 2i\zeta u^2 & 4u\zeta^2 + 2i\zeta u_x - u_{xx} \mp 2u^3 \\ \mp 4u\zeta^2 \pm 2i\zeta u_x \pm u_{xx} + 2u^3 & 4i\zeta^3 \mp 2i\zeta u^2 \end{bmatrix} \begin{bmatrix} v_1 \\ v_2 \end{bmatrix} \qquad (7.5.36)$$

for which the integrability condition is

$$u_t \pm 6u^2 u_x + u_{xxx} = 0 \qquad (7.5.37)$$

In this case, soliton solutions can be found for either sign.

A version of the Gelfand–Levitan–Marchenko equation for these systems has been derived, although its solution is highly nontrivial. In addition, the basic eigenvalue problem (7.5.27) can, unlike the Schrödinger equation (7.4.2), have solutions whose eigenvalues form complex conjugate pairs. This leads to an oscillatory type of soliton solution known as *breathers* or *bions*.

7.6 HAMILTONIAN STRUCTURE OF INTEGRABLE SYSTEMS

One of the most important properties of soliton equations is that they are integrable Hamiltonian systems. Here we give a simple account of this which, in turn, provides a link between the properties of the finite-degree-of-freedom integrable systems described in Chapters 2 and 3 and their continuum counterparts discussed here.

7.6.a The Functional Derivative

An important mathematical technique that we require is the *variational* or *functional derivative*. This is easily understood by recalling the variational principle used in Chapter 2. Consider some *functional*, denoted by $F[u]$, of the form

$$F[u] = \int_{x_1}^{x_2} f(x, u, u_x) \, dx \qquad (7.6.1)$$

in which f is some function of x, u, and u_x with $u = u(x)$ and $u_x = du/dx$. An obvious example of (7.6.1) is the action integral where f is the Lagrangian and where $u(x)$ is interpreted as $q(t)$. An ordinary derivative (of, for example, $g(x)$) is evaluated by determining the effect of adding a small deviation to the argument of the function, that is, $g(x + \Delta x)$; the functional derivative is evaluated by determining the effect of a small deviation to the *function* $u(x)$, that is, $u(x) + \delta u(x)$ in $f(x, u, u_x)$. In this way we can evaluate the *first variation* of $F[u]$ in the usual way, that is,

$$\delta F[u] \doteq \int_{x_1}^{x_2} f(x, u + \delta u, (u + \delta u)_x) \, dx - \int_{x_1}^{x_2} f(x, u, u_x) \, dx \quad (7.6.2)$$

where the variation $\delta u(x)$ is assumed to vanish at the end points x_1 and x_2. Expanding the first integrand to first order in δu, one obtains

$$\delta F[u] = \int_{x_1}^{x_2} \left[\frac{\partial f}{\partial u} \delta u + \frac{\partial f}{\partial u_x} (\delta u)_x \right] dx \quad (7.6.3)$$

where $(\delta u)_x = d(\delta u)/dx$. Evaluating the second term by parts (assuming that $\delta u(x_1) = \delta u(x_2) = 0$) gives the standard result

$$\delta F[u] = \int_{x_1}^{x_2} \delta u \left(\frac{\partial f}{\partial u} - \frac{d}{dx} \left(\frac{\partial f}{\partial u_x} \right) \right) dx \quad (7.6.4)$$

The integrand is termed the *variational derivative* and we denote it by $\delta F / \delta u$, that is,

$$\frac{\delta F}{\delta u} = \frac{\partial f}{\partial u} - \frac{d}{dx} \left(\frac{\partial f}{\partial u_x} \right) \quad (7.6.5)$$

Now consider the more general case in which f is a function of any number of derivatives of u, that is,

$$F[u] = \int_{x_1}^{x_2} f(x, u, u_x, u_{xx}, \ldots, u_{nx}) \, dx \quad (7.6.6)$$

where $u_{nx} = d^n u / dx^n$. In this case the variational derivative is easily found to be

$$\frac{\delta F}{\delta u} = \sum_{m=0}^{n} (-1)^m \frac{d^m}{dx^m} \left(\frac{\partial f}{\partial u_{mx}} \right) \quad (7.6.7)$$

where the alternating signs have come from repeated integrations by parts to bring $(\delta u)_{mx}$ down to δu. Some simple examples are as follows: For

$$F[u] = \int_{-\infty}^{\infty} (\tfrac{1}{2} u^2) \, dx \quad (7.6.8)$$

we obtain

$$\frac{\delta F}{\delta u} = u$$

and for

$$F[u] = \int_{-\infty}^{\infty} (\tfrac{1}{2} u_x^2) \, dx \tag{7.6.9}$$

we obtain

$$\frac{\delta F}{\delta u} = - u_{xx}$$

A less trivial example is the conserved density in the KdV equation (7.2.27), that is,

$$F[u] = \int_{-\infty}^{\infty} (u^3 + \tfrac{1}{2} u_x^2) \, dx \tag{7.6.10}$$

for which

$$\frac{\delta F}{\delta u} = 3 u^2 - u_{xx} \tag{7.6.11}$$

Thus the KdV equation can be expressed as

$$u_t = \frac{\partial}{\partial x} \left(\frac{\delta F}{\delta u} \right) \tag{7.6.12}$$

where F is the functional given in (7.6.10).

Functional derivatives can also be defined in the following way. Consider the functional (7.6.1) in which u is also a function of some parameter α (i.e., $u = u(x; \alpha)$), so that

$$F[u(x, \alpha)] = \int_{x_1}^{x_2} f(x, u(x, \alpha), u_x(x, \alpha)) \, dx \tag{7.6.13}$$

The derivative of F with respect to α, using the chain rule, is

$$\frac{dF}{da} = \int_{x_1}^{x_2} \left(\frac{\partial f}{\partial u} \frac{\partial u}{\partial \alpha} + \frac{\partial f}{\partial u_x} \frac{\partial u_x}{\partial \alpha} \right) dx \tag{7.6.14}$$

where $\partial u_x / \partial \alpha = d(\partial u / \partial \alpha)/dx$. Again integrating by parts (and assuming vanishing end-point contributions) gives

$$\frac{dF}{d\alpha} = \int_{x_1}^{x_2} \frac{\partial u}{\partial \alpha} \left(\frac{\partial f}{\partial u} - \frac{d}{dx} \left(\frac{\partial f}{\partial u_x} \right) \right) dx \tag{7.6.15}$$

This can be written as

$$\frac{dF}{d\alpha} = \int_{x_1}^{x_2} \frac{\partial u}{\partial \alpha} \frac{\delta F}{\delta u} \, dx \tag{7.6.16}$$

and can be taken as the definition of $\delta F / \delta u$. (This is obviously generalizable to (7.6.6).)

7.6.b Hamiltonian Structure of the KdV Equation

The Hamiltonian nature of (7.6.12) was first demonstrated by Gardner (1971), whose approach we follow here. In his derivation the solution $u = u(x, t)$ to the KdV equation is assumed to be periodic in the interval $(0, 2\pi)$. In this case, u can be represented by the Fourier series

$$u(x, t) = \sum_{k=-\infty}^{\infty} u_k e^{ikx} \tag{7.6.17}$$

where $u_k = u_k(t)$ are a set of complex coefficients. Thinking of the function $F[u]$ given in (7.6.10) as a function of the set of "parameters" u_k, we can use (7.6.16) to write

$$\begin{aligned}
\frac{\partial F}{\partial u_k} &= \int_0^{2\pi} \frac{\delta F}{\delta u} \frac{\partial u}{\partial u_k} \, dx \\
&= \int_0^{2\pi} \frac{\delta F}{\delta u} e^{ikx} \, dx
\end{aligned} \tag{7.6.18}$$

where we have used (7.6.17). From this relation we obtain the Fourier representation of $\delta F / \delta u$, that is,

$$\frac{\delta F}{\delta u} = \frac{1}{2\pi} \sum_{k=-\infty}^{\infty} \frac{\partial F}{\partial u_{-k}} e^{ikx} \tag{7.6.19}$$

Using (7.6.12), the equations of motion for the individual u_k are just

$$\frac{du_k}{dt} = \frac{ik}{2\pi} \frac{\partial F}{\partial u_{-k}} \tag{7.6.20}$$

By defining, for $k > 0$, the variables

$$q_k = \frac{u_k}{k}, \qquad p_k = u_{-k}, \qquad H = \frac{i}{2\pi} F \tag{7.6.21}$$

(7.6.20) is exactly in the form of the Hamiltonian system

$$\frac{dq_k}{dt} = \frac{\partial H}{\partial p_k}, \qquad \frac{dp_k}{dt} = -\frac{\partial H}{\partial q_k} \qquad (7.6.22)$$

Using the above definitions, one can then go on to define the *Poisson bracket* of two functionals, F and G, as

$$
\begin{aligned}
[F, G] &= \frac{i}{2\pi} \sum_{k=1}^{\infty} \left(\frac{\partial F}{\partial q_k} \frac{\partial G}{\partial p_k} - \frac{\partial F}{\partial p_k} \frac{\partial G}{\partial q_k} \right) \\
&= \frac{i}{2\pi} \sum_{k=1}^{\infty} \left(k \frac{\partial F}{\partial u_k} \frac{\partial G}{\partial u_{-k}} - k \frac{\partial F}{\partial u_{-k}} \frac{\partial G}{\partial u_k} \right) \\
&= \frac{i}{2\pi} \sum_{k=-\infty}^{\infty} k \frac{\partial F}{\partial u_k} \frac{\partial G}{\partial u_{-k}} \qquad (7.6.23)
\end{aligned}
$$

Using (7.6.19), one can then show that (7.6.23) can be re-expressed as

$$[F, G] = \int_0^{2\pi} \frac{\delta F}{\delta u} \frac{\partial}{\partial x} \left(\frac{\delta G}{\delta u} \right) dx \qquad (7.6.24)$$

which can be taken as the definition of the Poisson bracket $[F, G]$. Such a bracket can be shown to satisfy the Jacobi indentity.

We recall from Section 7.2 that the KdV equation has an infinite number of conserved densities of the form

$$F_n[u] = \int T_n \, dx \qquad (7.6.25)$$

where the range of integration is $(0, 2\pi)$ for periodic systems or $(-\infty, \infty)$ for systems on the infinite domain. Since the F_n are conserved (i.e., are constants of motion), their Poisson bracket with the Hamiltonian H (defined by (7.6.21) and (7.6.10)) must vanish, that is,

$$\frac{dF_n}{dt} = [F_n, H] = \int \frac{\delta F_n}{\delta u} \frac{\partial}{\partial x} \left(\frac{\delta H}{\delta u} \right) dx = 0 \qquad (7.6.26)$$

One can further go on to show that all the F_n commute with each other, that is,

$$[F_n, F_m] = 0 \qquad (7.6.27)$$

for all n and m. This is the continuum analogue of the property (2.5.11) of finite-degree-of-freedom integrable Hamiltonians. Thus the KdV equation can be thought of as a completely integrable, infinite-degree-of-freedom Hamiltonian system. Continuing the analogy with finite systems, the result (7.6.27) furthermore suggests that the KdV flow must, it some sense, be confined to an infinite-dimensional torus. Further, remarkable work by Zakharov and Faddeev (1971) identified the canonical transformation of H to action-angle variables in which these variables are expressible in terms of the IST scattering data.

7.6.c Hamiltonian Structure of the NLS Equation

All the soliton equations discussed in this chapter can be shown to be Hamiltonian systems with an associated Poisson bracket (the original restriction of $(0, 2\pi)$ is not required). To conclude we just mention the case of the NLS equation (7.5.26), which we write in the form

$$iu_t = u_{xx} + 2u^2v \qquad (7.6.28a)$$

$$- iv_t = v_{xx} + 2v^2u \qquad (7.6.28b)$$

where $v = u^*$. In this case the Hamiltonian is

$$H = - i \int (u^2v^2 - u_xv_x) \, dx \qquad (7.6.29)$$

and Eqs. (7.6.28) are immediately given by the canonical relations

$$u_t = \frac{\delta H}{\delta v}, \qquad v_t = -\frac{\delta H}{\delta u} \qquad (7.6.30)$$

7.7 DYNAMICS OF NONINTEGRABLE EVOLUTION EQUATIONS

Integrable partial differential equations exhibiting soliton solutions arise surprisingly often in the derivation of realistic physical models of various wave phenomena occurring in one dimension. (For an introductory review see the article by Gibbon (1985).) Equally important are the host of closely related nonlinear evolution equations which are not integrable and have no IST solution. These equations can exhibit behavior ranging from finite time singularities ("blowup") to spatial chaos. It seems likely that an improved understanding of spatiotemporal chaos (and perhaps even fluid dynamical turbulence) will be provided by the study of some of these model equations. This is an enormous topic, but to complete our picture of chaos and integrability (in a sense, these two opposing concepts are brought together in this context) in dynamical systems we briefly mention some of the "canonical" models and their associated behaviors.

7.7.a Self-Focusing Singularities

The nonlinear Schrödinger equation in two dimensions, namely,

$$iu_t + \Delta u + u|u|^2 = 0 \tag{7.7.1}$$

where Δ denotes the two-dimensional Laplacian, is not soluble by IST and, furthermore, its solutions can exhibit a finite time blowup or a "self-focusing" singularity. This singularity is usually studied with (7.7.1) cast in radially symmetric coordinates, that is,

$$iu_t + \frac{1}{r}\frac{\partial}{\partial r}\left(r\frac{\partial u}{\partial r}\right) + u|u|^2 = 0 \tag{7.7.2}$$

with $u = u(r, t)$. That a singularity is possible can be seen from quite simple mechanical principles (see, for example, the paper by Berkshire and Gibbon (1983)). For Eq. (7.7.2), one can write down integrals corresponding to conservation of mass and energy, respectively, namely,

$$M = 2\pi \int_0^\infty |u|^2 r\, dr \tag{7.7.3}$$

and

$$E = 2\pi \int_0^\infty \frac{1}{2}\left(\frac{1}{2}\left|\frac{\partial u}{\partial r}\right|^2 - \frac{1}{4}|u|^4\right) r\, dr \tag{7.7.4}$$

The moment of inertia can also be defined, that is,

$$I = 2\pi \int_0^\infty |u|^2 r^3\, dr \tag{7.7.5}$$

which, in turn, can be shown to be related to the energy integral through

$$\frac{\partial^2 I}{\partial t^2} = 4E \tag{7.7.6}$$

For suitable choice of initial conditions $u(r, 0)$, one can have $E < 0$ with the consequence that $\ddot{I} < 0$. This leads to a vanishing moment of inertia in finite time, that is, collapse (blowup).

The precise nature of the singularity is difficult to determine and has stimulated a lot of theoretical work—much of it initiated by Zakharov and co-workers (see, for example, Zakharov and Synakh (1976)). This earlier work suggested that the singularity (in time) was algebraic; that is, the solution behaved like $(t - t_*)^{2/3}$ as it approached the blowup time t_*. Subsequent work suggests that the singularity has a much more complicated logarithmic structure (see, for example, the paper by McLaughlin et al. (1986)). Singularities in the NLS equation are not confined to the two-dimensional case. Indeed they can occur for the general equation

$$iu_t + \Delta_d u + u|u|^{2\sigma} = 0 \qquad (7.7.7)$$

where Δ_d is the d-dimensional Laplacian and σ is the order of the nonlinearity. For each dimension d there will be a σ for which a self-focusing singularity can be found. For example, in one dimension, the quartic NLS equation

$$iu_t + u_{xx} + u|u|^4 = 0 \qquad (7.7.8)$$

exhibits blowup. Early work by Zakharov and Synakh (1976) suggested an algebraic blowup going as $(t - t_*)^{4/7}$—but again it seems that the singularity is more complicated than this.

7.7.b The Zakharov Equations

In many physical contexts, one is interested in modeling the interaction of long waves with short waves. For example, in the theory of Langmuir waves in plasma physics the interaction between a rapidly oscillating electric field (denoted by u) and a slowly varying ion density (denoted by v) takes the form

$$iu_t + u_{xx} + uv = 0 \qquad (7.7.9a)$$

$$v_{xx} - \frac{1}{C^2} v_{tt} = -\beta(|u|^2)_{xx} \qquad (7.7.9b)$$

These are known as the *one-dimensional Zakharov equations*. Equations such as (7.7.9) also arise in models of excitations in idealized DNA chains proposed by Davydov (1979). An important feature of Eqs. (7.7.9) is the limit of large C in (7.7.9b). In this case, one has $v_{xx} = -\beta(|u|^2)_{xx}$; in other words, v is directly proportional to u and (7.7.9a) reduces to the integrable, soliton-bearing, NLS equation. However, the full equations are not integrable and cannot be solved by IST. Numerical studies show them to be capable of very complicated behavior. By contrast, if Eqs. (7.7.9) are reduced to a "one-wave" form by factorizing the operator $(\partial^2/\partial x^2 - (1/C^2)(\partial^2/\partial t^2))$ into $(\partial/\partial x + (1/C)(\partial/\partial t))$ $(\partial/\partial x - (1/C)(\partial/\partial t))$, one obtains

$$iu_t + u_{xx} + uv = 0 \qquad (7.7.10a)$$

$$v_x + \frac{1}{C} v_t = -\beta(|u|^2)_x \qquad (7.710b)$$

which are integrable and have an IST solution. Particularly important is the physically more realistic, two-dimensional version of (7.7.9), namely,

$$iu_t + \Delta u + uv = 0 \qquad (7.7.11a)$$

$$\Delta v - \frac{1}{C^2} v_{tt} = -\beta\Delta(|u|) \tag{7.7.11b}$$

In this case the large C limit reduces (7.7.11a) to the 2-D NLS equation which can display blowup. Numerical simulations of (7.7.11) display a rich behavior including near blowup followed by "burnout." (A good review of these phenomena, in the plasma context, is given by Goldman (1984).)

7.7.c Coherence and Chaos

An important model system for the study of spatiotemporal chaos is the damped and driven one-dimensional Sine–Gordon equation, namely,

$$u_{tt} - u_{xx} + \sin u = \Gamma\cos(\omega t) - \alpha u_t \tag{7.7.12}$$

This equation is typically studied with periodic boundary conditions, that is, $u(x + L, t) = u(x, t)$. Detailed numerical studies of (7.7.12) have been carried out by Bishop et al. (1983) and others. A rich range of behaviors is found as the driving and damping parameters, Γ and α, respectively, are varied. Typically, though, the spatial structure of the solutions tends to be quite coherent; that is, they just exhibit a few well-defined spatial modes. What is especially striking is that this spatial coherence can be maintained even when the temporal evolution becomes chaotic. It seems clear that the soliton structure of the unperturbed ($\Gamma = \alpha = 0$) system can be quite robust.

In a variety of problems in statistical mechanics and fluid dynamics, two equations frequently occur—these are the Ginzburg–Landau and closely related Newell–Whitehead equations, respectively. A typical form for these equations (in one dimension) is

$$u_t = \alpha u_{xx} + \beta u - \gamma|u|^2 u \tag{7.7.13}$$

where u usually represents the (complex) amplitude of some unstable mode and α, β, γ are adjustable parameters that can be complex. This equation, which can be generalized to higher dimensions, can exhibit an enormous variety of behaviors, ranging from the coherent to the chaotic, depending on the choice of parameter values. The properties of such equations are a most active area of research.

SOURCES AND REFERENCES

Texts and General Review Articles

Ablowtiz, M. J., and H. Segur, *Solitons and the Inverse Scattering Transform*, SIAM, Philadelphia, 1981.

Dodd, R. K., J. C. Eilbeck, J. D. Gibbon, and H. C. Morris, *Solitons and Nonlinear Wave Equations*, Academic, London, 1982.

Drazin, P. G., *Solitons*, London Mathematical Society Lecture Notes, Vol. 8, Cambridge University Press, Cambridge, 1983.

Lamb, G. L., Jr., *Elements of Soliton Theory*, Wiley, New York, 1980.

Miura, R. M., The Korteweg–de Vries equation, a survey of results, *SIAM Rev.* **18**, 412 (1976).

Section 7.1

Fermi, E., J. Pasta, S. Ulam, Studies in nonlinear problems, I, Los Alamos report LA 1940 (1955). Reproduced in (A. C. Newell, Ed.), *Nonlinear Wave Motion*, American Mathematical Society, Providence, RI, 1974.

Korteweg, D. J., and G. de Vries, On the change of form of long waves advancing in a rectangular canal, and on a new type of long stationary waves, *Phil. Mag.*, **39**, 422 (1895).

Russell, J. Scott, Report on waves, *Report of the 14th Meeting of the British Association for the Advancement of Science*, John Murray, London (1844).

Zabusky, N. J., and M. D. Kruskal, Interaction of solitons in a collisionless plasma and the recurrence of initial states, *Phys. Rev. Lett.*, **15**, 240 (1965).

Section 7.2

Carrier, G. F., and Carl E. Pearson, *Partial Differential Equations*, Academic, New York, 1976.

Miura, R. M., Korteweg–de Vries equation and generalizations, I. A remarkable explicit nonlinear transformation, *J. Math. Phys.* **9**, 1202 (1968).

Mirua, R. M., C. S. Gardner, and M. D. Kruskal, Korteweg–de Vries equation and generalizations, II. Existence of conservation laws and constants of motion," *J. Math. Phys.*, **9**, 1204 (1968).

Section 7.3

Gardner, C. S., J. M. Greene, M. D. Kruskal, and R. M. Miura, Method for solving the Korteweg–de Vries equation, *Phys. Rev. Lett.*, **19**, 1905 (1967).

Gardner, C. S., J. M. Greene, M. D. Kruskal, and R. M. Miura, The Korteweg–de Vries equation and generalizations, VI. Methods of exact solution, *Commun. Pure Appl. Math.*, **27**, 97 (1974).

Gelfand, I. M., and B. M. Levitan, On the determination of a differential equation from its spectral function, *Am. Math. Soc. Transl. Ser 2*, **1**, 259 (1955).

Marchenko, V. A., On the reconstruction of the potential energy from phases of the scattered waves, *Doklady Akad. Nauk.*, **104**, 695 (1955).

Section 7.4

Lax, P. D., Integrals of nonlinear evolution equations and solitary waves, *Commun. Pure Appl. Math.*, **21**, 467 (1968).

Zabusky, N. J., and M. D. Kruskal, Interaction of solitons in a collisionless plasma and the recurrence of initial states, *Phys. Rev. Lett.*, **15**, 240 (1965).

Section 7.5

Ablowitz, M. J., D. J. Kaup, A. C. Newell, and H. Segur, The inverse scattering transform—Fourier analysis for nonlinear problems, *Stud. Appl. Math.*, **53**, 249 (1974)

Perring, K. K., and Skyrme, T. H. R., A model unified field equation, *Nucl. Phys.*, **31**, 550 (1962).

Zakharov, V. E., and P. B. Shabat, A scheme for integrating the nonlinear equations of mathematical physics by the method of the inverse scattering problem, I, *Funct. Anal. Appl.*, **8**, 226 (1974); **13**, 166 (1979).

Section 7.6

Gardner, C. S., The Korteweg–de Vries equation and generalizations, IV. The Korteweg–de Vries equation as a Hamiltonian system, *J. Math. Phys.*, **12**, 1548 (1971).

Zakharov, V. E., and L. D. Faddeev, Korteweg–de Vries equation, a completely integrable system, *Funct. Anal. Appl.*, **5**, 280 (1971).

Section 7.7

Berkshire, F. H., and J. D. Gibbon, Collapse in the n-dimensional nonlinear Schrödinger equation—a parallel with Sundman's results in the N-body problem, *Stud. Appl. Math.*, **69**, 229 (1983).

Bishop, A. R., K. Fesser, P. S. Lomdahl, and S. E. Trullinger, Influence of solitons in the initial state on chaos in the driven, damped Sine–Gordon system, *Physica*, **7D**, 259 (1983).

Davydov, A. S., Solitons in molecular systems, *Phys. Scr.*, **20**, 387 (1979).

Gibbon, J. D., A survey of the origins and physical importance of soliton equations, *Philos. Trans. R. Soc. London*, **A315**, 335 (1985).

Goldman, M. V., Strong turbulence of plasma waves, *Rev. Mod. Phys.*, **56**, 709 (1984).

McLaughlin, D. W., G. C. Papanicolaou, C. Sulem, and P. L. Sulem, Focusing singularity of the cubic Schrödinger equation, *Phys. Rev. A*, **34**, 1200 (1986).

Newell, A. C., and J. A. Whitehead, Finite amplitude, finite band width convection, *J. Fluid Mech.* **38**, 279 (1969).

Zakharov, V. E., Collapse of Langmuir waves, *Sov. Phys. JETP*, **72**, 908 (1972).

Zakharov, V. E., and V. S. Synakh, The nature of the self-focusing singularity, *Sov. Phys. JETP*, **41**, 465 (1976).

8

ANALYTIC STRUCTURE OF DYNAMICAL SYSTEMS

8.1 THE QUEST FOR INTEGRABLE SYSTEMS

A recurrent theme of the preceding chapters has been the distinction between integrable and nonintegrable systems. The latter can exhibit chaotic behavior, whereas the former are distinguished by the existence of a full complement of integrals and exhibit stable, multiply-periodic behavior. Integrable systems form the "building blocks" about which perturbation theories can be developed. In the case of finite-dimensional Hamiltonian systems, this ultimately led to the KAM theorem concerning the preservation of phase space tori under perturbation. In the case of infinite-degree-of-freedom systems (i.e. p.d.e.'s), we have seen that, when integrable, they can exhibit soliton solutions.

Despite all the advances in nonlinear dynamics, a fundamental question still remains, namely: Given a system of equations, how can one tell a priori whether or not they are integrable? For example, given the Henon–Heiles system with adjustable coefficients, that is,

$$H = \tfrac{1}{2}(p_x^2 + p_y^2 + Ax^2 + By^2) + Dx^2 y - \tfrac{1}{3}Cy^3 \qquad (8.1.1)$$

are there any combinations of the four parameter values A, B, C, D for which it is integrable? As it turns out, there are just four such cases. These are:

(a) $D/C = 0$, any A, B;
(b) $D/C = -1$, $A/B = 1$;

(c) $D/C = -\frac{1}{6}$, any A, B;

(d) $D/C = -\frac{1}{16}$, $A/B = \frac{1}{16}$.

Of these, case (a) is trivial and case (b) is not too difficult to spot since the equations of motion separate under a simple change of variables. However, cases (c) and (d)—particularly case (d)—are far from obvious.

One approach to identifying the integrable cases of such a system is to find the integrals of motion. In general, this is a most difficult undertaking, requiring a combination of genius, luck, and prayer. For certain special classes of system, namely, two-degree-of-freedom Hamiltonians with simple algebraic integrals (i.e., integrals that are polynomial in the canonical p and q), algorithms based on Bertrand's method (devised in 1852 and discussed in Whittaker (1959)) can be used to find the integrals. For non-Hamiltonian systems such as the Lorenz equations

$$\dot{x} = \sigma(y - x)$$
$$\dot{y} = -xz + Rx - y \qquad (8.1.2)$$
$$\dot{z} = xy - Bz$$

determination of integrability requires one to identify, if they exist, time-dependent integrals (see Section 1.6). As we shall see, there are, in fact, a few special combinations of the parameter values σ, R, B for which these equations are integrable. Although there are also procedures that can sometimes find the corresponding integrals, they, like Bertrand's method for Hamiltonian systems, have a restricted range of applicability and are computationally cumbersome.

The situation for partial differential equations is no less happy. For example, given the following set of nonlinear p.d.e.'s

$$u_t + 6uu_x + u_{xxx} = 0 \qquad (8.1.3a)$$
$$u_t + 6u^2 u_x + u_{xxx} = 0 \qquad (8.1.3b)$$
$$u_t + 6u^3 u_x + u_{xxx} = 0 \qquad (8.1.3c)$$

how can one determine which ones are integrable and capable of exhibiting n-soliton solutions? Of course, from Chapter 7 we identify the first two as the KdV and modified KdV equations, respectively, for which the inverse scattering transforms are known and explicit n-soliton solutions have been constructed. But what about the last equation? In lieu of finding its inverse scattering transform (not known), one could attempt to construct conservation laws. So far only three have been found, but there is no proof that there are no more. Another approach is to study the system numerically by colliding solitary waves to see if they have soliton-like properties, that is, retain their shape and speed after collision. For this system, soliton behavior

is not observed numerically. Although this is a good indication of nonintegrability, it is clearly not a proof.

Overall, our aim is to find a simple analytical test that can identify
integrable cases of both o.d.e.'s and p.d.e.'s—be they Hamiltonian or not.
The answer appears to lie in the complex plane, that is, in the types of
singularities exhibited by the solutions when analytically continued into the
complex domain of their independent variables. Although, at first, this
might sound rather intimidating, such an analysis is straightforward to
implement and, in fact, only depends on the properties of the given
differential equation(s)—it does not need the explicit construction of solutions. This idea is, in fact, quite old and has its origins in the classic work of
the great Russian mathematician Sofya Kovalevskaya.†

8.1.a The Work of Kovalevskaya

Her famous work, for which she won the Bordin prize of the Paris Academy
of Science in 1888, concerned the solution of the Euler–Poisson equations,
which describe the motion of a heavy top rotating about a fixed point. They
are a system of six first-order, nonlinear, coupled ordinary differential
equations of the form

$$A \frac{dp}{dt} = (B - C)qr - \beta z_0 + \gamma y_0 \tag{8.1.4a}$$

$$B \frac{dq}{dt} = (C - A)pr - \gamma x_0 + \alpha z_0 \tag{8.1.4b}$$

$$C \frac{dr}{dt} = (A - B)pq - \alpha y_0 + \beta x_0 \tag{8.1.4c}$$

$$\frac{d\alpha}{dt} = \beta r - \gamma q \tag{8.1.4d}$$

$$\frac{d\beta}{dt} = \gamma p - \alpha r \tag{8.1.4e}$$

$$\frac{d\gamma}{dt} = \alpha q - \beta p \tag{8.1.4f}$$

where p, q, r are the components of angular velocity and α, β, γ are the
direction cosines that define the orientation of the top. The sets of variables
A, B, C and x_0, y_0, z_0 are the moments of inertia and the position

†The life of Kovalevskaya (1850–1891) is as fascinating as her work. Her autobiography, *A
Russian Childhood*, is well worth reading. She had to overcome enormous obstacles to pursue
her mathematical studies in an age when such a career (if any) was deemed totally "unsuitable"
for a woman.

coordinates of the center of gravity, respectively. These are the adjustable parameters of the system, for different values of which the system may or may not be integrable.

In Kovalevskaya's day, solutions to (8.1.4) had only been found for a few special cases and it was not known if it was possible to solve the equations for general A, B, C and x_0, y_0, z_0. The system (8.1.4) has three "classical" first integrals for all parameter values. These are

$$I_1 = Ap^2 + Bq^2 + Cr^2 - 2(x_0\alpha + y_0\beta + z_0\gamma) \qquad (8.1.5a)$$

$$I_2 = Ap\alpha + Bq\beta + Cr\gamma \qquad (8.1.5b)$$

$$I_3 = \alpha^2 + \beta^2 + \gamma^2 = 1 \qquad (8.1.5c)$$

The first and second of these are the total energy and angular momentum, respectively; the the third represents a simple geometric constraint. The key to solving the Eqs. (8.1.4) is to find a fourth integral, thereby reducing the system to a second-order form that can then be integrated by quadratures.

The cases for which a fourth integral was known were:

(i) Euler's case, in which $x_0 = y_0 = z_0$, that is, the center of gravity coincides with the fixed point. Here it is not difficult to verify that the fourth integral is

$$I_4 = A^2p^2 + B^2q^2 + C^2r^2 \qquad (8.1.6)$$

(ii) Lagrange's case, in which $A = B$, $x_0 = y_0 = z_0 = 0$, that is, a symmetric top with center of gravity along the z-axis. In this case, Eq. (8.1.4c) becomes trivial and the fourth integral is just

$$I_4 = r \qquad (8.1.7)$$

(iii) The completely symmetric case, in which $A = B = C$.

These three cases can all be integrated in terms of Jacobi elliptic functions.

Kovalevskaya's approach to this mechanical problem was completely new and involved the ostensibly unphysical techniques of complex variable theory. Apparently motivated by the work of Fuchs on the properties of first-order differential equations in the complex plane, she set out to determine the types of singularities that Eqs. (8.1.4) could exhibit. Her aim was to find the conditions under which the only *movable singularities* exhibited by the solutions in the complex time plane were ordinary poles.

At this point we must digress in order to understand what is meant by a movable singularity. For linear ordinary differential equations the singularities are determined by the coefficients in the equation and are at fixed locations in the complex domain. For example, the equation

$$\frac{d}{dz}f(z) + \frac{1}{z^2}f(z) = 0 \tag{8.1.8}$$

has a fixed singularity at the point $z = 0$. In this case the solution is just $f(z) = ce^{1/z}$ and we see that the singularity at $z = 0$ is, in fact, an essential singularity. (Essential singularities will be discussed in more detail in Section 8.2.a.) In contrast to linear equations, nonlinear differential equations can exhibit singularities that are movable, that is, whose location depends on the initial conditions. For example, the equation

$$\frac{d}{dz}f(z) + f^2(z) = 0 \tag{8.1.9}$$

has the solution

$$f(z) = \frac{1}{z + z_0} \tag{8.1.10}$$

where $z_0 = 1/f(0)$. Thus $f(z)$ has a simple *pole* at $z = -z_0$, where z_0 is determined by the initial value $f(0)$. The equation

$$\frac{d}{dz}f(z) + f^3(z) = 0 \tag{8.1.11}$$

has the solution

$$f(z) = \frac{1}{\sqrt{2z + z_0)}} \tag{8.1.12}$$

where now $z_0 = 1/\sqrt{f(0)}$. So in this case the equation exhibits a movable *branch point*.

Kovalevskaya found that there were just four cases for which the Euler–Poisson equations exhibited only movable poles. These were the three cases already known and a new one (Kovalevskaya's case) in which $A = B = 2C$, $z_0 = 0$. For this case she found the fourth integral to be

$$I_4 = \left(p^2 - q^2 - \frac{x_0}{C}\alpha\right)^2 + \left(2pq - \frac{x_0}{C}\beta\right) \tag{8.1.13}$$

and integrated the equations of motion by means of some virtuoso work involving hyperelliptic functions.

At the time, it was not understood why such an approach should work, that is, why a particular singularity structure in the complex domain could determine the (real time) integrability of a mechanical system. Her result was thought to be a special property of the rigid-body problem, and there do not seem to have been any other applications of this idea to mechanical systems. It is only in recent years that the generality of this approach has been appreciated and that some insights into why it works have been obtained.

8.1.b The Work of Painlevé

Although Kovalevskaya's work was apparently not pursued outside the top problem, a major activity in late nineteenth-century mathematics was the classification of ordinary differential equations according to the types of singularities that their solutions could exhibit. The most extensive of these investigations was carried out by the French mathematician Paul Painlevé.† Following the earlier work of Fuchs and others on the classification of first-order equations, he studied the class of second-order equations

$$\frac{d^2 y}{dx^2} = F\left(\frac{dy}{dx}, y, x\right) \tag{8.1.14}$$

where F is analytic in x and rational in y and dy/dx. Within this class of equations he found just 50 types whose only movable singularities are ordinary poles. This special analytic property is now frequently referred to as the *Painlevé property*. Of these 50, he found that 44 could be integrated in terms of "known" functions (Ricatti equations, elliptic functions, etc.). The remaining six, now termed the *Painlevé transcendents*, do not have algebraic integrals and cannot be integrated by quadratures. Painlevé himself only found the first two of the transcendents (which we denote as PI and PII, respectively), namely,

$$\text{PI:} \quad \frac{d^2 y}{dx^2} = 6y^2 + x \tag{8.1.15}$$

$$\text{PII:} \quad \frac{d^2 y}{dx^2} = 2y^3 + xy + \alpha \tag{8.1.16}$$

where α is an arbitrary parameter. The remaining four transcendents were found by some of his pupils; the sixth, due to Gambier, contains the other five as limiting cases.

Although the Painlevé transcendents can only be represented in the form of convergent local expansions they are related, asymptotically, to some of

---

†Painlevé was both an important mathematician and politician—serving as French defense minister in World War I. He was an aviation enthusiast and, as a friend of the Wright brothers, became aviation's first passenger (nonpaying).

the "known" functions of analysis. For example, in the case of PI the so-called Boutroux transformation

$$y = z^{1/2}\omega, \qquad t = \tfrac{4}{5}z^{5/4}$$

transforms it into

$$\frac{d^2\omega}{dz^2} = 6\omega^2 - \frac{1}{t}\frac{d\omega}{dt} + \frac{4}{25t^2}\omega$$

which, in the limit $t \to \infty$, becomes a special case of the Weierstrass elliptic function.

Painlevé's work seemed to have little relevance to physical problems and quickly disappeared into the mathematical literature. However, in the last 10 years or so there has been a remarkable resurgence of interest in the work of Kovalevskaya and Painlevé (and others), and their ideas have been found to play a central role in determining and understanding the integrability of dynamical systems. Before we describe this work, we devote the next section to a brief review of the relevant properties of differential equations in the complex domain.†

8.2 ORDINARY DIFFERENTIAL EQUATIONS IN THE COMPLEX DOMAIN

8.2.a Local Representations

In the case of very simple nonlinear differential equations such as (8.1.9) and (8.1.11), the nature of their movable singularities is revealed by the form of the exact solution. Although exact solutions are typically not obtainable, the nature of the movable singularities can be determined rather easily by studying the "local" properties of the solutions.

Given an nth-order ordinary differential equation of the form

$$\frac{d^n y}{dz^n} = F\left(\frac{d^{n-1}y}{dz^{n-1}}, \ldots, \frac{dy}{dz}, y, z\right) \tag{8.2.1}$$

where F is analytic in the independent variable z and rational in the remaining arguments, the behavior of its solution(s) *at* a movable singularity is determined by a *leading-order* analysis. Here one makes the ansatz

$$y(z) = a(z - z_0)^\alpha \tag{8.2.2}$$

---

†An excellent account of this topic is given in *Ordinary Differential Equations in the Complex Domain* by Einar Hille (1976).

where a and α are to be determined and z_0 is an arbitrary location in the complex z-plane (i.e., the location of the movable singularity). Substitution of (8.2.2) into (8.2.1) and balancing the most singular terms determines a and α. For example, consider the second-order equation

$$\frac{d^2y}{dz^2} = 6y^2 + Ay \tag{8.2.3}$$

which we know from Chapter 1 has an exact solution in terms of Weierstrass elliptic functions. The substitution (8.2.2) gives

$$a\alpha(\alpha - 1)(z - z_0)^{\alpha-2} = 6a^2(z - z_0)^{2\alpha} + Aa(z - z_0)^\alpha$$

The most singular terms, often referred to as the *dominant balance* terms (i.e., the second derivative and the $6y^2$), must balance at the singularity z_0. Equating the exponents (i.e., $\alpha - 2 = 2\alpha$) gives $\alpha = -2$. Similarly, the coefficients must balance (i.e., $a\alpha(\alpha - 1) = 6a^2$) from which we deduce that $a = 1$. Thus at z_0 the solution to (8.2.3) behaves as $y(z) = (z - z_0)^{-2}$ (i.e., a second-order pole). Notice that the lower-order term in this equation (i.e., Ay) does not affect the behavior at z_0. For higher nonlinearities in (8.2.3), the order of the singularity changes. Thus the equation

$$\frac{d^2y}{dz^2} = 2y^3 + Ay \tag{8.2.4}$$

which we know is soluble in terms of Jacobi elliptic functions, is easily shown to behave at a movable singularity as $y(z) = (z - z_0)^{-1}$ (i.e., a first-order pole). Higher-order nonlinearities in (8.2.4) are easily seen to result in movable branch point behavior.

The leading-order analysis only tells us about the behavior of a solution *at* a singularity. In order to determine the behavior in the *neighborhood* of the singularity, a local expansion must be constructed. If the singularity is indeed a movable pole, then this expansion will be a simple *Laurent series*. Consider, for example, Eq. (8.2.3). We already know that at leading order it behaves as a second-order pole, so a Laurent series, if it is valid, should take the form

$$y(z) = \sum_{j=0}^{\infty} a_j(z - z_0)^{j-2} \tag{8.2.5}$$

This ansatz can be tested for self-consistency by direct substitution into the equation. This leads to the relationship

$$\sum_{j=0}^{\infty} a_j(j - 2)(j - 3)(z - z_0)^{j-4} = 6 \sum_{j=0}^{\infty} \sum_{k=0}^{\infty} a_j a_k(z - z_0)^{j+k-4} + A \sum_{j=0}^{\infty} a_j(z - z_0)^{j-2}$$

which can be simplified to give the *recursion relations* for the a_j, namely,

$$a_j(j+1)(j-6) = 6 \sum_{l=1}^{j-1} a_{j-l}a_l + Aa_{j-2} \tag{8.2.6}$$

where we have used $a_0 = 1$. Solving recursively for the a_j, one finds that

$$j = 1: \quad a_1 = 0$$
$$j = 2: \quad a_2 = -A/12$$
$$j = 3: \quad a_3 = 0$$
$$j = 4: \quad a_4 = -A^2/24$$
$$j = 5: \quad a_5 = 0$$

At $j = 6$, one has the relationship

$$0 \cdot a_6 = 12a_4a_2 + Aa_4 \tag{8.2.7}$$

and, by virtue of the values of a_2 and a_4, the right-hand side vanishes. This means that a_6 is *arbitrary*. Thus the Laurent series (8.2.5) is seen to have *two* arbitrary parameters, namely, a_6 and z_0—the latter just being the arbitrariness of the pole position. We know that since (8.2.3) is a second-order o.d.e., its *general solution* is characterized by two arbitrary parameters. The above analysis shows that these are manifested in the (local) Laurent expansion (8.2.5) by the arbitrariness of a_6 and z_0. Thus we may conclude that in the neighborhood of a movable singularity z_0, the general solution to (8.2.3) does indeed behave as a second-order pole. We will often refer to this result by saying that Eq. (8.2.3) has the Painlevé property.† It is a straightforward exercise to show that Eq. (8.2.4) has the Painlevé property with $y(z)$ having the local expansion

$$y(z) = \sum_{j=0}^{\infty} a_j(z - z_0)^{j-1} \tag{8.2.8}$$

in which a_4 and z_0 are the arbitrary parameters.

The powers of $(z - z_0)$ at which the arbitrary coefficient appear are often termed the *resonances*.‡ A simple analysis can be performed to determine the resonances without having to grind through the full recursion relations.

---

†We also often say that a *solution* has the Painlevé property if it has a local single-valued expansion. It is possible for a differential equation to exhibit more than one singularity type. These lead to different solution branches. When all solution branches have the Painlevé property, we say that the differential *equation* has the Painlevé property.
‡Resonances are sometimes termed *Kovalevskaya exponents*.

This is carried out by making the substitution

$$y(z) = a(z - z_0)^\alpha + p(z - z_0)^{r+\alpha} \tag{8.2.9}$$

where a and α have been determined from a leading-order analysis. From this, one can set up a linear equation in p and determine the powers, r, at which p is arbitrary. So, for example, in the case of Eq. (8.2.3), one makes the substitution

$$y(z) = (z - z_0)^{-2} + p(z - z_0)^{r-2} \tag{8.2.10}$$

Working with the dominant balance terms, one has

$$6(z - z_0)^{-4} + p(r-2)(r-3)(z - z_0)^{r-4} = 6[(z - z_0)^{-2} + p(z - z_0)^{r-2}]^2$$

and equating terms linear in p yields

$$p(r+1)(r-6) = 0 \tag{8.2.11}$$

Thus for p to be arbitrary, one requires that either $r = -1$ or $r = +6$. The root $r = 6$ tells us that the coefficient a_6, corresponding to the power $(z - z_0)^4$ in (8.2.5), should be arbitrary. The root $r = -1$ does not correspond to the power $(z - z_0)^{-3}$ (this is not supported by the equation); instead, it corresponds to the arbitrariness of z_0.

The resonance analysis only tells us which coefficients *should* be arbitrary, and this has to be verified by checking the full recursion relations. At the resonances, one usually finds some condition, termed a *compatibility condition*, that has to be satisfied in order to secure arbitrariness of the coefficient. For example, in Eq. (8.2.3), the relevant compatibility condition is $12a_4a_2 + Aa_4 = 0$. For this problem, it is satisfied for all values of A. For more complicated problems, one will often find that arbitrariness is only obtained for special values of the system parameters. For example, if (8.2.4) is modified to include a first-derivative term, that is,

$$\frac{d^2y}{dz^2} + B\frac{dy}{dz} - Ay - 2y^3 = 0 \tag{8.2.12}$$

the coefficient a_4 in the series (8.2.8) is only arbitrary if $A = -2B^2/9$. If this condition is not satisfied, the simple Laurent series can no longer be a valid local representation of the general solution. In this case the series has to be generalized in such a way so as to "recapture" an arbitrary coefficient at $j = 4$. The new series, termed a *psi series*, reveals that the singularity is no longer an ordinary pole but, instead, has a complicated logarithmic multivaluedness. This analysis will be described in Section 8.2.c.

In all the discussion so far, we have assumed only integer leading orders and resonances. It is quite possible, depending on the order and non-linearities of the equation, for one or both of these quantities to be non-integer (e.g., irrational or complex). In these cases the singularities are clearly no longer ordinary poles, and the system in question does not have the Painlevé property. Since, as suggested by Kovalevskaya's work, the Painlevé property is a hallmark of integrability, we see that even at the level of a leading order and resonance analysis, one has a simple analytical "hint" about a system's nonintegrability.

We conclude this subsection by making a few remarks about essential singularities. These are singularities whose local expansions have an infinite number of negative powers. A simple example is the fixed essential singularity in the solution $f(z) = ce^{1/z}$ to the linear, first-order equation (8.1.8). Here the local expansion is just

$$e^{1/z} = \sum_{n=0}^{\infty} \frac{1}{n! z^n} \qquad (8.2.13)$$

For nonlinear first-order equations of the form

$$\frac{dy}{dz} = F(y, z) \qquad (8.2.14)$$

it turns out that for a fairly wide class of $F(y, z)$, these equations can only exhibit movable poles or algebraic branch points—but only fixed essential singularities. However, if the nonlinear o.d.e. is of the second order, it is possible for the essential singularities to become movable. A simple example is the o.d.e.

$$\left(\frac{d}{dz}\left(\frac{y'}{y}\right)\right)^2 + 4\left(\frac{y'}{y}\right)^3 = 0 \qquad (8.2.15)$$

where y' denotes dy/dz, which has the general solution

$$y(z) = c_1 e^{1/(z-c_2)} \qquad (8.2.16)$$

exhibiting a movable essential singularity at $z = c_2$. In addition to movable essential singularities, second-order nonlinear o.d.e.'s can also exhibit movable logarithmic and transcendental branch points. Although the latter types of singularities can be identified by local analysis, essential singularities are difficult to identify unless the solution is known explicitly. At this time, little is known about the role of movable essential singularities in determining the integrability of a given system. We also comment that as differential equations become of higher order, ever more complicated types of movable singularity can appear.

8.2.b General and Singular Solutions

So far we have concentrated on local representations of general solutions, that is, expansions with as many arbitrary parameters as the order of the equation. It is important to recognize that solutions with less than the full complement of arbitrary parameters can also exist; these are termed *singular solutions*. As a simple example, consider the first-order nonlinear o.d.e.

$$\frac{df(z)}{dz} = 1 + f^2(z) \tag{8.2.17}$$

The general solution is $f(z) = \tan(z - c)$, where c is the one arbitrary parameter. This solution exhibits a series of movable poles at $z = c + (2n + 1)\pi/2$. However, (8.2.17) also has the singular solutions $f(z) = \pm i$ which have no arbitrary parameters.

It turns out that singular solutions, if they exist, play an important and rather subtle role in determining the integrability of a differential equation. What these solutions in fact represent are the *envelopes* of families of general solutions. We illustrate this idea with an example of Clairaut's equation.† Consider the first-order equation

$$f = zf' - (f')^2 \tag{8.2.18}$$

where f' denotes df/dz. The general solution is

$$f_g(z) = cz - c^2 \tag{8.2.19}$$

where c is the one arbitrary parameter. If Eq. (8.2.18) is differentiated with respect to z, one obtains

$$f''(z - 2f') = 0 \tag{8.2.20}$$

This equation has two solutions, namely, (a) $f'' = 0$ and (b) $z = 2f'$. One integration of solution (a) gives $f' = c$, which, on substitution into (8.2.18), yields the one-parameter general solution (8.2.19). From solution (b) we have $f' = z/2$, and substitution of this into (8.2.18) yields the (no-parameter) singular solution

$$f_s(z) = z^2/4 \tag{8.2.21}$$

If we plot $f_s(z)$ and a family (i.e., different c values) of $f_g(z)$, we see that the former is the envelope of the latter (see Figure 8.1).

---

†Clairaut's equation takes the general form $f = zf' + F(f')$, where F is an analytic function of f'. The reader will observe the parallel between what follows and the discussion of Legendre transformations in Appendix 2.1 of Chapter 2.

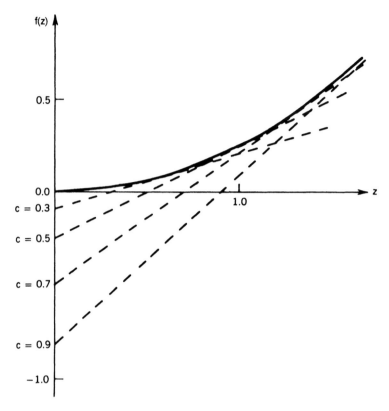

Figure 8.1 General and singular solutions of Clairaut's equation (8.2.18). Dashed lines represent one-parameter family of general solutions $f_g(z) = cz - c^2$ for different c values; solid line is singular solution $f_s(z) = z^2/4$ giving envelope of general solutions.

8.2.c Psi Series

Returning to Eq. (8.2.12), we know from the leading-order and resonance analyses that at a singularity the solution behaves as a first-order pole and that the resonances occur at $j = 4$ and $j = -1$. Direct substitution of the ansatz (8.2.8) into (8.2.12) yields the recursion relations

$$a_j(j+1)(j-4) = -Ba_{j-1} + Aa_{j-2} + 2\sum_{k=1}^{j-1}\sum_{l=1}^{k} a_{j-k-l}a_k a_l \qquad (8.2.22)$$

From this, one obtains, using $a_0 = 1$,

$$j = 1: \quad a_1 = -B/6$$
$$j = 2: \quad a_2 = -(A + B^2/6)/6$$

$$j = 3: \quad a_3 = -B(A/2 + B^2/9)/6$$
$$j = 4: \quad 0 \cdot a_4 = B^2(A + 2B^2/9)/3$$

The last of these gives the compatability condition that ensures the arbitrariness of a_4. This will only be satisfied if $A = -2B^2/9$.

If this condition is not satisfied, the arbitrariness of a_4 can be recaptured by modifying the ansatz (8.2.8). If one now tries the modified expansion

$$y(z) = \sum_{j=0}^{\infty} a_j(z - z_0)^{j-1} + b(z - z_0)^3 \ln(z - z_0) \qquad (8.2.23)$$

the same results are obtained at $j = 1$, 2, and 3; however, at $j = 4$ (i.e., at order $(z - z_0)^3$), one obtains

$$0 \cdot a_4 = \frac{B^2}{3}\left(A + \frac{2B^2}{9}\right) + 5b$$

Thus if one sets $b = -B^2(A + 2B^2/9)/15$, a_4 becomes arbitrary. Of course the additional term in (8.2.23) introduces various powers and combinations of terms involving $\ln(z - z_0)$. In order to make a self-consistent expansion in which all these additional terms balance, the series (8.2.23) must be generalized to

$$y(z) = \sum_{j=0}^{\infty} \sum_{k=0}^{\infty} a_{jk}(z - z_0)^{j-1}((z - z_0)^4 \ln(z - z_0))^k \qquad (8.2.24)$$

This is known as a (logarithmic) *psi series*. It is the local representation of the general solution to (8.2.12) about a movable singularity at $z = z_0$. Now the arbitrary parameters are z_0 and a_{40}. From (8.2.24) we see that the singularity is no longer a movable pole but is, instead, a movable logarithmic branch point. Obviously, the equation (8.2.12) does not have the Painlevé property unless $A = -2B^2/9$.

Other types of psi series are possible. Suppose we had a second-order o.d.e. whose leading order was, say, $\alpha = -1$ but with a resonance at some irrational power of $(z - z_0)$, say β. In this case the local expansion takes the form

$$y(z) = \sum_{j=0}^{\infty} \sum_{k=0}^{\infty} a_{jk}(z - z_0)^{j-1}\psi^k \qquad (8.2.25)$$

where $\psi = (z - z_0)^\beta$. The psi series (8.2.24) and (8.2.25) indicate that the solutions of the associated equations have a very complicated multisheeted structure in the complex plane. However, it turns out that a detailed analysis of these series is still possible and that some remarkable insight into the nature of the solutions can be obtained (see Section 8.3.e).

8.2.d Elliptic Functions and Algebraic Curves*

In Chapter 1 we introduced the idea of elliptic functions and wrote the differential equation for Weierstrass elliptic functions in the standard form

$$\left(\frac{dx}{dt}\right)^2 = 4x^3 - g_2 x - g_3 \tag{8.2.26}$$

The quadrature is thus

$$t - t_0 = \int^x \frac{dx'}{\sqrt{4(x' - e_1)(x' - e_2)(x' - e_3)}} \tag{8.2.27}$$

where the cubic form has been factorized. The roots e_1, e_2, e_3 are related to g_2 and g_3 according to the standard formulae given in Eq. (1.2.20). As it stands, (4.2.27) gives t as an infinitely multivalued function of x. The inverse (i.e., $x = \mathscr{P}(t)$) is just the Weierstrass elliptic function. As already indicated in Chapter 1, $\mathscr{P}(t)$ is a doubly periodic function of complex t in which the singularities are organized in a regular latticelike structure. Using the local analysis techniques described above, the singularities of $\mathscr{P}(t)$ are easily seen to be second-order poles.

Further insights can be obtained by examining the integrand in (8.2.27) in the complex x-plane. Clearly, it has three square root branch points at $x = e_i$, $i = 1, 2, 3$. By pairing up two of these branch points and the third one with the point at infinity, the complex plane acquires two branch cuts. Standard "cutting and pasting" along these cuts shows that the associated Riemann surface can be folded into a two-dimensional torus with a single hole. We term this a surface of *genus one*. This takes us into the realms of algebraic geometry. Here one can associate with a given Riemann surface a certain irreducible polynomial which, in turn, defines an *algebraic curve*.†
For surfaces of genus one, the canonical form of the associated curve is

$$r^2 = 4s^3 - g_2 s - g_3 \tag{8.2.28}$$

If we identify $r = d\mathscr{P}/dt$ and $s = \mathscr{P}(t)$ we see that this curve corresponds exactly to Eq. (8.2.26). Curves such as (8.2.28) are termed *elliptic curves* and they have the important property that the coordinates r and s are meromorphic functions of some parameter t. From the results of our local analysis this is exactly what we would expect since we have demonstrated that $P(t)$ only exhibits movable poles.

The Weierstrass and Jacobi elliptic functions are both meromorphic

---

†A good source for these concepts is *Analytic Function Theory* by E. Hille (1982).

functions (of t) and are easily associated with elliptic curves. Consider now Eq. (8.2.26) in which the right-hand side is a polynomial of order 5, denoted as $f_5(x)$. Local analysis tells us that the movable singularities are now square root branch points. The corresponding quadrature is now the hyperelliptic integral

$$t - t_0 = \int^x \frac{dx'}{\sqrt{f_6(x')}} \tag{8.2.29}$$

where f_6 is a sixth-order polynomial obtained from f_5. The corresponding Riemann surface is now a surface of genus two—a torus with two holes. The associated algebraic curve, a hyperelliptic curve, can no longer be parameterized in terms of meromorphic functions. However, Jacobi observed that certain symmetrical combinations of hyperelliptic integrals do have meromorphic inverses.

The theory of hyperelliptic integrals and their inverses can be cast in a very general form and generalized to many dimensions. The integrals (8.2.27) and (8.2.29) are special cases of what are termed *abelian integrals*.† The Riemann surfaces associated with these integrals are termed *abelian varieties*, and the right combinations of variables, termed *abelian functions*, are meromorphic functions. It now seems that these concepts play a crucial role in determining the integrability of dynamical systems and can help explain why the Painlevé property—as first used by Kovalevskaya—can provide a test for integrability.

8.3 INTEGRABLE SYSTEMS OF ORDINARY DIFFERENTIAL EQUATIONS

8.3.a The Henon–Heiles System

We now demonstrate how the integrable cases of the Henon–Heiles Hamiltonian (8.1.1) can be identified by means of the Painlevé property. Here it is convenient to write the equation of motion in Newtonian form, that is,

$$\ddot{x} = -Ax - 2Dxy \tag{8.3.1a}$$

$$\ddot{y} = -By - Dx^2 + Cy^2 \tag{8.3.1b}$$

where "$\dot{\ }$" denotes d/dt. The first step is to carry out a leading-order analysis by setting

$$x = a(t - t_0)^\alpha, \qquad y = b(t - t_0)^\beta \tag{8.3.2}$$

---

†Named after the great Norwegian mathematician Niels Abel (1802–1829), who died in poverty and relative obscurity before his 27th birthday.

where both α and β are assumed < 0. The balance at the singularity t_0 is not affected by the linear terms, so one considers the equations

$$a\alpha(\alpha - 1)(t - t_0)^{\alpha - 2} = -2Dab(t - t_0)^{\alpha + \beta} \tag{8.3.3a}$$

$$b\beta(\beta - 1)(t - t_0)^{\beta - 2} = -Da^2(t - t_0)^{2\alpha} + Cb^2(t - t_0)^{2\beta} \tag{8.3.3b}$$

These equations immediately reveal that two different types of leading order are possible. These are, where $\lambda = D/C$,

$$\text{case (i):} \quad \alpha = -2, \quad a = \pm\frac{3}{D}\sqrt{2 + 1/\lambda}$$

$$\beta = -2, \quad b = -\frac{3}{D}$$

and, if one assumes that α is less negative than β,

$$\text{case (ii):} \quad \alpha = \tfrac{1}{2} \pm \tfrac{1}{2}\sqrt{1 - 48\lambda}, \quad a = \text{arbitrary}$$

$$\beta = -2, \qquad\qquad b = \frac{6}{C}$$

where the above form of α has come from equating the coefficients in (8.3.3a) (β and b having already been determined from (8.3.3b)). In case (ii), α has two roots. However, since the most singular behavior that can be supported by the equations of motion is $(t - t_0)^{-2}$, both roots can only exist for $\lambda > -\tfrac{1}{2}$.

The leading-order analysis is already yielding valuable information. If we are interested in the Painlevé property (i.e., that the movable singularities be poles), this immediately restricts λ to those values for which the case (ii) α is integer. Typically, α is irrational and for $\lambda > 1/48$, α becomes complex. We note that for the original Henon–Heiles system (i.e., in which $A = B = C = D = 1$), $\alpha = (1 \pm i\sqrt{47})/2$. As it turns out, these complex singularities lead to some remarkably complicated structures in the complex t-plane.

The two different solution *branches*, case (i) and case (ii), must both be tested for the Painlevé property. At this stage of the analysis it is not clear which one corresponds to the general solution and which one corresponds to a singular solution—if it should exist. The next step is to carry out a resonance analysis. For the case (i) branch, we set

$$x = \pm\frac{3}{D}\sqrt{2 + 1/\lambda}\, t^{-2} + pt^{r-2}$$

$$y = -\frac{3}{D}t^{-2} + qt^{r-2}$$

where p and q are arbitrary parameters. Setting up the linear equations for p and q from the dominant balance terms in Eqs. (8.3.1), we find that the resonances occur at

$$r = -1, 6, \tfrac{5}{2} \pm \tfrac{1}{2}\sqrt{1 - 24(1 + 1/\lambda)} \qquad (8.3.4)$$

Since Eqs. (8.3.1) are fourth order, there must be four arbitrary parameters in the local expansion of the general solution. One of these is just the singularity position t_0 (corresponding to the resonance at $r = -1$), the other is at $r = 6$, but the remaining two are determined by the value of λ. In order that both of these be greater than zero we must have $\lambda > 0$ or $\lambda < -\tfrac{1}{2}$. In the range $-\tfrac{1}{2} < \lambda < 0$ there are only three positive resonances and the case (i) branch must then correspond to a singular solution. Furthermore, in order to have only integer resonances (in the hope of obtaining a solution with the Painlevé property), Eq. (8.3.4) imposes further restrictions on allowable λ values. In fact it is not difficult to determine that these λ values are (a) $\lambda = -1$, $r = -1, 2, 3, 6$, (b) $\lambda = -\tfrac{1}{2}$, $r = -1, 0, 5, 6$, and (c) $\lambda = -\tfrac{1}{6}$, $r = -3$, $-1, 6, 8$, with the last of these cases corresponding to a singular solution.

For a resonance analysis of the case (ii) branch, we set

$$x = at^{\alpha_\pm} + pt^{r+\alpha_\pm}$$

$$y = \frac{6}{C}t^{-2} + qt^{r-2}$$

where $\alpha_\pm = \tfrac{1}{2} \pm \tfrac{1}{2}\sqrt{1 - 48\lambda}$ and a is already arbitrary. We also note that the dominant balance terms for this analysis are just

$$\ddot{x} = -2Dxy \qquad (8.3.5a)$$

$$\ddot{y} = Cy^2 \qquad (8.3.5b)$$

In this case the resonances are found to be at

$$r = -1, 0, 6, \mp\sqrt{1 - 48\lambda} \qquad (8.3.6)$$

where the resonance at $r = 0$ corresponds to the arbitrariness of the leading-order coefficient a and the $\mp$ sign in the last of (8.3.6) corresponds to the choice α_+ and α_-, respectively. Now, four-parameter solutions only exist for $\lambda > -\tfrac{1}{2}$. Again, integer resonances will only occur for special λ values.

The combination of leading-order and resonance analyses has already narrowed down considerably the λ values for which *both* case (i) and case (ii) solutions have integer leading orders and resonances. These are just $\lambda = -1$, $-\tfrac{1}{2}$, and $-\tfrac{1}{6}$. The associated resonance structure is tabulated below.

	Case (i)		Case (ii)	
λ	Leading Orders	Resonances	Leading Orders	Resonances
-1	$\beta = -2,$ $\alpha = -2$	$r = -1, 2, 3, 6$	$\beta = -2,$ $\alpha_- = -3$	$r = -1, 0, 6, 7$
			$\beta = -2,$ $\alpha_+ = 4$	$r = -7, -1, 0, 6$
$-\frac{1}{2}$	$\beta = -2,$ $\alpha = -2$	$r = -1, 0, 5, 6$	$\beta = -2,$ $\alpha_- = -2$	$r = -1, 0, 5, 6$
			$\beta = -2,$ $\alpha_+ = 3$	$r = -5, -1, 0, 6$
$-\frac{1}{6}$	$\beta = -2,$ $\alpha = -2$	$r = -3, -1, 6, 8$	$\beta = -2,$ $\alpha_- = -1$	$r = -1, 0, 3, 6$
			$\beta = -2,$ $\alpha_+ = 2$	$r = -3, -1, 0, 6$

The reader may have noticed something peculiar about the case $\lambda = -\frac{1}{2}$. The case (i) resonance at $r = 0$ corresponds to the vanishing of the leading-order coefficient $a = \pm\frac{3}{2}\sqrt{2 + 1/\lambda}$. Since the original ansatz was that $x(t) = a(t - t_0)^{-2}$ at leading order, there is clearly a contradiction. As it turns out, the leading order is not -2 but involves logarithmic terms.† Thus there are now only two candidates for the Painlevé property. The first is $\lambda = -1$, with the case (i) singularities corresponding to the general solution and the case (ii) singularities ($\beta = -2$, $\alpha_+ = 4$) corresponding to the singular solution. The other candidate is $\lambda = -\frac{1}{6}$ with case (i) now being the singular solution and case (ii) ($\beta = -2$, $\alpha_- = -1$) being the general solution.

In order to verify the single-valuedness of the solutions, the full recursion relations must now be tested for self-consistency at the resonances. For $\lambda = -1$ we make the ansatz

$$x(t) = \sum_{j=0}^{\infty} a_j t^{j-2} \tag{8.3.7a}$$

$$y(t) = \sum_{j=0}^{\infty} b_j t^{j-2} \tag{8.3.7b}$$

for the general solution (setting $t_0 = 0$ for notational convenience). The

---

†The analysis here is somewhat tricky, but one may show that at leading order (setting $t_0 = 0$ for notational convenience) we have

$$x = \left(\frac{15}{2D^2}\right)^{1/2} t^{-2}(\ln t)^{-1/2} \quad \text{and} \quad y = -\frac{3}{D} t^{-2} - \left(\frac{5}{4D}\right) t^{-2}(\ln t)^{-1}$$

recursion relations are (setting $C = -D$ in Eq. (8.3.1b))

$$
\begin{bmatrix} (j-2)(j-3) - 6 & \pm 6 \\ \pm 6 & (j-2)(j-3) - 6 \end{bmatrix} \begin{bmatrix} a_j \\ b_j \end{bmatrix}
$$
$$
= \begin{bmatrix} -Aa_{j-2} - 2D \sum\limits_{l=1}^{j-1} a_{j-l}b_l \\ -Bb_{j-2} - D \sum\limits_{l=1}^{j-1} (a_{j-l}a_l + b_{j-l}b_l) \end{bmatrix} \tag{8.3.8}
$$

Given that $a_0 = \pm 3/D$ and $b_0 = -3/D$ we find

$$j = 1: \quad a_1 = b_1 = 0$$

$$j = 2: \quad \begin{bmatrix} -6 & \pm 6 \\ \pm 6 & -6 \end{bmatrix} \begin{bmatrix} a_2 \\ b_2 \end{bmatrix} = \begin{bmatrix} \mp 3A/D \\ +3B/D \end{bmatrix}$$

The determinant of the matrix on the left-hand side vanishes corresponding to the resonance at $j = 2$. The *null vector* of this matrix is just $[1, \pm 1]$, so in order that a_2, b_2 be arbitrary we require that

$$[1, \pm 1] \begin{bmatrix} \mp 3A/D \\ -3B/D \end{bmatrix} = 0$$

This is the compatability condition and is only satisfied if $A = B$. If this is not the case, then logarithmic terms must be included in the series in order to recapture the arbitrariness of these coefficients. The recursion relations must be checked up to $j = 6$ to ensure that the compatibility conditions are satisfied at the remaining resonances $j = 3$ and $j = 6$. That this is so is fairly easy to demonstrate. The singular solution (i.e., the case (ii) branch) may similarly be shown to be single-valued. Thus, overall, one concludes that the Henon–Heiles system has the Painlevé property for $\lambda = -1$ provided that $A = B$. In this case the equations of motion are

$$\ddot{x} = -Ax - 2Dxy \tag{8.3.9a}$$

$$\ddot{y} = -Ay - D(x^2 + y^2) \tag{8.3.9b}$$

and, on introducing the new variables $u = x + y$, $v = x - y$, these equations separate into the pair

$$\ddot{u} = -Au - 2Du^2 \tag{8.3.10a}$$

$$\ddot{v} = -2v + 2Dv^2 \tag{8.3.10b}$$

which are both easily integrated in terms of elliptic functions.

In the case $\lambda = -\frac{1}{6}$ the general solution corresponds to the case (ii) branch and one has to test the series

$$x(t) = \sum_{j=0}^{\infty} a_j t^{j-1} \tag{8.3.11a}$$

$$y(t) = \sum_{j=0}^{\infty} b_j t^{j-2} \tag{8.3.11b}$$

for consistency at the resonances $r = 3$ and 6, with the resonance $r = 0$ already being guaranteed at leading order. In this case the recursion relations are found to be self-consistent without any restrictions. A similar result is found for the case (i) singular branch. Thus for $\lambda = -\frac{1}{6}$ we find that the Henon–Heiles system has the Painlevé property and that the Hamiltonian

$$H = \tfrac{1}{2}(p_x^2 + p_y^2 + Ax^2 + By^2) + Dx^2 y + 2Dy^3 \tag{8.3.12}$$

is integrable for all A and B. Verification of this prediction is not as easy as before but is provided by the discovery of the second integral, which takes the form

$$F = x^4 + 4x^2 y^2 - 4\dot{x}(\dot{x}y - \dot{y}x) + 4Ax^2 y + (4A - B)(\dot{x}^2 + Ax^2) \tag{8.3.13}$$

(It also turns out that the equations of motion separate in parabolic coordinates.)

8.3.b Integrable Systems with Movable Branch Points

The requirement that a system has the Painlevé property means that its solution(s) must live on a single Riemann surface with an arbitrary number of isolated movable poles. It is possible that this condition is too restrictive for determining integrability, and we now relax this condition and allow for movable *rational* branch points. We illustrate this idea by considering the Henon–Heiles system in the range $-\frac{1}{2} < \lambda < 0$. In this case it is only the case (ii) singularities that have four arbitrary parameters. In order that the associated resonances be rational numbers, we deduce from (8.3.6) that

$$\lambda = \frac{1}{48}\left(1 - \left(\frac{m}{n}\right)^2\right), \qquad 1 < \frac{m}{n} < 5 \tag{8.3.14}$$

which results in the leading orders

$$\alpha = \frac{1}{2}\left(1 - \frac{m}{n}\right), \qquad \beta = -2 \tag{8.3.15}$$

and resonances

$$r = -1, 0, \frac{m}{n}, 6 \qquad (8.3.16)$$

Included in this set is, of course, the known integrable case $\lambda = -\frac{1}{6}$, corresponding to $m/n = 3$. For these special λ values (8.3.14) we can then make the ansatz that about a movable singularity (set at $t_0 = 0$) the expansions take the form

$$x(t) = t^{-(m-n)/2n} \sum_{j=0}^{\infty} a_j t^{j/n} \qquad (8.3.17a)$$

$$y(t) = t^{-2} \sum_{j=0}^{\infty} b_j t^{j/n} \qquad (8.3.17b)$$

In order for these expansions to be valid the compatibility conditions at the resonances at $j = m$ and $j = 6n$ must be satisfied (the resonance at $r = 0$ being trivially satisfied). It turns out that both the compatibility conditions are only satisfied for $m/n = 2$, corresponding to $\lambda = -\frac{1}{16}$, provided that $B = 16A$ (this condition appearing at the first resonance). Thus, for this case, the local expansions take the form

$$x(t) = t^{-1/2} \sum_{j=0}^{\infty} a_j t^j \qquad (8.3.18a)$$

$$y(t) = t^{-2} \sum_{j=0}^{\infty} a_j t^j \qquad (8.3.18b)$$

Furthermore, the associated three-parameter solution corresponding to the case (i) singularities is single-valued with resonances at $r = -7, -1, 6, 12$. The general solution (8.3.18a) has isolated movable square-root branch points, and under the "local" change of variable $\tau^2 = t$ the expansion becomes single-valued in τ. However, in this particular case we can also make the "global" change of dependent variable $X = x^2$ and hence make the solutions single-valued in the variables (X, y). These results led to the prediction that the Hamiltonian

$$H = \tfrac{1}{2}(p_x^2 + p_y^2 + Ax^2 + 16Ay^2) + x^2 y + \tfrac{16}{3} y^3 \qquad (8.3.19)$$

is integrable. This result was independently confirmed by Hall (1983), who found the second integrable to be

$$G = \tfrac{1}{4}p_x^4 + (\tfrac{1}{2}x^2 + 4x^2 y)p_x^2 - \tfrac{4}{3}x^3 p_x p_y + \tfrac{1}{4}x^4 - \tfrac{4}{3}x^4 y - \tfrac{8}{9}x^6 - \tfrac{16}{3}x^4 y^2 \qquad (8.3.20)$$

The idea that systems with rational branch points—and these do not have to be "locally" or "globally" transformable to single-valued solutions—can sometimes be integrable has been termed the *weak Painlevé property*. Many examples of integrable systems with the weak Painlevé property have now been found. A brief discussion of why the Painlevé and weak Painlevé properties can indicate integrability will be given in Section 8.3.d.

8.3.c The Lorenz System

We now turn to the non-Hamiltonian dynamical system (8.1.2) and demonstrate that the Painlevé property can still be used to identify integrable cases. At leading order we set

$$x(t) = a_0(t - t_0)^\alpha, \qquad y(t) = b_0(t - t_0)^\beta, \qquad z(t) = c_0(t - t_0)^\gamma$$

and easily determine that

$$\alpha = -1, \qquad \beta = -2, \qquad \gamma = -2$$
$$a_0 = \mp 2i, \qquad b_0 = \pm 2i/\sigma, \qquad c_0 = -2/\sigma$$

A resonance analysis reveals that the resonances occur at

$$r = -1, 4, 6$$

Thus for the Lorenz system the leading orders and resonances do not depend on the system parameters. On making the ansatz

$$x(t) = \sum_{j=0}^{\infty} a_j t^{j-1} \tag{8.3.21a}$$

$$y(t) = \sum_{j=0}^{\infty} b_j t^{j-2} \tag{8.3.21b}$$

$$z(t) = \sum_{j=0}^{\infty} c_j t^{j-2} \tag{8.3.21c}$$

the recursion relations are found to be

$$\begin{bmatrix} (j-1) & -\sigma & 0 \\ -2/\sigma & (j-2) & 2i \\ 2i/\sigma & -2i & (j-2) \end{bmatrix} \begin{bmatrix} a_j \\ b_j \\ c_j \end{bmatrix} = \begin{bmatrix} -\sigma a_{j-1} \\ Ra_{j-2} - b_{j-1} - \sum_{l=1}^{j-1} a_{j-l}c_l \\ -Bc_{j-1} + \sum_{l=1}^{j-1} a_{j-l}b_l \end{bmatrix} \tag{8.3.22}$$

At $j = 1$, one finds

$$a_1 = -\frac{i}{3}(1 + 2B - 3\sigma), \qquad b_1 = 2i, \qquad c_1 = \frac{2}{3\sigma}(3\sigma + 1 - B) \qquad (8.3.23)$$

and at $j = 2$ we have

$$\begin{bmatrix} 1 & -\sigma & 0 \\ -2/\sigma & 0 & 2i \\ 2i/\sigma & -2i & 0 \end{bmatrix} \begin{bmatrix} a_2 \\ b_2 \\ c_2 \end{bmatrix} = \begin{bmatrix} -\sigma a_1 \\ Ra_0 - b_1 - a_1 c_1 \\ -Bc_1 + a_1 b_1 \end{bmatrix}$$

The determinant of the left-hand matrix vanishes corresponding to the resonance appearing at this order. Multiplying through by the null vector $[2i/\sigma, 0, -1]$ yields the compatibility condition

$$B(B - 1) - 6\sigma^2 + 2\sigma + B\sigma = 0 \qquad (8.3.24)$$

which will be satisfied if

$$B = 2\sigma \quad \text{or} \quad B = 1 - 3\sigma \qquad (8.3.25)$$

The analysis at $j = 3$ and $j = 4$ becomes more complex, but the final result is a compatibility condition at $j = 4$ of the form

$$a_3 c_1 + a_2 c_2 + a_1 c_3 - Ra_2 + b_3 + ia_3 b_1 + ia_2 b_2 + ia_1 b_3 - iBc_3 = 0 \qquad (8.3.26)$$

which, on substitution of the explicit forms of the a_i, b_i, c_i ($i = 1, 2, 3$), gives the compatibility condition in terms of the system parameters B, σ, R. Note, however, that (8.3.26) contains terms involving the a_2, b_2, c_2 which are arbitrary. This has the result of splitting the condition (8.3.26) into two parts—one group of terms independent of the arbitrary parameter introduced at $j = 2$ and another group of terms all multiplied by that parameter. In order for (8.3.26) to be satisfied, both groups of terms must vanish independently. (For full details, see the paper by Tabor and Weiss (1981).) Both compatibility conditions are only satisfied simultaneously for three sets of parameter values (excluding the trivial case $\sigma = 0$)

$$\text{(a)} \quad \sigma = \tfrac{1}{2}, \quad B = 1, \quad R = 0 \qquad (8.3.27a)$$

$$\text{(b)} \quad \sigma = 1, \quad B = 2, \quad R = \tfrac{1}{9} \qquad (8.3.27b)$$

$$\text{(c)} \quad \sigma = \tfrac{1}{3}, \quad B = 0, \quad R \text{ arbitrary} \qquad (8.3.27c)$$

For the first of these cases the integration of the equations is fairly straightforward and provides a nice exercise in the use of time-dependent integrals. Writing the equations as

$$\dot{x} + \tfrac{1}{2}x = \tfrac{1}{2}y \tag{8.3.28a}$$

$$\dot{y} + y = -xz \tag{8.3.28b}$$

$$\dot{z} + z = xy \tag{8.3.28c}$$

we notice that multiplying Eq. (8.3.28a) through by $2x$ and subtracting the result from (8.3.28c) yields

$$\frac{d}{dt}(z - x^2) = -(z - x^2)$$

Thus one obtains the time-dependent integral

$$I_1 = z - x^2 = Ae^{-t} \tag{8.3.29}$$

Furthermore, if we multiply (8.3.28b) through by y and (8.3.28c) by z, the resulting sum yields the second integral

$$I_2 = y^2 + z^2 = Be^{-2t} \tag{8.3.30}$$

These two integrals can now be used to reduce the original third-order system to a single quadrature. Introducing the new variables

$$u = xe^{t/2}, \quad v = ye^t, \quad w = ze^t$$

the two integrals become time independent, that is,

$$I_1 = w - u^2 = A \tag{8.3.31a}$$

$$I_2 = v^2 + w^2 = B \tag{8.3.31b}$$

Elimination w between the two of these gives

$$v = ((B - A^2) - 2A^2u^2 - u^4)^{1/2} \tag{8.3.32}$$

In the new variables the original system of equations (8.3.28) becomes

$$\dot{u} = \tfrac{1}{2}ve^{-t/2} \tag{8.3.33a}$$

$$\dot{v} = -uwe^{-t/2} \tag{8.3.33b}$$

$$\dot{w} = -uve^{-t/2} \tag{8.3.33c}$$

Inserting the expression for v from (8.3.32) into Eq. (8.3.33a) gives the quadrature

$$\int^u \frac{du'}{\sqrt{(B-A^2)-2A^2u'^2-u'^4}} = 2\int^t e^{-t'/2}\,dt' \qquad (8.3.34)$$

which can be solved in terms of Jacobi elliptic functions. The evolution of the remaining variables follows trivially from (8.3.31a) and (8.3.32a).

The integration of the Lorenz equations for the other two sets of parameter values (8.3.27b) and (8.3.27c) is a little more involved, but Segur (1980) has shown how this can be accomplished in terms of the second and third Painlevé transcendents, respectively.

It is interesting to note that for the parameter values $B = 2\sigma$, R arbitrary, one time-dependent integral, namely,

$$I_1 = 2\sigma z - x^2 = Ae^{-2\sigma t} \qquad (8.3.35)$$

can always be found. Except for the cases $R = 0$ and $R = \frac{1}{9}$, the system does *not* have the Painlevé property. However, the existence of this one integral is enough to eliminate the possibility of chaotic behavior even if it is not enough to permit a complete integration by quadratures. A similar situation arises for $B = 1$, $R = 0$, σ arbitrary. In this case the one integral is

$$I_1 = (y^2 + z^2) = e^{-2t} \qquad (8.3.36)$$

8.3.d Why Does the Painlevé Property Work?

The Painlevé property (and the "weak" Painlevé property) apparently provides a direct test of integrability. However, a proof of why this should be so is far from easy. In the past few years, some progress in this direction has been made, although these results require the sophisticated concepts of algebraic geometry introduced in Section 8.2.d. However, a simple argument that may be helpful is as follows. An important theorem due to Liouville states that if $f(z)$ is an entire bounded function of z, then it can only take on one value, that is,

$$f(z) = c \qquad (8.3.37)$$

In other words, $f(z)$ must be a constant. Now since (time-independent) integrals of the motion are constant in time, it follows that they are entire functions of complex t. For a rather narrow class of integrable Hamiltonian systems, termed *algebraically integrable*, the integrals are polynomial functions of the canonical variables, that is, they take the form

$$I = \sum_{n,m} c_{nm} \prod_{i=1}^{N} p_i^{n_i} q_i^{m_i} \qquad (8.3.38)$$

where **n**, **m** are the N-dimensional vectors $(n_1, \ldots, n_N)$, $(m_1, \ldots, m_N)$ and N is the number of degrees of freedom. The individual p_i and q_i will have movable singularities of various types. However, in order that I be constant we require that all the singular terms in the sum (8.3.38) must cancel at any singularity position t_0. This will only be possible for polynomials of the above form if the local expansions for the p_i and q_i correspond to poles or, perhaps, rational branch points. Clearly, if the expansions for the canonical variables have irrational, complex, or other multivalued powers of $(t - t_0)$, it would (typically) not be possible to construct a polynomial function of them in which these powers cancel. (Note, of course, that the Hamiltonian function itself will always be an entire (i.e., constant) function.)

Clearly, such an argument is only valid for polynomial integrals. It suggests that for Hamiltonians with integrals that are irrational or transcendental functions of the canonical variables the Painlevé property would not be an appropriate test for integrability. The analytic structure of such systems is, to date, an explored area.

8.3.e Singularity Structure of Nonintegrable Systems*

Although they are outside the scope of these lectures, we briefly mention some of the immensely complicated structures that can appear in the complex plane of nonintegrable systems. One example is the Henon–Heiles system when $\lambda = 1$ (i.e., its original form). In this case, one finds for the case (i) singularities the leading orders

$$\alpha = -2, \qquad \beta = -2$$

and resonances

$$r = -1, \tfrac{5}{2} \pm \frac{i}{2}\sqrt{47}, 6$$

and for the case (ii) singularities the leading orders are

$$\alpha = \tfrac{1}{2} \pm \frac{i}{2}\sqrt{47}, \qquad \beta = -2$$

$$r = -1, 0, \mp i\sqrt{47}, 6$$

Thus both types of singularity can appear in the general (four-parameter) solution. Notice that the case (ii) leading order coincides with the case (i) resonance. As it happens, this can only occur for the λ values $+1$ and $-\tfrac{1}{2}$, as may be verified by solving $\sqrt{1 - 48\lambda} = \sqrt{1 - 24(1 + 1/\lambda)}$. Numerical studies by Chang et al. (1982) show that in the complex t-plane the singularities arrange themselves in *self-similar* spirals. This leads to an immensely complicated, fractal-like distribution of singularities that cluster on ever-finer scales. This recursive clustering of singularities appears to lead to the formation of a *natural* boundary in the complex t-plane past which the solution cannot be analytically continued. In the neighborhood of a given complex singularity, the local expansion can be expressed formally as a psi series of the form (for the case (i) singularities)

$$x(t) = \sum_{j=0}^{\infty} \sum_{k=0}^{\infty} a_{jk} t^{j-2} \tau^k + \sum_{j=0}^{\infty} \sum_{k=1}^{\infty} \bar{a}_{jk} t^{j-2} \bar{\tau}^k \qquad (8.3.39a)$$

$$y(t) = \sum_{j=0}^{\infty} \sum_{k=0}^{\infty} b_{jk} t^{j-2} \tau^k + \sum_{j=0}^{\infty} \sum_{k=1}^{\infty} \bar{b}_{jk} t^{j-2} \bar{\tau}^k \qquad (8.3.39b)$$

where

$$\tau = t^{\alpha}, \qquad \alpha = \tfrac{1}{2} + \tfrac{1}{2}\sqrt{1 - 24(1 + 1/\lambda)}$$

$$\bar{\tau} = t^{\bar{\alpha}}, \qquad \bar{\alpha} = \tfrac{1}{2} - \tfrac{1}{2}\sqrt{1 - 24(1 + 1/\lambda)}$$

Similar formal expansions can be written for the case (ii) singularities. We emphasize that the expansions (8.3.39) are "formal" since, in view of the recursive singularity clustering, it is not clear if they have a finite radius of convergence. Nonetheless, it turns out that by carefully analyzing these expansions it is possible to determine the precise geometry of the observed singularity structures.

For the Lorenz equations the nonintegrable cases are characterized by logarithmic series of the form

$$x(t) = \sum_{j=0}^{\infty} \sum_{k=0}^{\infty} a_{jk} t^{j-1} \tau^k \qquad (8.3.40a)$$

$$y(t) = \sum_{j=0}^{\infty} \sum_{k=0}^{\infty} b_{jk} t^{j-2} \tau^k \qquad (8.3.40b)$$

$$z(t) = \sum_{j=0}^{\infty} \sum_{k=0}^{\infty} c_{jk} t^{j-2} \tau^k \qquad (8.3.40c)$$

where

$$\tau = t^2 \ln t \qquad (8.3.41)$$

if the compatibility conditions are not satisfied at the first resonance and

$$\tau = t^4 \ln t \qquad (8.3.42)$$

if the compatibility conditions first fail at the second resonance. Numerical studies of the Lorenz equation and of the closely related Duffing equation by Levine et al. (1988) show that now the singularities recursively cluster in the complex t-plane in the form of n-armed stars, where $n = 2$ if the psi series involves (8.3.41) and $n = 4$ if it involves (8.3.42). These structures can again be predicted analytically. Furthermore, very detailed analysis of the logarithmic psi series is possible, leading to some remarkable analytical insights into the properties of the associated nonintegrable motion.

8.4 PAINLEVÉ PROPERTY FOR PARTIAL DIFFERENTIAL EQUATIONS*

In view of the insights that can be obtained from studying the singularity structure of ordinary differential equations, it is natural to ask if a similar approach can be taken with partial differential equations. In particular, it would be extremely useful if some version of the Painlevé property for o.d.e.'s could be developed to test for integrability in p.d.e.'s. An obvious problem here is that the solutions to p.d.e.'s are functions of at least two independent variables and that the analytic continuation of such functions is far less straightforward than for functions of a single complex variable. The resolution of this issue will be described shortly—for the meantime we consider a simpler approach.

In the discussion of soliton equations in Chapter 7, considerable emphasis (and use) was placed on the various reductions of those p.d.e.'s to o.d.e.'s. Thus, for example, the KdV equation reduces to an elliptic o.d.e. under a traveling wave transformation and to a Painlevé transcendent under a similarity transformation. Indeed, the reader may have realized by now that for all the integrable equations considered there, the reductions always seemed to lead to o.d.e.'s with the Painlevé property. It was this observation that lead Ablowitz, Ramani, and Segur (ARS) (1980) to conjecture that "a nonlinear p.d.e. is solvable by an inverse scattering transform only if every nonlinear o.d.e. obtained by exact reduction has the Painlevé property." (The conjecture allows for changes of variable in the o.d.e. when testing for the Painlevé property.) Put another way, this conjecture suggests that a test for integrability of a p.d.e. (in the sense that an IST exists) is to test *all* its reductions to o.d.e.'s for the Painlevé property. Unfortunately, it does not seem possible to ever know what all these reductions are. Nonetheless, the ARS conjecture does suggest that some type of meromorphicity property of the p.d.e. is associated with its integrability. Naturally, though, one would like to have a direct means of testing a p.d.e. for this property without having to make reductions to o.d.e.'s.

8.4.a A Generalized Laurent Expansion

A major difference between analytic functions of one and several complex variables is that the singularities of the latter cannot be isolated. It turns out that if $f(z_1, \ldots, z_n)$ is a meromorphic function of the n complex variables $z_i (i = 1, \ldots, n)$, the singularities of f lie on analytic manifolds of dimension $2n - 2$. These manifolds, which we term *singular manifolds*, are determined by conditions of the form

$$\phi(z_1, \ldots, z_n) = 0 \qquad (8.4.1)$$

where ϕ is an analytic function in the neighborhood of the manifold defined by (8.4.1).

The existence of these singular manifolds suggests a way of generalizing the concept of a Laurent series for functions of one complex variable to functions of many complex variables. Thus for an analytic function $u = u(z_1, \ldots, z_n)$, Weiss et al. (1983) have proposed a generalized Laurent expansion of the form

$$u(z_1, \ldots, z_n) = \frac{1}{\phi^\alpha} \sum_{j=0}^{\infty} u_j \phi^j \qquad (8.4.2)$$

where

$$\phi = \phi(z_1, \ldots, z_n) \qquad (8.4.3)$$

and

$$u_j = u_j(z_1, \ldots, z_n) \qquad (8.4.4)$$

are analytic functions of $z_1, \ldots, z_n$ in the neighborhood of the manifold (8.4.1) and α is an integer. The analogy with traditional Laurent series in the one-variable case is easily seen by setting $\phi(z) = (z - z_0)$ and recognizing the u_j as the (now constant) expansion coefficients. If the validity of the expansion (8.4.2) can be demonstrated, then we can claim that u is "single-valued" about the arbitrary movable singular manifold (8.4.1). In this way we can use the expansions (8.4.2) to define a "Painlevé property" for the solutions to p.d.e.'s.

An important (and useful) property of the function ϕ is that its gradients should not vanish on the singular manifold itself. Consider, for example, ϕ as function of the two variables x and t. The singular manifold is thus defined by

$$\phi(x, t) = 0 \qquad (8.4.5)$$

In the neighborhood of this surface the implicit function theorem tells that ϕ can be represented as

$$\phi(x, t) = x - \psi(t) \qquad (8.4.6)$$

where

$$\phi(\psi(t), t) = 0 \qquad (8.4.7)$$

provided that

$$\phi_x(x, t) \neq 0 \qquad (8.4.8)$$

on (8.4.5). Similarly, if we expand $\phi(x, t)$ as $\phi = t - \theta(x)$ in the neighborhood of (8.4.5), then we require $\phi_t(x, t) \neq 0$ in order that $\phi(x, \theta(x)) = 0$.

To actually test a given (nonlinear) p.d.e. for the generalized Painlevé property, one proceeds analogously to ordinary differential equations. Substitution of the ansatz (8.4.2) into the given p.d.e. determines the possible values of α and the recursion relations for the u_j. Note that since the u_j are functions of the z_i, these recursion relations will now take the form of coupled p.d.e.'s involving the u_j and ϕ. As with o.d.e.'s, there will also be powers of ϕ for which the associated u_j are arbitrary (i.e., resonances). These resonances are a manifestation of the Cauchy–Kovalevskaya theorem for p.d.e.'s, which tells us that a local expansion of the general solution must have as many *arbitrary functions* as the order of the equation. Failure for a u_j to be arbitrary at a resonance tells us that the ansatz (8.4.2) is not valid and that some version of a generalized psi series is required to recapture the required arbitrariness. The initial stages of the analysis may be simplified by first carrying out the analogue of the leading order and resonance analyses. These quickly reveal the different singularity types and which ones correspond to general and singular solutions. After this the analysis becomes more difficult than for o.d.e.'s since the recursion relations for the u_j are systems of coupled p.d.e.'s. However, this difficulty can be overcome by using the implicit function theorem. Thus, for the two-dimensional problem $u(x, t)$, we just use $\phi = x - \psi(t)$ and the expansion coefficients $u_j(x, t)$ reduce to functions of t only (i.e., $u_j = u_j(t)$). This reduction renders the recursion relations much more tractable. However, as we shall see later, use of the full expansions yields far more information than just a test for single-valuedness.

8.4.b Examples of the Painlevé Property for p.d.e.'s

A simple illustration of the technique is provided by Burgers' equation

$$u_t + u u_x = u_{xx} \qquad (8.4.9)$$

Although this equation does not exhibit soliton solutions, it can be thought of as "integrable" since it can be exactly linearized (and hence solved) by means of the Hopf–Cole transformation.† The leading-order test is per-

---

†On making the substitution $u = -2\psi_x/\psi$, (8.4.9) reduces to

$$\psi_t = \psi_{xx}$$

which is the exactly soluble heat equation. Although this method of solution is credited to Hopf (1950) and Cole (1951), it was given as an exercise in Forsyth's *Theory of Differential Equations, Part IV* published at the turn of the century.

formed by setting

$$u = u_0 \phi^\alpha \qquad (8.4.10)$$

and balancing the dominant terms uu_x and u_{xx}. This easily gives

$$\alpha = -1 \quad \text{and} \quad u_0 = -2\phi_x \qquad (8.4.11)$$

A resonance analysis is also easily performed by setting

$$u = u_0\phi^{-1} + p\phi^{r-1} \qquad (8.4.12)$$

which yields resonances at $r = -1$ and 2. The first resonance just corresponds to the arbitrariness of ϕ itself. One now tests the ansatz.

$$u = \sum_{j=0}^{\infty} u_j(x, t)\phi^{j-1} \qquad (8.4.13)$$

by developing the recursion relations for the u_j and checking that u_2 is an arbitrary function. Since both the u_j and ϕ are functions of x and t we note that

$$u_x = \sum_{j=0}^{\infty} u_{j,x}\phi^{j-1} + (j-1)\phi^{j-2}(u_j\phi_x)$$

$$u_{xx} = \sum_{j=0}^{\infty} u_{j,xx}\phi^{j-1} + (j-1)\phi^{j-2}(2u_{j,x}\phi_x + u_j\phi_{xx}) + (j-1)(j-2)\phi^{j-3}(u_j\phi_x^2)$$

$$u_t = \sum_{j=0}^{\infty} u_{j,t}\phi^{j-1} + (j-1)\phi^{j-2}(u_j\phi_t)$$

The full recursion relations take the form

$$(j+1)(j-2)\phi_x^2 u_j = u_{j-2}\phi_t - u_{j-2,xx}$$
$$+ (j-2)[u_{j-1}\phi_t - 2u_{j-1,x}\phi_x - u_{j-1}\phi_{xx}]$$
$$+ \sum_{k=1}^{j-1} [u_{j-k}u_{k,x} + (k-1)u_ju_{j-k}\phi_x] \qquad (8.4.14)$$

From these one finds that

$$j = 0: \quad u_0 = -2\phi_x \qquad (8.4.15a)$$

$$j = 1: \quad \phi_t + u_1\phi_x = \phi_{xx} \qquad (8.4.15b)$$

$$j = 2: \quad \frac{\partial}{\partial x}(\phi_t + u_1\phi_x - \phi_{xx}) = 0 \tag{8.4.15c}$$

By (8.4.15b) the compatibility condition (8.4.15c) is satisfied identically and the arbitrariness of u_2 is confirmed. Thus the generalized Laurent series (8.4.13) is a valid "local" representation of the general solution to Burgers' equation in the neighborhood of a movable singular manifold $\phi(x, t) = 0$. Therefore we can say that the equation has the Painlevé property.

A more important example is the KdV equation

$$u_t + 6uu_x + u_{xxx} = 0 \tag{8.4.16}$$

In this case a leading-order analysis gives $\alpha = -2$ and $u_0 = -2\phi_x^2$ and a resonance analysis gives $r = -1, 4,$ and 6. For this equation the recursion relations become more complicated and it is convenient to test for the Painlevé property in the "reduced" framework, namely, by setting $\phi(x, t) = x - \psi(t)$ and $u_j = u_j(t)$. Now the recursion relations take the much simpler form

$$(j+1)(j-4)(j-6)u_j = -u_{j-3,t} - (j-4)u_{j-2}\phi_t - 6\sum_{k=1}^{j-1} u_{j-k}u_k(k-2) \tag{8.4.17}$$

from which one finds

$$j = 0: \quad u_0 = -2$$

$$j = 1: \quad u_1 = 0$$

$$j = 2: \quad u_2 = -\phi_t/6$$

$$j = 3: \quad u_3 = 0$$

$$j = 4: \quad 0 . u_4 = 0$$

$$j = 5: \quad u_5 = -\phi_{tt}/36$$

$$j = 6: \quad 0 . u_6 = -2u_4\phi_t - 12u_2u_4 = 0$$

Thus the expansion

$$u(x, t) = \sum_{j=0}^{\infty} u_j\phi^{j-2} \tag{8.4.18}$$

is a valid representation of the general solution, and we can say that the KdV equation also has the Painlevé property.

For the higher-order KdV equation

$$u_t = 30u^2 u_x + 20u_x u_{xx} + 10uu_{xxx} + u_{xxxxx} \qquad (8.4.19)$$

one finds that there are now two different leading-order behaviors, namely, $\alpha = -2$ with $u_0 = -2\phi_x^2$ and $\alpha = -2$ with $u_0 = -6\phi_x^2$. The former case has resonances at $r = -1, 2, 5, 6, 8$ and corresponds to the general solution, whereas the second case is a singular solution with resonances at $j = -3$, $-1, 6, 8, 10$. In this case, both branches may be shown to have single-valued expansions (of the form (8.4.18)), and the equation is again seen to have the Painlevé property.

8.4.c Lax Pairs and Auto-Bäcklund Transformations

Although the reduced framework provides the quickest way to determine whether or not a system has the Painlevé property, it turns out that the full expansions contain a remarkable amount of additional information. For example, if one develops the full recursion relations for the KdV equation, one finds that

$$j = 0: \quad u_0 = -2\phi_x^2 \qquad (8.4.20a)$$

$$j = 1: \quad u_1 = 2\phi_{xx} \qquad (8.4.20b)$$

$$j = 2: \quad \phi_x \phi_t + 4\phi_x \phi_{xxx} - 3\phi_{xx}^2 + 6\phi_x^2 u_2 = 0 \qquad (8.4.20c)$$

$$j = 3: \quad \phi_{xt} + 6\phi_{xx}u_2 + \phi_{xxxx} - 2\phi_x^2 u_3 = 0 \qquad (8.4.20d)$$

$$j = 4: \quad \frac{\partial}{\partial x}(\phi_{xt} + \phi_{xxxx} + 6\phi_{xx}u_2 - 2\phi_x^2 u_3) = 0 \qquad (8.4.20e)$$

the last of which demonstrates that u_4 is arbitrary by virtue of (8.4.20d). The relations at $j = 5$ and $j = 6$ are rather complicated but still self-consistent. Thus the full, single-valued series may be written as

$$u(x, t) = \frac{-2\phi_x^2}{\phi^2} + \frac{2\phi_{xx}}{\phi} + u_2 + u_3\phi + u_4\phi^2 + \cdots$$

$$= 2\frac{\partial^2}{\partial x^2}\log\phi + \sum_{j=2}^{\infty} u_j\phi^{j-2} \qquad (8.4.21)$$

By examining the recursion relations, one finds that the expansion (8.4.21) can be consistently *truncated* at $O(\phi^0)$ by

(i) setting the arbitrary functions u_4 and u_6 equal to zero,

(ii) requiring that $u_3 = 0$, and

(iii) requiring that u_2 itself satisfies the KdV equation.

If these three conditions are satisfied, it is easily demonstrated that

$$u_j = 0, \qquad j \geq 3 \tag{8.4.22}$$

This being so, one now obtains from (8.4.20) and (8.4.21) the system of equations

$$u(x, t) = 2 \frac{\partial^2}{\partial x^2} \log \phi + u_2 \tag{8.4.23a}$$

$$\phi_x \phi_t + 4 \phi_x \phi_{xxx} - 3 \phi_{xx}^2 + 6 u_2 \phi_x^2 = 0 \tag{8.4.23b}$$

$$\phi_{xt} + \phi_{xxxx} + 6 u_2 \phi_{xx} = 0 \tag{8.4.23c}$$

$$u_{2,t} + 6 u_2 u_{2,x} + u_{2,xxx} = 0 \tag{8.4.23d}$$

The three equations (8.4.23b)–(8.4.23d) would appear to be an overdetermined system for the variables $\phi(x, t)$ and $u_2(x, t)$. However, they are entirely self-consistent, with (8.4.23d) being the solvability condition for (8.4.23b) and (8.4.23c). This is demonstrated by means of the substitution

$$\phi_x(x, t) = \psi^2(x, t) \tag{8.4.24}$$

which transforms (8.4.23b) and (8.4.23c) into

$$\psi_{xx} + (\lambda + u_2(x, t))\psi = 0 \tag{8.4.25a}$$

$$\psi_t + 4 \psi_{xxx} + 6 u_2(x, t)\psi_x + 3 u_x(x, t)\psi = 0 \tag{8.4.25b}$$

which is precisely the Lax pair for the KdV equation (8.4.23d)! The truncated expansion (8.4.23a) provides what is termed an *auto-Bäcklund transformation* for the KdV equation.† It provides the means by which new solutions to the KdV equation can be created from old ones. Thus if one starts off with u_2 being the one-soliton solution, the pair (8.4.25) can be solved to give the corresponding scattering eigenfunction. Through (8.4.24) this gives ϕ which, together with the original u_2 in (8.4.23a), gives a new solution—the two-soliton solution in fact—to the KdV equation.

The above example demonstrates how the singular manifold expansions not only can provide a simple "Painlevé test" for p.d.e.'s but also how they

---

†More about Bäcklund transformations can be found in the books by Drazin (1983) and by Ablowitz and Segur (1981).

can be used to actually find the associated Lax pair and auto-Bäcklund transformation. Recent research has shown this to be a rather general procedure. Indeed, the method can also be applied to o.d.e.'s for which "Lax pairs" can also be defined. In these cases the derived Lax pair yields the integrals of motion and algebraic curve for the equation as well.

SOURCES AND REFERENCES

Texts and General Review Articles

Golubev, V. V., *Lectures on Integration of the Equations of Motion of a Rigid Body About a Fixed Point*, State Publishing House, Moscow, 1953.

Hille, E., *Analytic Function Theory*, Vols. I and II, Chelsea, New York, 1982.

Hille, E., *Ordinary Differential Equations in the Complex Plane*, Wiley, New York, 1976.

Section 8.1

Kovalevskaya, S., *A Russian Childhood* (B. Stillman, Ed.), Springer-Verlag, New York, 1978.

Kovalevskaya, S., Sur le probleme de la rotation d'un corps solide autour d'un point fixe, *Acta Math.*, **12**, 177 (1889); **14**, 81 (1889).

Oeuvres de Paul Painlevé, Vol. I, Centre National de la Research Scientifique, Paris, 1973.

Tabor, M., Modern dynamics and classical analysis, *Nature*, **310**, 277 (1984).

Whittaker, E. T., *A Treatise on the Analytical Dynamics of Particles and Rigid Bodies*, 4th ed., Cambridge University Press, Cambridge, 1959.

Section 8.3

Bountis, T., H. Segur, and F. Vivaldi, Integrable Hamiltonian systems and the Painlevé property, *Phys. Rev. A*, **25**, 1257 (1982).

Chang, Y. F., M. Tabor, and J. Weiss, Analytic structure of the Henon–Heiles Hamiltonian in integrable and nonintegrable regimes, *J. Math. Phys.*, **23**, 531 (1982).

Chang, Y. F., J. M. Greene, M. Tabor, and J. Weiss, The analytic structure of dynamical systems and self-similar natural boundaries, *Physica*, **8D**, 183 (1983).

Hall, L. S., A theory of exact and approximate configurational invariants, *Physica*, **8D**, 90 (1983).

Levine, G., J.-D. Fournier, and M. Tabor, Singularity clustering in the Duffing oscillator, *J. Phys. A*, **21**, 33 (1988).

Ramani, A., B. Dorizzi, and B. Grammaticos, Painlevé conjecture revisited, *Phys. Rev. Lett.*, **49**, 1539 (1982).

Segur, H., Solitons and the inverse scattering transform, lectures given at the International School of Physics, "Enrico Fermi," Varenna, Italy, 1980.

Tabor, M., and J. Weiss, Analytic structure of the Lorenz system, *Phys. Rev. A*, **24**, 2157 (1981).

Further insights into the workings of the Painlevé property can be found in:

Adler, M., and P. van Moerbeke, "The algebraic integrability of geodesic flow on SO(4), *Invent. Math.*, **67**, 297 (1982).

Ercolani, N., and E. Siggia, Painlevé property and integrability, *Phys. Lett. A*, **119**, 112 (1986).

Haine, L., The algebraic complete integrability of geodesic flow on SO(*N*), *Commun. Math. Phys.*, **94**, 271 (1984).

Yoshida, H., Necessary condition for the existence of algebraic first integrals, *Celest. Mech.*, **31**, 363 (1983); **31**, 381 (1983).

Section 8.4

Ablowitz, M. J., and H. Segur, *Solitons and the Inverse Scattering Transform*, SIAM, Philadelphia, 1981.

Ablowitz, M. J., A. Ramani, and H. Segur, A connection between nonlinear evolution equations and o.d.e.'s of *P*-type, *J. Math. Phys.*, **21**, 715 (1980).

Cole, J. D., On a quasi-linear parabolic equation occurring in aerodynamics, *Quart. Appl. Math.* **9**, 225 (1951).

Drazin, P. G., *Solitons*, London Mathematical Society Lecture Notes, Vol. 85, Cambridge University Press, Cambridge, 1983.

Forsyth, A. R., *Theory of Differential Equations*. Part IV. *Partial Differential Equations*. Vol. VI, Cambridge University Press, Cambridge, 1906 (republished by Dover, New York, 1959).

Hopf, E., The partial differential equation $u_t + uu_x = \mu u_{xx}$, *Comm. Pure Appl. Math.*, **3**, 201 (1950).

Newell, A. C., M. Tabor, and Y. B. Zeng, A unified approach to Painlevé expansions, *Physica*, **29D**, 1 (1987).

Osgood, W. F., *Topics in the Theory of Functions of Several Complex Variables*, Dover, New York, 1966.

Weiss, J., M. Tabor, and G. Carnevale, The Painlevé property for partial differential equations, *J. Math. Phys.*, **24**, 522 (1983).

INDEX